T0269677

LONDON MATHEMATICAL SOCIETY LECTURE NOTE SERIES

Managing Editor: Professor J.W.S. Cassels, Department of Pure Mathematics and Mathematical Statistics, 16 Mill Lane, Cambridge CB2 1SB, England

The books in the series listed below are available from booksellers, or, in case of difficulty, from Cambridge University Press.

4 Algebraic topology, J.F.ADAMS
5 Commutative algebra, J.T.KNIGHT
8 Integration and harmonic analysis on compact groups, R.E.EDWARDS
11 New developments in topology, G.SEGAL (ed)
12 Symposium on complex analysis, J.CLUNIE & W.K.HAYMAN (eds)
13 Combinatorics, T.P.McDONOUGH & V.C.MAVRON (eds)
16 Topics in finite groups, T.M.GAGEN
17 Differential germs and catastrophes, Th.BROCKER & L.LANDER
18 A geometric approach to homology theory, S.BUONCRISTIANO, C.P.ROURKE & B.J.SANDERSON
20 Sheaf theory, B.R.TENNISON
21 Automatic continuity of linear operators, A.M.SINCLAIR
23 Parallelisms of complete designs, P.J.CAMERON
24 The topology of Stiefel manifolds, I.M.JAMES
25 Lie groups and compact groups, J.F.PRICE
26 Transformation groups, C.KOSNIOWSKI (ed)
27 Skew field constructions, P.M.COHN
29 Pontryagin duality and the structure of LCA groups, S.A.MORRIS
30 Interaction models, N.L.BIGGS
31 Continuous crossed products and type III von Neumann algebras,A.VAN DAELE
32 Uniform algebras and Jensen measures, T.W.GAMELIN
34 Representation theory of Lie groups, M.F. ATIYAH et al.
35 Trace ideals and their applications, B.SIMON
36 Homological group theory, C.T.C.WALL (ed)
37 Partially ordered rings and semi-algebraic geometry, G.W.BRUMFIEL
38 Surveys in combinatorics, B.BOLLOBAS (ed)
39 Affine sets and affine groups, D.G.NORTHCOTT
40 Introduction to Hp spaces, P.J.KOOSIS
41 Theory and applications of Hopf bifurcation, B.D.HASSARD, N.D.KAZARINOFF & Y-H.WAN
42 Topics in the theory of group presentations, D.L.JOHNSON
43 Graphs, codes and designs, P.J.CAMERON & J.H.VAN LINT
44 Z/2-homotopy theory, M.C.CRABB
45 Recursion theory: its generalisations and applications, F.R.DRAKE & S.S.WAINER (eds)
46 p-adic analysis: a short course on recent work, N.KOBLITZ
47 Coding the Universe, A.BELLER, R.JENSEN & P.WELCH
48 Low-dimensional topology, R.BROWN & T.L.THICKSTUN (eds)
49 Finite geometries and designs,P.CAMERON, J.W.P.HIRSCHFELD & D.R.HUGHES (eds)
50 Commutator calculus and groups of homotopy classes, H.J.BAUES
51 Synthetic differential geometry, A.KOCK
52 Combinatorics, H.N.V.TEMPERLEY (ed)
54 Markov process and related problems of analysis, E.B.DYNKIN
55 Ordered permutation groups, A.M.W.GLASS
56 Journêes arithmêtiques, J.V.ARMITAGE (ed)
57 Techniques of geometric topology, R.A.FENN
58 Singularities of smooth functions and maps, J.A.MARTINET
59 Applicable differential geometry, M.CRAMPIN & F.A.E.PIRANI
60 Integrable systems, S.P.NOVIKOV et al
61 The core model, A.DODD
62 Economics for mathematicians, J.W.S.CASSELS
63 Continuous semigroups in Banach algebras, A.M.SINCLAIR

64 Basic concepts of enriched category theory, G.M.KELLY
65 Several complex variables and complex manifolds I, M.J.FIELD
66 Several complex variables and complex manifolds II, M.J.FIELD
67 Classification problems in ergodic theory, W.PARRY & S.TUNCEL
68 Complex algebraic surfaces, A.BEAUVILLE
69 Representation theory, I.M.GELFAND et al.
70 Stochastic differential equations on manifolds, K.D.ELWORTHY
71 Groups - St Andrews 1981, C.M.CAMPBELL & E.F.ROBERTSON (eds)
72 Commutative algebra: Durham 1981, R.Y.SHARP (ed)
73 Riemann surfaces: a view towards several complex variables, A.T.HUCKLEBERRY
74 Symmetric designs: an algebraic approach, E.S.LANDER
75 New geometric splittings of classical knots, L.SIEBENMANN & F.BONAHON
76 Linear differential operators, H.O.CORDES
77 Isolated singular points on complete intersections, E.J.N.LOOIJENGA
78 A primer on Riemann surfaces, A.F.BEARDON
79 Probability, statistics and analysis, J.F.C.KINGMAN & G.E.H.REUTER (eds)
80 Introduction to the representation theory of compact and locally compact groups, A.ROBERT
81 Skew fields, P.K.DRAXL
82 Surveys in combinatorics, E.K.LLOYD (ed)
83 Homogeneous structures on Riemannian manifolds, F.TRICERRI & L.VANHECKE
84 Finite group algebras and their modules, P.LANDROCK
85 Solitons, P.G.DRAZIN
86 Topological topics, I.M.JAMES (ed)
87 Surveys in set theory, A.R.D.MATHIAS (ed)
88 FPF ring theory, C.FAITH & S.PAGE
89 An F-space sampler, N.J.KALTON, N.T.PECK & J.W.ROBERTS
90 Polytopes and symmetry, S.A.ROBERTSON
91 Classgroups of group rings, M.J.TAYLOR
92 Representation of rings over skew fields, A.H.SCHOFIELD
93 Aspects of topology, I.M.JAMES & E.H.KRONHEIMER (eds)
94 Representations of general linear groups, G.D.JAMES
95 Low-dimensional topology 1982, R.A.FENN (ed)
96 Diophantine equations over function fields, R.C.MASON
97 Varieties of constructive mathematics, D.S.BRIDGES & F.RICHMAN
98 Localization in Noetherian rings, A.V.JATEGAONKAR
99 Methods of differential geometry in algebraic topology, M.KAROUBI & C.LERUSTE
100 Stopping time techniques for analysts and probabilists, L.EGGHE
101 Groups and geometry, ROGER C.LYNDON
102 Topology of the automorphism group of a free group, S.M.GERSTEN
103 Surveys in combinatorics 1985, I.ANDERSEN (ed)
104 Elliptical structures on 3-manifolds, C.B.THOMAS
105 A local spectral theory for closed operators, I.ERDELYI & WANG SHENGWANG
106 Syzygies, E.G.EVANS & P.GRIFFITH
107 Compactification of Siegel moduli schemes, C-L.CHAI
108 Some topics in graph theory, H.P.YAP
109 Diophantine analysis, J.LOXTON & A.VAN DER POORTEN (eds)
110 An introduction to surreal numbers, H.GONSHOR
111 Analytical and geometrical aspects of hyperbolic space, D.B.A.EPSTEIN (ed)
112 Low dimensional topology and Kleinian groups, D.B.A.EPSTEIN (ed)
113 Lectures on the asymptotic theory of ideals, D.REES
114 Lectures on Bochner-Riesz means, P.A.TOMAS & Y-C.CHANG
115 An introduction to independence for analysts, H.G.DALES & W.H.WOODIN
116 Representations of algebras, P.J.WEBB (ed)
117 Homotopy theory, E.REES & J.D.S.JONES (eds)
118 Skew linear groups, M.SHIRVANI & B.WEHRFRITZ

London Mathematical Society Lecture Note Series. 118

Skew Linear Groups

M. SHIRVANI and B.A.F. WEHRFRITZ
School of Mathematical Sciences, Queen Mary College, London

CAMBRIDGE UNIVERSITY PRESS
Cambridge
London New York New Rochelle
Melbourne Sydney

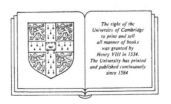

CAMBRIDGE UNIVERSITY PRESS
Cambridge, New York, Melbourne, Madrid, Cape Town, Singapore, São Paulo

Cambridge University Press
The Edinburgh Building, Cambridge CB2 8RU, UK

Published in the United States of America by Cambridge University Press, New York

www.cambridge.org
Information on this title: www.cambridge.org/9780521339254

First published 1986
Re-issued in this digitally printed version 2008

A catalogue record for this publication is available from the British Library

ISBN 978-0-521-33925-4 paperback

CONTENTS

FOREWORD

Our aim in writing this book is to give an up-to-date account of the group-theoretic properties of groups of invertible matrices, where the entries of the matrices lie in a division ring. Our knowledge of this branch of algebra has expanded very rapidly over the last decade. The published part of this material exists only as research papers, and quite a bit, although well-known to the small group of initiates, exists only as private notes. The situation has been reached where, when writing research papers, it is very difficult, or at least very cumbersome, to give the reader adequate references. We hope our book will solve this problem both by providing a textbook for those wishing to study the subject in a systematic way and by providing a convenient reference for research workers. In quite a number of places we have included substantial improvements of both statements and proofs of published theorems.

The importance of linear groups in many branches of mathematics, and even in some parts of physics and chemistry, is well-established. By contrast the much more recent theory of skew linear groups has yet to prove itself. Our final chapter indicates some applications to the theory of group algebras, but even there the intervention of skew linear groups is far from decisive. Thus we feel the need to say a few words here to justify our subject. Our hope of course is that in time many more applications will arise.

Matrix groups over fields are now very well understood (see Wehrfritz [2]) and even if one works over a commutative ring, a very great deal is known (e.g. Wehrfritz [2] Chapter 13 and Wehrfritz [5] Chapter 2). An obvious question is, what happens over a non-commutative ring? One of the first cases to consider is that of a division ring. From the ring-theoretic point-of-view, work of P. M. Cohn especially, shows that, unlike fields, division rings cannot be understood without considering the behaviour of matrices. Thirdly, for group theory, Schur's Lemma shows that every irreducible module over a group G gives rise to a representation of G, usually of infinite degree, over some division ring. Thus representations of groups over division rings are more natural objects of study than representations of groups over fields. The problem is that division rings are usually, or at least presently, too difficult to handle.

PREREQUISITES

We give full, and often alternative, references throughout the book to results that we require from elsewhere, but we indicate here in general terms what we assume of the reader. (We advise readers to start reading the book and to look up what they need as they go along, and not to try and learn the following first.)

We certainly assume a good understanding of algebra up to masters level. Beyond that the requisites for each chapter are very different. Moreover Chapters 2 to 5 inclusive are largely independent of each other.

The first three sections of Chapter 1 are widely used throughout the book. The fourth and final section is used intermittently, mainly to construct counterexamples. Section 1.1 requires little knowledge beyond the Artin-Wedderburn theorems, as does the earlier and main part of Section 1.2. The second half of Section 1.2 uses some Noetherian ring theory. Section 1.3 is in the main very elementary, the exception being 1.3.8 which requires a deep result of H. Heineken and R. H. Bruck. We do make use, in Section 1.3 and elsewhere in the book, of P. Hall's calculus of group classes (see the opening pages of Robinson [1], Vol. 1.). It makes many results very simple to state and clarifies many proofs. The main techniques used in Section 1.4 are Ore domains and Goldie's Theorem. The special case of Goldie's theorem used there we prove in detail.

The remaining chapters make heavier demands on the reader. Chapter 2 utilizes some finite group theory, including a little representation theory, a great deal of number theory in Section 2.1, and a weak version of the classification of the finite simple groups in Sections 2.4 and 2.5. Chapter 3 depends very heavily on the theory of soluble and nilpotent linear groups (Wehrfritz [2]), and to a much lesser extent on the general theory of solubility properties of groups (Robinson [1]). Chapter 4 requires much more ring theory than the preceding chapters. In particluar the structure of group algebras over polycyclic groups, and of modules over such algebras, plays a decisive role throughout the chapter. Our main references here are Passman [2], and for the group theory side, Segal [3]. The very short final Section 4.6 requires some knowledge of Lie algebras. The needs of Chapter 5 are again mainly ring-theoretic, but of a different flavour to those of Chapter 4, and much of what we need we develop in Sections 5.2 and 5.3. In addition some facts about periodic linear groups (see Wehrfritz [2]) are required, as well as the weak version of the classification of finite simple groups. The very brief Chapter 6 shows how some of the earlier results, especially those of Chapter 5, shed light on certain aspects of the theory of group algebras.

The first author wishes to thank Queen Mary College for their hospitality during his stay there. Further, we would like to thank Angela Ridealgh, Marie Fairbrass and Ann Cook, for their cheerful and uncomplaining struggles with illegible manuscripts.

London,
June 1986.

1 BASIC CONCEPTS

By a ***skew linear group*** of degree n, a positive integer, we mean a subgroup of the general linear group GL(n,D) for some division ring D. Certain aspects of the theories of skew linear and of linear groups are very similar and in this chapter we concentrate on some of these. Other aspects are very different indeed, or at least require very different proofs.

Throughout this chapter the symbols n and D have the above designation. In the first three sections we investigate how much of the linear theory of irreducibility, absolute irreducibility, and unipotence can be extended to cover skew linear groups. Intentionally these sections are in the main more elementary than the rest of the book and we hope the reader will find them comparatively easy reading. They are also fundamental for much that follows. In the fourth and final section of Chapter 1 we construct, for later use, a wide range of examples of groups with faithful skew linear representations. This section may be omitted from a first reading.

1.1 IRREDUCIBILITY

Let G be a subgroup of GL(n,D) and set $V = D^n$, the space of row n-vectors over D. Then V is a D-G bimodule in the obvious way. We say that G is an ***irreducible*** (resp. ***reducible, completely reducible***) subgroup of GL(n,D) whenever V is irreducible (resp. reducible, completely reducible) as D-G bimodule.

Viewing the elements of V as column vectors instead of row vectors, we can regard V as a G-D bimodule. It is easy to see that V is irreducible (reducible, completely reducible) as a G-D bimodule precisely when it has the property as a D-G bimodule. For let $H = \mathrm{Hom}_D(V,D)$, where

V, D are left D-spaces. Then the right GL(n,D)-action on V and the right D-action on D make H into a GL(n,D)-D bimodule. Standard duality theory shows that V and H are isomorphic as GL(n,D)-D bimodule. If $V = V_1 \oplus V_2$ as D-G bimodule then $H \cong \mathrm{Hom}_D(V_1,D) \oplus \mathrm{Hom}_D(V_2,D)$ as G-D bimodule. Then V is D-G irreducible whenever V is G-D irreducible and the converse follows by symmetry. The claim concerning complete reducibility can be proved in a similar way. The following exercise is the basis of a less conceptual proof of these facts.

1.1.1. V *is reducible as* D-G *bimodule if and only if for some* x *in* GL(n,D) *the* (1,n)-*entry of* g^x *is* 0 *for each* $g \in G$, *if and only if* V *is reducible as* G-D *bimodule.* □

By a *representation* of a group G of degree n over D we mean a homomorphism ρ of G into GL(n,D). We say that ρ is irreducible (similarly with the other adjectives) whenever Gρ is irreducible, etc. We usually (but not always) put scalars on the left and mappings on the right.

Let G be a subgroup of GL(n,D) acting on $V = D^n$. Suppose there exists a series of left D-submodules of V of the form:

$$\{0\} = V_0 \le V_1 \le \ldots \le V_r = V \qquad (*)$$

such that G acts trivially on each factor V_i/V_{i-1} (i.e. for each i, $[V_i,G] \subseteq V_{i-1}$). Then we say that G stabilizes the series (*), and we call G a ***stability subgroup*** of GL(n,D). G is, of course, a stability group in the usual group- theoretical sense. Obvious examples are the groups $Tr_1(n,D)$ (resp. $Tr^1(n,D)$) of all lower (resp. upper) unitriangular matrices in GL(n,D).

Given such a series (*) for a stability group G, choose a left D-basis of V that contains a basis of each V_i. With a suitable ordering of this basis, if $x \in GL(n,D)$ is the change-of-basis matrix from the standard basis to this one, then $G^x \subseteq Tr_1(n,D)$. Hence G is ***unitriangularizable*** over D. Clearly G is unipotent. The converse question of whether unipotence implies stability is discussed in Section 1.3.

Let G be a subgroup of GL(n,D) and set $V = D^n$. There exists a D-G composition series

$$\{0\} = V_0 \le V_1 \le \ldots \le V_r = V$$

of V. Then $W = \oplus_i V_i/V_{i-1}$ is a completely reducible D-G bimodule that is

D-isomorphic to V. The kernel of the action of G on W clearly stabilizes the composition series. We have therefore proved:

1.1.2. *Let* G *be a subgroup of* GL(n,D). *Then there exists a completely reducible representation* ρ *of* G *into* GL(n,D) *whose kernel is a stability subgroup.* □

Note that the specific construction of ρ will frequently be used below.

We now consider analogues of Maschke's, Schur's and Clifford's theorems. Care is required as certain parts of the classical arguments decidedly fail in this more general situation.

1.1.3. *Let* G *be a group and* H *a subgroup of* G *of finite index* m. *Let* D *be a division ring of characteristic* 0 *or prime to* m *and let* V *be a* D-G *bimodule which is* D-H *completely reducible. Then* V *is* D-G *completely reducible.*

Proof: Let V_1 be a D-G submodule of V. We have to show that V_1 is a D-G direct summand of V. There exists a D-H submodule U of V with $V = V_1 \oplus U$. Let μ be the natural projection of V onto U with kernel V_1 and let T be a right transversal of H to G. Set

$$\Phi = \frac{1}{m} \Sigma_{t \in T} \; t^{-1}\mu t \in \mathrm{Hom}(V,V)$$

It is easy to check that Φ is a D-G map, so $V_2 = V\Phi$ is a D-G submodule of V. Also $V_1\mu = \langle 0 \rangle$ and $V_1 G \subseteq V_1$, so $V_1\Phi = \langle 0 \rangle$. Further $V(\mu - 1) \subseteq V_1$, so $V(\Phi - 1) \subseteq V_1$. It follows that $V = V_1 \oplus V_2$. □

By setting $H = \langle 1 \rangle$ in 1.1.3 we obtain the following well-known result.

1.1.4. (H. Maschke). *If* G *is a finite subgroup of* GL(n,D) *such that* charD *does not divide the order of* G, *then* G *is completely reducible.* □

So far this is exactly as for linear groups, but now something different happens: Schur's theorem (Wehrfritz [2], 1.6) does not extend.

For the construction of the following example see 1.4.24 below.

1.1.5. EXAMPLE. (Wehrfritz [13], Theorem). *Let* p *be zero or a prime. Then there exists a division ring* D, *locally finite-dimensional over its centre, and a periodic abelian subgroup* G *of* GL(2,D) *of rank* 1 *that is not completely reducible and contains no non-trivial elements of order* p. □

Obviously every subgroup of GL(1,D) = D* is irreducible. If the division ring is finite-dimensional over its centre then Schur's theorem does extend. The reader should check the proof of Schur's theorem given in Wehrfritz [2], 1.6 and spot the reason that the proof does not work more generally. The first half of Clifford's theorem, however, presents no difficulty.

1.1.6. (A. H. Clifford). *Let* G *be a group,* N *a normal subgroup of* G *and* V *an irreducible* D-G *bimodule of finite dimension over* D. *Then* V *is a direct sum of irreducible* D-N *bimodules of the same dimension over* D. *The homogeneous* D-N *components of* V *are permuted transitively by* G.

Proof: Let U be an irreducible D-N submodule of V; such exists as $\dim_D V$ is finite. Then Ug is an irreducible D-N submodule of V with $\dim_D U = \dim_D Ug$ for all $g \in G$. Also $\Sigma_{g\in G}$ Ug is a D-G submodule of V and therefore is V itself. This proves the first part.

Suppose that U_1 and U_2 are irreducible D-N submodules of V and that $\phi : U_1 \longrightarrow U_2$ is a D-N isomorphism. Then $g^{-1}\phi g : U_1 g \longrightarrow U_2 g$ is also a D-N isomorphism for all $g \in G$ and the second part follows. □

1.1.7. *A subnormal subgroup of a completely reducible skew linear group is completely reducible.* □

Just as in the linear case one can now define systems of imprimitivity, imprimitive groups and primitive groups and, for example, 1.9 and 1.10 of Wehrfritz [2] go through unchanged. We are unaware of any useful analogue of the main part of Clifford's theorem (Wehrfritz [2], 1.15) although much of the argument given there does have a skew linear version. We leave this to the reader.

The irreducibility or otherwise of a linear group is determined by a particular finitely generated subgroup of the group. This useful technique is not available for skew linear groups.

1.1.8. EXAMPLE. (Wehrfritz [13]). *Let* p *be a zero or a prime. Then there exists a locally finite-dimensional division ring* D *of characteristic* p *and abelian subgroups* G *and* H *of* GL(2,D) *such that* G *is not completely reducible but every finitely generated subgroup of* G *is completely reducible, and such that* H *is irreducible while every finitely generated subgroup of* H *is reducible.* □

In 1.1.8 the groups G and H can be chosen to be periodic or torsion-free. For a proof of 1.1.8 see 1.4.24 below.

If F is a field and D is a finite-dimensional F-algebra then there is no problem. For if G is a subgroup of GL(n,D) there is a finite subset X_o of G such that every element of G is an F-linear combination of elements of X_o. Set $X = \langle X_o \rangle$. Then if G is respectively irreducible, reducible or completely reducible (as a subset of GL(n,D)) then so is X, and if X is irreducible (resp. reducible, completely reducible) and if $X \leq Y \leq G$ then Y is irreducible etc, since F[G] = F[X] = F[Y]. It is tempting to try and repeat the above argument for any division ring D using linear dependence over D instead of over F, but 1.1.8 has already ruled that out. Some weakened versions, however, exist and we conclude this section with an account of them. Before that, however, we make a general but very useful remark.

1.1.9. *The matrix ring* $D^{n\times n}$ *containes at most* n *non-zero pairwise orthogonal idempotents.*

Proof: Let E be a (possibly infinite) set of non-zero pairwise orthogonal idempotents of $D^{n\times n}$. Let $V = D^n$. It is easy to see that $\Sigma_{e\in E}\, Ve = \oplus_{e\in E}\, Ve \subseteq V$, and this decomposition is as left D-module. Clearly $e \neq 0$ implies that $Ve \neq 0$. Dimension theory now implies that the cardinality of E is at most n. □

1.1.10. *Let* F *be a field and* R *an* F-*algebra that is locally semisimple Artinian (meaning that every finite subset of* R *lies in a semisimple Artinian*

F-subalgebra of R). *If for some integer* n *there are at most* n *pairwise orthogonal idempotents in* R, *then* R *is semisimple Artinian.*

Proof: If X is any semisimple F-subalgebra of R then X contains at most n pairwise orthogonal idempotents. Thus X_X has composition length at most n. Suppose $M_0 < M_1 < \ldots < M_{n+1}$ is a chain of distinct right ideals of R. For $0 \leq i \leq n$ pick $x_i \in M_{i+1} \setminus M_i$. By hypothesis there exists a semisimple F-subalgebra X of R containing $\{x_0, \ldots, x_n\}$. It clearly follows that $M_0 \cap X < M_1 \cap X < \ldots < M_{n+1} \cap X$, which contradicts the above. Thus R is certainly Artinian. If $\underline{n}(R)$ denotes the nilradical of R then $\underline{n}(R) \cap X \subseteq \underline{n}(X) = \langle 0 \rangle$ for any semisimple subalgebra X of R, and so $\underline{n}(R) = \langle 0 \rangle$. Therefore R is semisimple. □

1.1.11. (Wehrfritz [13]). *Let* G *be a subgroup of* GL(n,D) *and set* $V = D^n$.

a) *If* G *is completely reducible then there exists a finitely generated subgroup* X *of* G *such that* Y *is completely reducible for every finitely generated subgroup* Y *of* G *containing* X.

b) *If* G *is irreducible then there is a finitely generated subgroup* X *of* G *such that for every finitely generated subgroup* Y *of* G *containing* X *the* D-Y-*bimodule* V *is completely reducible and homogeneous, and every irreducible* D-Y *submodule of* V *is* D-X *irreducible.*

The converse of Part a) is false, even in the locally finite-dimensional case; the group G of 1.1.8 shows this. Trivially the converse of Part b) is false, even if G is linear, as is shown by the group of scalar matrices for $n > 1$. The group H of 1.1.8 shows that we cannot strengthen Part b) to make X (and hence Y) irreducible, even when D is locally finite-dimensional.

Proof: Assume first that G is irreducible. Pick a finitely generated subgroup X_0 of G such that the D-X_0 composition length of $V = D^n$ is minimal. Let Y be a finitely generated subgroup of G containing X_0. Then any D-Y composition series of V, being also a series of D-X_0 submodules, is also a D-X_0 composition series for V. In particular the irreducible D-Y submodules of V are also D-X_0 irreducible. Let S_Y denote the D-Y-socle of V. The above shows that $S_Y \subseteq S_{X_0}$. Of all possible X_0

choose one X with $\dim_D S_X$ minimal. Then $S_X = S_Y$ for every finitely generated $Y \supseteq X$. Thus S_X is actually a D-G submodule of V and so $S_X = V$. Therefore for any finitely generated subgroup $Y \supseteq X$ we have $V = S_Y$; that is, Y is completely reducible. Clearly the result remains true when G is completely reducible. Part a), therefore, is established.

Finally let $X \leq Y \leq Z \leq G$ with Y, Z finitely generated. Suppose $V = V_1 \oplus ... \oplus V_r$, where each V_i is D-Z irreducible. By the minimality of the composition length of V as D-X bimodule, each V_i is D-X, and hence D-Y, irreducible. If H is any non-trivial homogeneous D-Y component of V then H is a sum of certain of the V_i and as such is a Z-module. This is for all such Z, so H is a non-trivial D-G submodule of V. Therefore H = V and V is D-Y homogeneous. Part b) follows. □

1.1.12. (Wehrfritz [13]). *Let* F *be a field,* D *a locally finite-dimensional division* F-*algebra and* G *a subgroup of* GL(n,D). *Set* $R = F[G] \subseteq D^{n\times n}$.

a) *If* G *is completely reducible then* R *is semisimple Artinian.*

b) *If* G *is irreducible then* R *is simple Artinian.*

If G is the group of 1.1.8, then by 1.1.12 a) and 1.1.10 the ring R is semisimple Artinian. Thus the converse of 1.1.12 a) is false. In fact the ring R is simple. For if e is a central idempotent of R then $V = D^2 = Ve \oplus V(1-e)$, and so if R is not simple then G is completely reducible, which is not true. Hence R is simple, which shows that the converse of 1.1.12 b) is also false. A less interesting counter-example to the converse of 1.1.12 b) is furnished by the group $G = \langle \operatorname{diag}(\alpha,1) : \alpha \in D^* \rangle$ for any division ring D. For $R \cong D$ is simple and yet G is reducible.

Note also that 1.1.12 is not true for division rings in general. For suppose that D is a division ring with a central subfield F and that $x \in D$ is not algebraic over F. Set $G = \langle x \rangle \leq D^* = GL(1,D)$. Then G is clearly irreducible and yet $F[G] \cong F[x,x^{-1}]$ is not Artinian.

Proof: a) First assume that G is finitely generated and completely reducible. Then R has finite dimension over F. Let W be an irreducible D-G submodule of $V = D^n$ and pick $w \in W\backslash\langle 0\rangle$. Then wR contains an irreducible R-module W_1, say.

Clearly $W = DW_1$ and $dW_1 \cong_R W_1$ for all $d \in D^*$. Thus W is completely reducible as R-module. But V is D-G completely reducible, so V is completely reducible as R-module. Thus so too is $D^{n\times n}$ as a right R-module, and it follows that R is semisimple Artinian.

In the general case choose the finitely generated subgroup X of G as in 1.1.11 a). Then by what we have already proved the set $\langle F[Y] : X \leq Y \leq G$ with Y finitely generated $\rangle$ is a local system of semisimple Artinian F-subalgebras of R. By 1.1.9 the ring R contains at most n orthogonal idempotents, so R is semisimple Artinian by 1.1.10. Part a) is, therefore, established.

b) By Part a) R is semisimple Artinian, so $V = D^n$ contains an irreducible R-module W, say. Clearly $V = DW$ and $dW \cong_R W$ for every $d \in D^*$. Thus V is homogeneous and faithful as R-module. It follows that R is simple. □

In the opposite direction we have.

1.1.13. (Wehrfritz [13]). *Let* F *be a field,* D *a division* F*-algebra and* G *a subgroup of* GL(n,D). *Set* $R = F[G] \subseteq D^{n\times n}$.

a) *If* R *is semiprime (e.g. if* R *is semisimple Artinian) then* G *is isomorphic to a completely reducible subgroup of* GL(n,D).

b) *If* R *is simple Artinian then for some* $m \leq n$ *the group* G *is isomorphic to an irreducible subgroup of* GL(m,D).

Proof: a) Let $\langle 0\rangle = V_0 < V_1 < \ldots < V_r = V$ be a D-G composition series of $V = D^n$. Set $H = \bigcap_i C_G(V_i/V_{i-1})$. As in 1.1.2 the group G/H is isomorphic to a completely reducible subgroup of GL(n,D). Also $H \subseteq 1 + \underline{n}$ where $\underline{n} = \langle x \in R : V_i x \subseteq V_{i-1}$ for each i $\rangle$. But $\underline{n}^r = \langle 0\rangle$ and so $\underline{n}$ is a nilpotent ideal of the semiprime ring R. Thus $\underline{n} = \langle 0\rangle$ and Part a) follows.

b) Let W be an irreducible D-G submodule of V. Since R is Artinian W contains an irreducible R submodule W_1, say. The simplicity of R implies that W_1, and thus also W, is a faithful R-module. The result follows with $m = \dim_D W$. □

As a weak generalization of 1.1.12 and a converse of 1.1.13 we have the following .

1.1.14. *Let* F, D, G *and* R *be as in* 1.1.13.

a) *If* G *is completely reducible then* R *is semiprime; more generally so is* F[N] *for every subnormal subgroup* N *of* G.

b) *If* G *is irreducible then* R *is prime.*

c) *If* G *is locally finite and completely reducible then* R *is semisimple Artinian.*

Proof: Let $V = D^n$. Since F is central the D-G and D-R submodules of V coincide.

a) Let $\underline{a}$ be an ideal of R with $\underline{a}^2 = \langle 0 \rangle$. By hypothesis $V = V\underline{a} \oplus W$ for some D-G submodule W of V. But then

$$V\underline{a} = V\underline{a}^2 \oplus W\underline{a} = W\underline{a} \subseteq V\underline{a} \cap W = \langle 0 \rangle$$

Consequently $\underline{a} = \langle 0 \rangle$ and R is semiprime. By Clifford's Theorem (1.1.7), N is also completely reducible, so F[N] is also semiprime.

b) Suppose $\underline{a}$ and $\underline{b}$ are ideals of R with $\underline{ab} = 0$. Then $V\underline{a}$ is a D-G submodule of V, and so is either V or $\langle 0 \rangle$. If $V\underline{a} = V$ then $V\underline{b} \subseteq V\underline{ab} = \{0\}$ and so $\underline{b} = \{0\}$. If $V\underline{a} = \{0\}$ then $\underline{a} = \{0\}$. Therefore R is prime.

c) By 1.1.11 there is a finite subgroup X of G such that every finite subgroup Y of G containing X is completely reducible. Then the subalgebra F[Y] of R is semiprime by Part a) and finite-dimensional. Thus each F[Y] is semisimple Artinian, and consequently so is R, by 1.1.9 and 1.1.10. □

Occasionally we need to deal with subrings of $D^{n \times n}$ which are not of the form F[G] for some $G \leq GL(n,D)$. They will, however, be normalized by G. The most common type to arise has the form K[N] where K is some non-central subfield of $D^{n \times n}$ normalized by G and N is a normal subgroup of G. Then following result will then be helpful.

1.1.15. *Let* G *be an irreducible subgroup of* GL(n,D) *and* S *a subring of* $D^{n \times n}$ *normalized by* G. *Then* S *is semiprime.*

S need not be prime, for let G be the full monomial group for $n > 1$ and S the subring of all diagonal matrices. Also if G is only completely reducible then S need not be semiprime, for example if $G = \langle 1 \rangle$ and S is the subring of lower triangular matrices for $n > 1$. (In the latter case S is semiprime if S = F[N] for some N normal in G, by 1.1.14.)

Proof: Let $\underline{n}$ be the upper nilradical of S and $V = D^n$. Then $\underline{n}$ is nilpotent (Chatters & Hajarnavis [1], 1.35, or see 1.3.9) and normalized by G. Suppose $\underline{n}^r \neq \underline{n}^{r+1} = \{0\}$. Then $V\underline{n}^r$ is a non-zero D-G submodule of V and hence is V. But then $\underline{n}$ annihilates V and so $\underline{n} = \{0\}$, as required. □

1.2 ABSOLUTE IRREDUCIBILITY

Suppose G is a subgroup of GL(n,F) for some field F. Then G is absolutely irreducible if one of the following hold: G is irreducible over the algebraic closure of F ; $F[G] = F^{n\times n}$; G is irreducible over every extension field of F; $C_{F^{n\times n}}(G) = F1_n$ (Wehrfritz [2], 1.18). The concept of absolute irreducibility for skew linear groups is much more important than that of irreducibility but it is perhaps not obvious what the correct definition should be.

Let D be a division ring with centre F. A subgroup G of GL(n,D) is ***absolutely irreducible over*** D if $F[G] = D^{n\times n}$. This clearly extends the definition for linear groups. Also the following is obvious.

1.2.1. *If* $G \leq GL(n,D)$ *is absolutely irreducible over* D *then* G *is irreducible over* D. □

Suppose that G is an absolutely irreducible subgroup of GL(n,D). Let F' be any extension field of F. Then $R' = F' \otimes_F D^{n\times n}$ is a simple F'-algebra (Cohn [2], p. 364). If $\dim_F D$ is finite then clearly $\dim_{F'} R'$ is finite and R' is Artinian. If $\dim_F F'$ is finite then R' is finitely generated as right $D^{n\times n}$-module and again R' is Artinian. If R' is Artinian then R' is a matrix ring of degree n', say, over some division F'-algebra D' by the Artin-Wedderburn theorem. Also clearly $G \cong 1 \otimes G \leq R'$ and $F'[G] = R'$. Hence we have the following result.

1.2.2. *Let* G *be an absolutely irreducible subgroup of* GL(n,D) *and let* F' *be an extension field of* F *with either* $\dim_F D$ *or* $\dim_F F'$ *finite. Then there is a natural way of regarding* G *as an absolutely irreducible subgroup of* $GL(n',D')$ *for some integer* n' *and division* F'*-algebra* D'. □

The proof of the following is obvious.

1.2.3. *Let* G *be a subgroup of* GL(n,D). *If* G *is absolutely irreducible then* $C_{D^{n\times n}}(G) = F1_n$. □

What sort of possible converses are there to 1.2.1? If G is any subgroup of D^* then G is irreducible over D and clearly G is seldom absolutely irreducible. The following is an immediate consequence of 1.1.12 and the Artin-Wedderburn Theorem.

1.2.4. *Let* D *be locally finite-dimensional over* F *and let* G *be a subgroup of* GL(n,D).

a) *If* G *is completely reducible then* G *is a subdirect product of a finite number of absolutely irreducible skew linear groups over locally finite-dimensional division* F*-algebras.*

b) *If* G *is irreducible then* G *is isomorphic to an absolutely irreducible skew linear group over a locally finite-dimensional division* F*-algebra.* □

We now consider analogues of Clifford's theorem. The following is a special case of 1.1.15.

1.2.5. *Let* G *be an absolutely irreducible subgroup of* GL(n,D), *let* N *be a normal subgroup of* G *and* K *a subring of* $R = D^{n\times n}$ *normalized by* G. *Then* S = K[N] *is semiprime.* □

Ideally in 1.2.5 we would like S to be semisimple Artinian, at least if K is. In general this is quite impossible. For example let F be any field, let R = F(X) be the field of rational functions in some set of variables X and set $G = R^*$, K = F and N = ⟨X⟩. Then S = K[N] is the group algebra over F of the free abelian group N. We can, however, show that S is semisimple Artinian in two special cases, namely 1.2.6 b) and 1.2.12 below.

1.2.6. a) *Let* S *be a semiprime locally Artinian ring with only a finite number of primitive pairwise orthogonal idempotents. Then* S *is semisimple Artinian.*

b) *Assume the notation of* 1.2.5. *If* N *is locally finite and* K *is Artinian then* S *is semisimple Artinian.*

(If K is central then 1.2.6 b) follows from 1.1.7, 1.1.14 c), and 1.2.1.)

Proof: Let $e_1,\dots,e_m$ be a maximal set of pairwise orthogonal idempotents of S. Suppose that $X \neq e_jS$ is an S-submodule of e_jS and pick $x \in X$ and $y \in e_jS\backslash X$. There exists an Artinian subring T of S containing $x, y, e_1,\dots, e_m$. Idempotents lift over the radical J of T (Passman [2], 2.3.7). Then $T = \oplus e_iT$, $J = \oplus e_iJ$, each e_iT/e_iJ is T-irreducible and each e_iJ is nilpotent and is the maximal submodule of e_iT (Passman [2], 2.4.10). But $y \in e_jT\backslash X$, so $e_jT \cap X$ is nilpotent. It follows that X is a nil right ideal of S, which is semiprime by hypothesis. Therefore $X = (0)$ by Chatters and Hajarnavis [1], 1.35, or see 1.3.9 for a more general result. Hence each e_jS is irreducible and S is semisimple Artinian.

b) This follows immediately from 1.1.9, 1.2.5, and Part a). □

Before we can prove a dual result we need some preliminary lemmas. The next three results hold for a finite normalizing extension R of a ring S, although we only prove them in the following special situation.

1.2.7. *Let* R *be a ring,* S *a subring of* R, G *a subgroup of the group of units of* R *normalizing* S *such that* R = S[G], *and* H *a normal subgroup of finite index in* G *such that* $H \subseteq S \cap G$.

1.2.8. (Formanek & Jategaonkar [1]). *Assume the notation of* 1.2.7 *and let* M *be a non-trivial right* R*-module.*

a) *If* N *is an essential* S*-submodule of* M *then* N *contains a non-zero* R*-submodule of* M.

b) M *is* R*-Noetherian if and only if* M *is* S*-Noetherian.*

c) M *has an* R*-composition series if and only if* M *has an* S*-composition series.*

d) R *is right Artinian (resp. right Noetherian) if and only if* S *is.*

The main part for us is d) and if the reader is prepared to accept this he can skip the proof below without harm.

Proof: a) Let $g \in G$. The mapping $x \mapsto xg$ of M into itself permutes the S-submodules of M since $Sg = gS$. Thus Ng is also an essential S-submodule of M. If T is a transversal of H to G then $N_o = \bigcap_{x \in T} Nx$ is non-zero and a G-submodule of M. The result follows.

b) Let M be R-Noetherian, and assume that M is not S-Noetherian. Pick an R-submodule N of M maximal subject to M/N not being S-Noetherian. Replacing M by M/N we may assume that every proper R-image of M is S-Noetherian.

Suppose $K = \oplus_{i=1}^{\infty} K_i \leq M$ where each K_i is a non-zero S-module. By Zorn's lemma there is an S-submodule L of M such that $K \oplus L$ is essential. By Part a) there exists $u \neq 0$ with $uR \subseteq K \oplus L$. Now R is finitely S-generated and hence so is uR. Thus $u \in (\oplus_{i \leq m} K_i) \oplus L$ for some m and so $\oplus_{i > m} K_i$ is isomorphic to an S-submodule of M/uR, which by assumption is S-Noetherian. This contradiction shows that M has finite uniform dimension as S-module.

Let $M_1 \subseteq M_2 \subseteq ..$ be an ascending chain of S-submodules. By the previous paragraph there exists n such that each M_i for $i \geq n$ has the same uniform S-dimension, s say. By Zorn's lemma there exists an S-submodule L of M with $M_n \oplus L$ essential in M. If for $i > n$ we have $M_i \cap L \neq \langle 0 \rangle$ then M_i has S-dimension at least s+1, so $M_i \cap L = \langle 0 \rangle$. By Part a) and our choice of M, $M/(M_n \oplus L)$ is S-Noetherian. Thus for some $i \geq n$ we have $M_i \oplus L = M_j \oplus L$ for all $j \geq i$. By the modular law $M_j = M_i + (M_j \cap L) = M_i$ and the result follows. The converse is trivial.

c) If M has an S-composition series it is Noetherian and Artinian as S-module and thus as R-module. Thus M has an R-composition series. Conversely suppose M is R-irreducible and let N be an S-submodule of M. By Zorn's lemma there exists an S-submodule L of M with $N \oplus L$ essential in M. By Part a) and the irreducibility of M we have $N \oplus L = M$. Thus M is completely reducible as S-module. By Part b) we have that M is S-Noetherian so M has only finitely many S-irreducible factors.

d) If R_R is Artinian it has an R-composition series (Kaplansky [1], Theorem 34). Hence R_S, and so S_S, has an S-composition series by Part c). That is, S is right Artinian. If R_R is Noetherian then R_S and hence S_S is Noetherian by Part b). The reverse implications are easy. □

1.2.9. (Lorenz [1]). *Assume the notation of* 1.2.7. *Then there exists an integer* $t \geq 1$ *such that for any right ideal* A *of* S *and elements* $r_1, \ldots, r_t$ *of*

AR *we have* $r_1 \dots r_t = \Sigma w_j$, *where each* w_j *is a product of elements of* $\{ r_1, \dots, r_t \} \cup A$ *with at least one factor from* A.

This is a special case of Theorem 1.3 of Lorenz [1].

1.2.10. (Snider [3]). *Assume the notation of* 1.2.7. *If* $x \in S$ *is such that* xR *is an* R-*direct summand of* R, *then* xS *is an* S-*direct summand of* S.

Proof: Let $R = xR \oplus K$ as R-modules, and let e be the component of 1 in xR. Then e is an idempotent, $xR = eR$ and $ey = y$ for all $y \in xR$. Apply 1.2.9 with $A = xS$ and $r_1 = \dots = r_t = e$. Then for some $t \geq 1$ we have $e = e^t = \Sigma w_j$ where each w_j is a product of e's and at least one element of xS. Since $ey = y$ for all $y \in xS$ we have $e = ee = ae$ for some $a \in xS$. Now $a = ea = aea = a^2$, so a is an idempotent and $S = aS \oplus (1-a)S$. Also $x = ex = aex = ax$, whence $xS \subseteq aS \subseteq xS$ and the result follows. □

1.2.11. *Let* R *be a ring such that* xR *is a direct summand of* R *for every* $x \in R$. *If* R *has either the maximal or the minimal condition on right annihilators then* R *is right Noetherian.*

Proof: Let M be a non-zero right ideal of R. Pick $x_1 \in M \backslash \langle 0 \rangle$. Then $R = x_1R \oplus K_1$ for some K_1. Set $L_1 = K_1$. Then $M = x_1R \oplus (L_1 \cap M)$. If possible pick $x_2 \in L_1 \cap M \backslash \langle 0 \rangle$. Then $R = x_2R \oplus K_2$ for some K_2, $L_1 = x_2R \oplus L_2$ for $L_2 = K_2 \cap L_1$, and $M = x_1R \oplus x_2R \oplus (L_2 \cap M)$. Repeat the process by choosing $x_3 \in L_2 \cap M \backslash \langle 0 \rangle$, and keep going. If $R = X \oplus Y$ as right R-modules and if $1 = e + f$ for $e \in X$ and $f \in Y$ then $X = r(f)$. Thus $\langle x_1R \oplus \dots \oplus x_iR \rangle$ is an ascending and $\langle L_i \rangle$ a descending chain of right annihilators. By hypothesis one of these chains terminates. Thus $L_n \cap M = \langle 0 \rangle$ for some n and $M = x_1R \oplus \dots \oplus x_nR$ is finitely generated. Therefore R is right Noetherian. □

The following "dual" of 1.2.6 b) generalizes the proposition of Wehrfritz [18].

1.2.12. THEOREM. (Snider [3]). *Assume the notation of* 1.2.5 *with* $F \subseteq K$. *If* G/N *is locally finite then* S *is semisimple Artinian.*

Proof: Let $x \in S$. Then xR is a direct summand of $R = K[G]$ and $xR = eR$ for some idempotent e. Choose y, z in R with $xy = e$ and $ez = x$. There exists a subgroup $G_1 \supseteq N$ of G with G_1/N finite and $y, z \in R_1 = K[G_1]$. Then $xR_1 = eR_1$ and since e is an idempotent, $R_1 = xR_1 \oplus (1-e)R_1$. By 1.2.10 it follows that xS is a direct summand of S. Trivially R, and hence S, has the maximal (and the minimal) condition on right annihilators. Therefore S is right Noetherian by 1.2.11.

Let E be the set of matrix units of R, so $D = C_R(E)$. There exists a finite subgroup G_2/N of G/N such that $E \subseteq R_2 = K[G_2]$. Suppose we can prove that R_2 is simple Artinian. Then S is Artinian by 1.2.8 d) and semiprime by 1.2.5, so S will be semisimple Artinian.

By Passman [2], 6.1.5 we have $R_2 = EC \cong C^{n\times n}$, where $C = C_{R_2}(E) \subseteq D$. Let $x \in C\backslash\{0\}$. There exists a finite subgroup G_x/N of G/N containing G_2/N with $x^{-1} \in K[G_x]$. Now $R_x = K[G_x]$ and R_2 are both finitely generated right S-modules and S is right Noetherian, so R_x is a right Noetherian R_2-module. As $x \in R_2$ and $x^{-1} \in R_x$, we have $R_2 \subseteq x^{-1}R_2 \subseteq \ldots \subseteq x^{-i}R_2 \subseteq \ldots \subseteq R_x$. Consequently $x^{-i}R_2 = x^{-i-1}R_2$ for some i and thus $x^{-1} \in R_2$. It follows that $x^{-1} \in R_2 \cap D = C$, so C is a division ring. Therefore R_2 is simple Artinian, as required. □

Of course in 1.2.6 b) and 1.2.12 there is no need for S to be simple Artinian, any more than a normal subgroup of an irreducible linear group need be irreducible. However a (normal) subgroup of finite index of a connected irreducible linear group is always irreducible and there is an analogue of this for skew linear groups which we shall find useful in Chapter 5.

Let G be an absolutely irreducible subgroup of GL(n,D). Say that G is ***persistent*** (short for persistently of simple Artinian type) if the F-subalgebra F[H] of $D^{n\times n}$ is simple Artinian for every subgroup H of G of finite index. Say that G is ***prime*** if $F[N] \leq D^{n\times n}$ is a prime ring for every normal subgroup N of G. (This terminology should probably be regarded as provisional. A number of minor, and for our purposes irrelevant, variations could be made, e.g. the absolute irreducibility condition could be weakened. Alternatively one might introduce notions such as subnormally (or normally) persistent by requiring H above to be subnormal (or normal) respectively. At this stage of the development of the subject of skew linear groups, the most useful formulation is not apparent.)

1.2.13. *Let* G *be an absolutely irreducible subgroup of* GL(n,D). *Then* G *has a normal subgroup* H *of finite index such that* H *is a subdirect product of groups* H_i, *where for each* i, H_i *is a persistent subgroup of* $GL(n_i,D_i)$ *for some integers* n_i *satisfying* $\Sigma\ n_i \leq n$ *and division* F*-algebras* D_i.

Proof: Let H_o be a subgroup of G of finite index such that the composition length of $V = D^n$ as D-H_o bimodule is maximal, and set $H = \bigcap_{x\in G} H_o^x$. Clearly H is normal of finite index in G. By 1.2.1 and 1.1.6 the D-H bimodule V is completely reducible, say $V = \oplus V_i$, where each V_i is D-H irreducible. Let $\pi_i : F[H] \longrightarrow End_D(V_i)$ be the obvious map, and set $H_i = H\pi_i$.

Each V_i is D-H irreducible and hence is D-H_i irreducible. Further F[H] is Artinian by 1.2.8 and hence $F[H_i] = F[H]\pi_i$ is also Artinian. Let U_i be an irreducible $F[H_i]$-submodule of V_i ; such must exist. Then $V_i = DU_i$ and $dU_i \cong_{F[H_i]} U_i$ for every $d \in D^*$. Thus ir fact $F[H_i]$ is primitive and hence simple Artinian, and is therefore a matrix ring of degree n_i say over some division F-algebra D_i. After the obvious identifications H_i is an absolutely irrreducible subgroup of $GL(n_i,D_i)$. Also $n_i \leq \dim_D V_i$ by 1.1.9 and so $\Sigma\ n_i \leq n$.

Let K be any subgroup of H_i of finite index. It remains only to show that $F_i[K]$ is simple Artinian, where F_i is the centre of D_i. By 1.2.8 the ring $F_i[K]$ is Artinian. By the choice of H_o we have that V_i is D-$K\pi_i^{-1}$ irreducible. Let W be an irreducible $F_i[K]$-submodule of V_i. Then V_i = DW and $dW \cong_{F_i[K]} W$ for all $d \in D^*$. Hence $F_i[K]$ is simple Artinian, as required. □

1.2.14. *Let* G *be an absolutely irreducible subgroup of* GL(n,D). *Then the following are equivalent:*

a) G *is prime.*

b) $F[H] \leq D^{n\times n}$ *is prime for every normal subgroup* H *of* G *of finite index.*

c) F[H] *is simple Artinian for every* H *as in* b), *i.e.* G *is normally persistent.*

Proof: Trivially a) implies b). If H is a normal subgroup of G of finite index then F[H] is semisimple Artinian by 1.2.8 and 1.2.5. Thus b) implies c). Assume that c) holds and let N be a normal subgroup of G. Then

F[N] is semiprime by 1.2.5. Also F[N] satisfies the maximal condition on right annihilators, since F[G] does, and hence F[N] has only a finite number of minimal prime ideals, $P_1,\ldots,P_n$ say, by Chatters & Hajarnavis [1], 1.16, and each P_i is an annihilator ideal. Clearly G permutes the P_i, so $H = \bigcap_i N_G(P_i)$ is a normal subgroup of G of finite index. By hypothesis F[H] is simple, and yet P_1H is an ideal of F[H]. If $P_1H = F[H]$ then $l_{F[N]}(P_1) = \{0\}$ and hence $P_1 = r_{F[N]}\ l_{F[N]}(P_1) = F[N]$. This is false. Therefore $P_1 = \{0\}$ and F[N] is prime. □

As an immediate consequence of 1.2.14 and the definition of persistent we have the following result.

1.2.15. *A persistent subgroup of* GL(n,D) *is prime.* □

Presumably not every absolutely irreducible prime group is subnormally persistent. If H is any subgroup of finite index of the subnormally persistent subgroup G of GL(n,D), then F[H] is always Artinian but need not be semisimple (Exercise). It is not known whether normal subgroups of prime groups are again prime.

1.3 UNIPOTENCE

An element g of a ring is called *unipotent* if g-1 is nilpotent. It follows that g is a unit, its inverse being $\Sigma_i\ (1-g)^i$. Let D be a division ring. A subgroup of GL(n,D) is unipotent if each of its elements is unipotent in the ring $D^{n\times n}$. A fundamental result in the theory of linear groups is the equivalence of the notions of unipotent and stability subgroups. Although the equivalence is unproved in general for skew linear groups, so many special cases are known that for practical purposes the equivalence holds. The following result is obvious.

1.3.1. *Every stability subgroup of* GL(n,D) *is unipotent.* □

1.3.2. *Let* $x \in$ GL(n,D). *The following are equivalent.*

a) x *is unipotent.*

b) $\langle x\rangle$ *is a stability subgroup of* GL(n,D).

c) $(x-1)^n = 0$.

Let charD = p ≥ 0. *If* p = 0 *and* x *is unipotent then* ⟨x⟩ *is torsion-free. If* p > 0 *then* a), b), c) *are equivalent to*

d) x *is a* p-*element.*

In case d) *the order of* x *is at most* $p^{-[-\log_p n]}$.

Proof:

a) implies b): If $(x-1)^m = 0$ then x stabilizes the series

$$D^n \supseteq D^n(x-1) \supseteq D^n(x-1)^2 \supseteq \ldots \supseteq D^n(x-1)^m = \langle 0 \rangle.$$

Thus b) holds.

b) implies c): If x stabilizes the series

$$D^n = V_0 \supseteq V_1 \supseteq \ldots \supseteq V_m = \langle 0 \rangle \qquad (*)$$

of left D-modules, where the V_i are distinct, then $m \leq \dim_D D^n = n$. Also $D^n(x-1)^m = \langle 0 \rangle$, so c) holds.

c) implies a) is trivial.

If (*) is a series of left D-modules stabilized by x then each V_i/V_{i+1} is torsion-free if p = 0 and of exponent p if p > 0. By elementary stability theory (Kegel & Wehrfritz [1], 1.C) ⟨x⟩ is torsion-free in the first case and has order dividing p^m in the second. Now suppose x is an element of order p^m, say. Then $(x-1)^{p^m} = 0$ and x is unipotent. Thus $(x-1)^n = 0$ by Part c), so $(x-1)^{p^k} = 0$ for any $p^k \geq n$, and $x^{p^k} = 1$. The result follows. □

In the following result, and indeed in much of this section, we make use of Hall's calculus of group classes. A list of the symbols we use can be found in the Notation Index. A detailed account of this calculus is given in the opening pages of Volume 1 of Robinson [1]. At this point we use the local, quotient, and poly-normal operators L, Q, and P_n.

1.3.3. *Let* U̲ *be the class of all groups such that every unipotent skew linear* U̲*-group is a stability subgroup. Then* U̲ *is* L*-closed and* $\underline{U}^Q$ *is* $\langle L, P_n \rangle$*-closed.*

Proof: Let G be a unipotent subgroup of GL(n,D). Assume that G ∈ LU̲. For V = D^n and H a finitely generated subgroup of G there exists a U̲-subgroup of G containing H and so we have

$$V \supseteq [V,H] \supseteq \ldots \supseteq [V,_nH] = \{0\}.$$

Thus $[V,_nG] = \{0\}$, as required. We have therefore shown that LU̲ ⊆ U̲, and

$QL \leq LQ$ implies that $QL\underline{U}^Q \leq LQ\underline{U}^Q \leq L\underline{U} \leq \underline{U}$, whence $L\underline{U}^Q \subseteq \underline{U}^Q$.

Suppose now that G is a $P_n\underline{U}^Q$-group. We induct on $n = \dim_D V$. If $n = 1$ then clearly $G = \langle 1 \rangle$. Suppose that $n > 1$ and $G \neq \langle 1 \rangle$. Then G has a non-trivial normal $\underline{U}$-subgroup N. Therefore $W = [V,N]$ is a proper D-G submodule of V. Also $P_n\underline{U}^Q$ is Q-closed. Thus by induction on n, G stabilizes a series of left D-submodules in W and V/W and so G is a stability subgroup of GL(n,D). □

1.3.4. $\langle L,P_n \rangle\, \underline{A}\, \underline{F} \subseteq \underline{U}$, *where* $\underline{A}$ *is the class of abelian groups and* $\underline{F}$ *that of finite groups. In particular* $\underline{U}$ *contains the classes of locally soluble groups, locally finite groups, and hyperabelian groups.*

Proof: $\underline{A}$ and $\underline{F}$ are both Q-closed, and $\underline{A} \subseteq LP_n\underline{G}_1$, where $\underline{G}_1$ is the class of cyclic groups. Now cyclic unipotent groups are stability groups by 1.3.2. In view of 1.3.3 it remains only to prove that $\underline{F} \subseteq \underline{U}$. Thus let G be a finite unipotent subgroup of GL(n,D). If F is the centre of D then $R = F[G]$ has finite dimension over F. By the linear case G acts as a stability subgroup on R. Let $\mathfrak{g}$ be the image of the augmentation ideal of FG in R. Then $\mathfrak{g}$ is nilpotent and G stabilizes $D^n \supseteq D^n\mathfrak{g} \supseteq D^n\mathfrak{g}^2 \supseteq \ldots$, as required. □

As an immediate corollary we obtain the following.

1.3.5. *If* D *is locally finite-dimensional over its centre and* G *is a unipotent subgroup of* GL(n,D) *then* G *is a stability group.*

Proof: G is locally linear and so locally nilpotent. The result now follows from 1.3.4. □

1.3.6. *If* G *is a unipotent subgroup of* GL(2,D) *then* G *is abelian and a stability subgroup.*

Proof: Let $x, y \in G\backslash\langle 1 \rangle$, with u,v eigenvectors of x, y respectively in $V = D^2$. If $v \notin Du$ then u,v is a left basis for V. Relative to this basis x and y have coordinate matrices

$$X = \begin{bmatrix} 1 & 0 \\ \xi & 1 \end{bmatrix} \quad , \quad Y = \begin{bmatrix} 1 & \eta \\ 0 & 1 \end{bmatrix}$$

where η and ξ are non-zero. But XY-1 is nilpotent since xy is unipotent. Thus $\begin{bmatrix} 0 & \eta \\ \xi & \xi\eta \end{bmatrix}^2 = 0$ which clearly yields the contradiction $\eta\xi = 0$.

It follows that Du is a D-G submodule of V and G stabilizes the series $V \supseteq Du \supseteq \{0\}$. □

1.3.7. *If* G *is a unipotent subgroup of* GL(n,D) *where* p = charD $\geq$ 0 *and* p + n $\leq$ 6 *then* G *is a stability subgroup.*

Proof: If p = 5 then n = 1 and the result is trivial. If p is 2 or 3 then G has exponent at most 4. By a theorem of Sanov (Magnus, Karrass, and Solitar [1], 5.25) G is locally finite, and the result follows from 1.3.4.□

By direct calculation Suprunenko [1] has shown that a stability subgroup of GL(n,D) is unipotent whenever n $\leq$ 5 and charD is 2 or 3. This may seem very special, but in fact the problem only remains unsolved for small positive characteristics.

1.3.8. (Mochizuki [1]) *Let* G *be a unipotent subgroup of* GL(n,D) *where* charD = 0 *or* charD $\geq$ (n-1)(n-[n/2]). *Then* G *is a stability subgroup.*

We need the following two results.

1.3.9. (J. Levitzki) *Let* S *be a multiplicative semigroup of nilpotent elements of* $D^{n\times n}$. *Then* $S^{(n)} = \{0\}$, *where for each positive integer* r *we set* $S^{(r)} = S...S$, *the product of* r *copies of* S.

Proof: We induct on n, the case n = 1 being trivial. Clearly $S = \bigcup_T T$ where T ranges over all the finitely generated sub-semigroups of S. Hence we may assume that S is finitely generated as a semigroup, by non-zero elements $s_1,...s_r$, say.

If $V = D^n$ is D-S reducible, that is if V has a proper non-zero left D-submodule W with $WS \subseteq W$, then we may apply the inductive hypothesis to W and V/W. Thus assume that V is D-S irreducible. Then since $S = \bigcup Ss_i$ we have $V = \Sigma Vs_i$. Suppose we have $i_1,...i_k \in \langle 1,...,r\rangle$ such that $s_{i_1}...s_{i_k} \neq 0$. Then we can find $i_0 \in \langle 1,..., r\rangle$ with $s_{i_0}s_{i_1}...s_{i_r} \neq 0$. For otherwise $s_{i_1}\dots s_{i_k}$ will annihilate $V = \Sigma\, Vs_i$. Inductively we may find $r(n-1)+1$ subscripts $i_o,\dots,i_{r(n-1)} \in \langle 1,\dots, r \rangle$ such that $\Pi\, s_{i_j} \neq 0$. At least n of these subscripts must be equal. We have therefore constructed $u_1,\dots, u_n \in S$ with $u_1s_iu_2s_i\dots u_ns_i \neq 0$, for some i.

Now Ss_i is a sub-semigroup of S. Also s_i is non-zero and nilpotent, so $V \supset Vs_i \supset \langle 0\rangle$ is a series of Ss_i-submodules of V. Thus V is D-Ss_i reducible and induction yields that $(Ss_i)^{(n)} = \langle 0\rangle$. But this contradicts the above paragraph. We have therefore shown that V is D-S reducible, and the proof is complete. □

1.3.10. (Heineken [1], Bruck [1]). *Let* H *be a finitely generated multiplicative subgroup of an algebra over a prime field of characteristic* $p \geq 0$ *and* n *a positive integer such that* $(h-1)^n = 0$ *for every* h *in* H. *If* $p = 0$ *or* $p > (n-1)(n-[n/2])$ *then there exists an integer* k, *depending on* n *and* H, *such that* $(H-1)^{(k)} = \{0\}$, *the notation being as in* 1.3.9. □

The original result, proved by Heineken, has the bound $p > ((n-1)^2)!$. The above improvement is due to Bruck. We omit the proof since it is long, does not involve skew linear groups, and we make no use of 1.3.8 in what follows.

Proof of 1.3.8: Let S be the multiplicative sub-semigroup of $D^{n\times n}$ generated by G-1. If $s = (g_1-1)(g_2-1)\dots(g_r-1) \in S$, where the $g_i \in G$ then s is nilpotent by 1.3.10 applied to $\langle g_1,\dots, g_r \rangle$. Then $S^{(n)} = \{0\}$ by 1.3.9. Hence $[D^n,{}_nG] = D^n.S^{(n)} = \{0\}$, so G is a stability subgroup. □

Let G be any subgroup of GL(n,D). If G has a unique maximal unipotent normal subgroup we denote it, as in the linear case, by u(G). An elementary application of Zorn's Lemma shows that G always has maximal unipotent normal subgroups. The uniqueness is far from clear and is related to the questions considered above.

1.3.11. *Let* G *be any subgroup of* GL(n,D). *Then:*

a) G *has a unique maximal normal stability subgroup* s(G).

b) *There is an exact sequence*

$$\langle 1\rangle \longrightarrow s(G) \longrightarrow G \longrightarrow GL(n,D).$$

where the image $G\rho$ *of* G *in* $GL(n,D)$ *is completely reducible.*

c) *If either* D *is locally finite-dimensional over its centre or* $G \in \underline{U}^{S_n}$ (e.g. *if* $G \in \langle L, P_n \rangle \, \underline{A}\, \underline{F}$) *then* $u(G)$ *exists and is equal to* $s(G)$.

Proof: a) A subgroup X of G is a stability subgroup if and only if $[D^n,_n X] = \{0\}$. This is clearly an inductive condition, so by Zorn's Lemma G contains a maximal normal stability subgroup, Y say. (Exercise: Prove the existence of such a Y without using the axiom of choice.)

To show that Y in fact contains every normal stability subgroup of G it is evidently sufficient to prove the following: if X_1, X_2 are two normal stability subgroups of G, then X_1X_2 is also a stability subgroup. Now every D-X_2 composition factor of D^n is X_2-trivial. Hence every D-X_2 composition factor of $[D^n,_{i-1}X_1]/[D^n,_i X_1]$ is trivial. But each of these factors is a D-X_1X_2 composition factor, so X_1X_2 is a stability group. Part a) follows.

b) Here we use the construction of 1.1.2. Let $V = D^n$ and let (V_i) be a D-G composition series of V. Then $W = \oplus V_i/V_{i-1} \cong_D V$ and is D-G completely reducible. Thus we have an exact sequence

$$\langle 1 \rangle \longrightarrow S \longrightarrow G \longrightarrow GL(n,D)$$

where $S = \ker\rho = \{ x \in G : V_i x \subseteq V_{i-1} \text{ for each } i \}$. By Part a) we have $S \subseteq s(G)$. By Clifford's Theorem (1.1.6) the module V_i/V_{i-1} is D-$s(G)$ completely reducible. But the D-$s(G)$ composition factors of V are $s(G)$-trivial. Thus each V_i/V_{i-1} is $s(G)$-trivial, whence $s(G) \subseteq S$, as required.

c) If D is locally finite-dimensional over its centre then $s(G) = u(G)$ by 1.3.5. Let $G \in \underline{U}^{S_n}$. If X is any unipotent normal subgroup of G then $X \in \underline{U}$, and so $X \subseteq s(G)$ by definition of $\underline{U}$ and Part a). The result follows. □

Everything we have done so far in this section suggests that unipotent skew linear groups look very much like unipotent linear groups, except that the proofs are harder. This conclusion is, however, unwarranted. A unipotent linear group, like all linear groups, satisfies the maximal condition on centralizers. Thus, for example, the direct product of infinitely many copies of the lower triangular group $Tr_1(3,\mathbb{Z})$ is a torsion-free nilpotent group of class 2 which is not isomorphic to any linear group. In contrast we have the following result.

1.3.12. THEOREM. (Hartley & Menal [1]). *Let* G *be any torsion-free nilpotent group of class* c. *Then there exists a division ring* D *such that* G *is isomorphic to a subgroup of* $Tr_1(c+1,D)$. □

Note that necessarily charD = 0 (by 1.3.2). Thus the class of skew linear stability groups of degree n and characteristic zero coincides with the class of torsion-free nilpotent groups of class less than n. The positive characteristic case seems to be unknown.

1.4 CONSTRUCTIONS

Clearly every linear group is a skew linear group, but many skew linear groups are very different from linear groups. In this section we present some general techniques for constructing examples of such groups. Frequently the example will be a subgroup of D^*, that is, of GL(1,D), for some division ring D. There are two very commonly used constructions for building the division ring. The first involves Ore domains and the application of Goldie's Theorem, and the second uses ordered groups and rings of formal infinite series.

Let R be a ring, S a subring of R and G a group of units of R normalizing S such that R = S[G]. Suppose that $N = S \cap G$ is a normal subgroup of G and $R = \oplus_{t\in T} tS$, where T is some transversal of N to G. Set H = G/N. We summarize the above by saying that (R, S, G, H) is a *crossed product*. The classical crossed product (Jacobson [5], Vol. 2, Chapter 8) is a special case of this. Sometimes we say that R is a *crossed product of* S *by* H; by H note, not by G.

Let (R, S, G, H) be a crossed product. Since G normalizes S we also have $R = \oplus_{t\in T} St$. Further if U is another transversal of N to G then for each $t \in T$ there is a unique $u_t \in U$ and $n_t \in N$ with $t = u_t n_t$ and $R = \oplus u_t S$. If $x \in R$ then $x = \Sigma t\mu_t$ for unique elements μ_t of S, almost all of which are zero. We call the non-zero μ_t the coefficients of x. They depend on T and are determined only to within elements of N; explicitly $x = \Sigma\, u_t(n_t\mu_t)$ in the above notation. However statements such as "the coefficients of x are regular" are well-defined. (A *regular* element of a ring is a non zero-divisor). If x is as above then the *support* of x is defined to be

$$\text{supp } x = \{\ tN : \mu_t \neq 0\ \}.$$

1.4.1. (G. Higman). *Let (R, S, G, H) be a crossed product, where H is such that every non-trivial finitely generated subgroup of H has an infinite cyclic image. If x is an element of R , all of whose coefficients are regular in S , then x is regular in R . In particular if S is a division ring then R is a domain.*

Proof: Let $y \in R$ with $xy = 0$. We prove by induction on $d = |\mathrm{supp}\, x| + |\mathrm{supp}\, y|$ that $y = 0$. We can pre-multiply x and post-multiply y by an element of G without affecting the problem. That is, we may assume that $1 \in \mathrm{supp}\, x \cap \mathrm{supp}\, y$ and seek a contradiction. Let $K = \langle \mathrm{supp}\, x, \mathrm{supp}\, y \rangle$. If $K = \langle 1 \rangle$ then x is regular in S and $y = 0$. Otherwise K has a normal subgroup L such that K/L is infinite cyclic.

Let M be the inverse image of L in G with $\langle g \rangle M$ the inverse image of K in G. Then (S[M], S, M, M/N) is also a crossed product, as is $(S[\langle g \rangle M], S[M], \langle g \rangle, \langle g \rangle)$. Then $x = \Sigma_{i=1}^{r} g^{m_i} \mu_i$ and $y = \Sigma_{j=1}^{s} g^{n_j} \rho_j$, where $m_1 < m_2 < \ldots < m_r = m$, $n_1 < n_2 < \ldots < n_s = n$, $|\mathrm{supp}\, x| = \Sigma |\mathrm{supp}\, \mu_i|$ and $|\mathrm{supp}\, y| = \Sigma |\mathrm{supp}\, \rho_j|$. Now $0 = xy = g^{m+n}.\tilde{\mu}\rho$ + lower terms, where $\tilde{\mu} = g^{-n} \mu g^{n}$. Hence $\tilde{\mu}.\rho = 0$. The coefficients of $\tilde{\mu}$, as conjugates of the coefficients of x relative to suitable transversals, are regular, and $\rho \neq 0$. Since $\mathrm{supp}\, x \cup \mathrm{supp}\, y \not\subseteq L$ and $1 \in \mathrm{supp}\, x \cap \mathrm{supp}\, y$, $|\mathrm{supp}\, \tilde{\mu}| + |\mathrm{supp}\, \rho| < d$ and induction yields the contradiction $\rho = 0$. Therefore x is right regular. The left regularity of x may be established in the same way. The result follows. □

It is easy to construct crossed products. Let S be a ring and G a group such that $N = G \cap S$ is a subgroup of both G and the group of units of S (with the same operations). Suppose that there is an action of G on S (which we denote by $s \mapsto s^x$ for $s \in S$ and $x \in G$) that extends the action of N on S by conjugation and the action of G on N by conjugation (so N is normal in G). Then there is a unique crossed product (R,S,G,G/N) determined by this information.

For let T be a transversal of N to G with f the corresponding factor set, so $t_1 t_2 = t_3.f(t_1,t_2)$ where the $t_i \in T$. Let R be the free right S-module on T and define a multiplication on R by extending linearly the product

$$(t_1\alpha)(t_2\beta) = t_3.f(t_1,t_2)\,\alpha^{t_2}\,\beta$$

where t_1, t_2, $t_3 \in T$ and α, $\beta \in S$. It is elementary to check that R becomes a ring with subring S and subgroup $G = TN \subseteq R$ such that (R, S, G, G/N) is a crossed product. The uniqueness is also easy, and shows incidentally that the construction does not depend on the choice of T .

There is another construction of R. Replace G by an isomorphic copy of itself such that $S \cap N = \langle 1 \rangle$, so that now we have an embedding, say $n \mapsto n'$ of N into the units of S such that $\alpha^n = (n')^{-1} \alpha n'$ and $(n^x)' = (n')^x$ for $\alpha \in S$, $n \in N$ and $x \in G$. Let R_0 be the skew group ring SG; that is R_0 is the free right S-module on G with multiplication given by the multiplications on S and G and the rule $\alpha x = x\alpha^x$. Let $\underline{n} = \Sigma_{n \in N} (n-n')R_0$ Then $\underline{n}$ is an ideal of R_0 and $R_0/\underline{n}$ is isomorphic to the crossed product R constructed above.

We seldom have cause to deal with a general crossed product (R, S, G, H). In the constructions below S is a division ring and often a field. Other common restrictions are that S should be central or that G = H (the case of skew group rings).

A ***right*** (resp. ***left***) ***Ore domain*** is a ring R with no zero divisors such that $aR \cap bR \neq \langle 0 \rangle$ (resp. $Ra \cap Rb \neq \langle 0 \rangle$) for every pair of non-zero elements of R. An *Ore domain* is a domain that is right and left Ore. By Ore's theorem every right (left) Ore domain has a clasical ring of right (left) quotients that is a division ring, and for an Ore domain the two coincide (e.g. Passman [2], 4.4.2 and 4.4.3). From our point of view Ore domains are simply a convenient class of rings that we can guarantee embed into division rings. The result that makes everything work is the following special case of Goldie's theorem.

1.4.2. (A. W. Goldie) *Let* R *be a domain that is locally right (resp. left) Noetherian, meaning that every finite subset of* R *lies in a right (left) Noetherian subring of* R. *Then* R *is right (left) Ore.*

Proof: We prove the right-handed version. Let a and b be non-zero elements of R. There exists a right Noetherian subring S of R containing a and b. Then $abS + a^2bS + a^3bS + \ldots$ is finitely S-generated, so for some $m > n \geq 0$ and some elements $s_i \in S$ with $s_n \neq 0$ we have

$$a^m b = a^n bs_n + a^{n+1}bs_{n+1} + \ldots + a^{m-1}bs_{m-1}$$

Set $t = a^{m-n-1}b - a^{m-n-2}bs_{m-1} - \ldots -bs_{n+1}$, so $a^n(at-bs_n) = 0$. Since R is a domain and a, b, and s_n are non-zero we obtain $at = bs_n \neq 0$. Thus $aR \cap bR \neq (0)$, as required. □

Let $\underline{O}$ be the class of all groups H such that every crossed product of a division ring by H is an Ore domain.

1.4.3. *The infinite cyclic group lies in $\underline{O}$, and $\underline{O}$ is locally and poly closed. Consequently, $\langle P,L\rangle C_\infty \subseteq \underline{O}$.*

Proof: Let (R, E, G, H) be a crossed product with E a division ring and H infinite cyclic. Then R is a domain by 1.4.1 and is Noetherian by a theorem of P. Hall (Passman [2], 10.2.6). Thus R is an Ore domain by 1.4.2, and so C_∞ belongs to $\underline{O}$.

Trivially being an Ore domain is a local property, so $\underline{O}$ is locally closed. Let (R, E, G, H) be a crossed product where E is a division ring and there exists $E \cap G \subseteq K \lhd G$ such that $K/(E \cap G)$ and G/K are both $\underline{O}$-groups. Clearly E[K] is a crossed product of E by $K/(E \cap G)$ and is therefore an Ore domain by hypothesis. Thus E[K] has a division ring D of quotients. The action of G on E[K] extends to one of G on D and we can form a crossed product (D[G], D, G, G/K). Since $G/K \in \underline{O}$ it follows that D[G] is an Ore domain.

Clearly $R \subseteq D[G]$ is a domain. Let a,b be non-zero elements of R. Then there exist non-zero elements c,d of D[G] with ac = bd. Let T be a transversal of K to G. Then $c = \Sigma tc_tq^{-1}$ and $d = \Sigma td_tq^{-1}$ for some c_t and d_t in E[K] and some q in $E[K]^*$. Thus a(cq) = b(dq) and cq and dq are non-zero elements of R. Consequently R is right Ore. Similarly it is left Ore and so $G/(E \cap G)$ belongs to $\underline{O}$. □

1.4.4. REMARKS. Clearly $P(\underline{A} \cap \underline{F}^{-S}) \subseteq PLP\, C_\infty$, so the group ring EG is an Ore domain for any division ring E and any poly(torsion-free abelian) group G. Thus for any charactersitic $p \geq 0$ and any such group G there is a division ring D of characteristic p such that G is a subgroup of D^*. This is Lemma 3 of Snider [1]. In particular G could be any torsion-free locally nilpotent group; we will see another proof of this fact later. It is elementary that $\acute{P} \leq \langle P,L\rangle$, so for example all hyper torsion-free locally nilpotent groups are in $\underline{O}$.

If G is a group with a soluble series whose factors involve torsion the situation is more complicated, and the embedding is indeed impossible in general. We will see in the next chapter that not every periodic soluble group is isomorphic to a skew linear group. Obviously if G_n is an extension of an $\underline{O}$-group by a finite group of order n then G_n is isomorphic to a skew linear group of monomial matrices of degree n. One further case where one can incorporate some torsion is given by the following result. The proof is harder than the constructions above and we refer the reader to 1.5 of Wehrfritz [17] for a proof.

1.4.5. *Let* $A \rightarrowtail G \twoheadrightarrow H$ *be a group extension, where* A *is abelian of finite exponent a power of a prime* p *and* H *is a finite extension of an* $\underline{O}$*-group* K. *Then there exists a division ring* $D = F(K)$ *of characteristic* p, *where* F *is a subfield of* D *normalized by* K, *and an integer* n *such that* G *is isomorphic to a subgroup of* $GL(n,D)$. □

An important open question, with profound implications for the theory of abstract soluble groups, is whether in 1.4.5 one can choose F to be central in D, at least if H is polycyclic. (It follows from 1.4.3 that all polycyclic groups lie in $\underline{O}\,\underline{F}$.)

Let E be a division ring and let G be an $\underline{O}$-group acting (not necessarily faithfully) on E as a group of automorphisms. Let R be the skew group ring of G over E; that is $R = \bigoplus_{x\in G} xE$ with multiplication given by the multiplications of E and G and by $\alpha x = x\alpha^x$ for $\alpha \in E$ and $x \in G$, where α^x is the image of α under x. Then (R, E, G, G) is a crossed product. Let D be the division ring of quotients of R, which exists since $G \in \underline{O}$.

1.4.6. *Assume* E, G, *and* D *are as above.*

a) *The split extension* $E^*] G$, *with* G *acting by conjugation, is isomorphic to a subgroup of* $D^* = GL(1,D)$.
b) *The split extension* $E] G$, *with* G *acting by conjugation, is isomorphic to a subgroup of* $GL(2,D)$.
c) *The split extension* $D] G$, *with* G *acting by right multiplication, is isomorphic to a subgroup of* $GL(2,D)$.

Proof: a) $E^*] G \cong \langle E^*, G \rangle \subseteq D^*$.

b) $E] G \cong \langle \begin{bmatrix} 1 & 0 \\ e & 1 \end{bmatrix}, \begin{bmatrix} g & 0 \\ 0 & g \end{bmatrix} : e \in E, g \in G \rangle$

c) $D] G \cong \langle \begin{bmatrix} 1 & 0 \\ d & 1 \end{bmatrix}, \begin{bmatrix} g & 0 \\ 0 & 1 \end{bmatrix} : d \in D, g \in G \rangle$. □

1.4.7. (G. Higman). *Let* F *be a free group and* R *a normal subgroup of* F. *Then* F/R' *is torsion-free.*

Proof: Let $x \in F$ and $r \geq 0$ with $x^r \in R'$. Then $E = \langle x \rangle R$ is a free group, so E/E' is free abelian. But $x^r \in R' \subseteq E' \subseteq R$. Thus $x \in E' \subseteq R$, whence $E = R$ and $x \in R'$, as required. □

1.4.8. *Let* F *be any field and* G *any torsion-free locally polycyclic-by-finite group. Then the group algebra* FG *is an Ore domain,* FG *has a division ring of quotients* D *and* G *is embedded in the multiplicative group of* D.

Proof: FG is a domain by results of Farkas and Snider (Passman [2], 13.4.18) and Cliff [1]. If X is any finitely generated subgroup of G then FX is Noetherian by P. Hall's theorem (Passman [2], 10.2.8) and so by 1.4.2 it is an Ore domain. Consequently FG is an Ore domain and the result follows from Ore's Theorem. □

1.4.9. THEOREM. (Lichtman [4]). *For any characteristic* $p \geq 0$ *there is a finitely generated skew linear group* G *of degree* 1 *and characteristic* p *such that* G *satisfies a non-trivial identity and yet* G *is not soluble-by-finite.*

Trivially every free subgroup of G is cyclic. Thus neither Platonov's Theorem (Wehrfritz [2], 10.15) nor Tits's Theorem (ibid. 10.16) extend from linear to skew linear groups.

Proof: Let S be any finite perfect simple group. By 1.4.7 there is a finitely generated torsion-free group T with an abelian normal subgroup A such that $T/A \cong S$. Let C be infinite cyclic, and form the restricted wreath product $G = T \wr C$. Let B be the base group of G. Trivially G is finitely generated and satisfies the identity $[[x_1,x_2]^s,[x_3,x_4]^s] = 1$, where s is the order of S. Suppose H is a soluble normal subgroup of G of finite index. We may assume that $H \supseteq A^G = \langle g^{-1}Ag : g \in G \rangle$. Now B/A^G is the direct product of infinitely many copies of S. By a well-known theorem of Remak (Robinson [1], 5.45) $(H \cap B)/A^G$ is also a direct product of copies of S. This contradicts either the solubility of H or the finiteness of $(G:H)$. Thus G is not soluble by finite.

Let F be any field of characteristic p. By 1.4.8 the group ring FB is an Ore domain; let E be its division ring of quotients. The action of C on B extends to one on E and we can form the skew group ring EC. Thus (EC, E, C, C) is a crossed product. Then EC is an Ore domain by 1.4.3 and so has a division ring of quotients D. Clearly $G = BC \subseteq D^*$. □

We will see in the next chapter that locally finite skew linear groups are very restricted. Abelian-by-(locally finite) linear groups are also very restricted. The following result is therefore quite surprising.

1.4.10. THEOREM. (Snider [4]) *Let* H *be any locally finite group and* p *any prime. Then there exists an absolutely irreducible skew linear group* G *of characteristic* p *and degree* 1 *and an abelian normal subgroup* A *of* G *such that* $G/A \cong H$.

Proof: Let F be any field of characteristic p. By 1.4.7 there exists a torsion-free group G_1 with an abelian normal subgroup A_1 such that $G_1/A_1 \cong H$. Let $D = F(G_1)$ be the division ring of quotients of FG_1; this exists by 1.4.8. Now G_1 normalizes the subfield $K = F(A_1)$ of D. Set $A = K^*$, $G = G_1A$ and $L = F[G]$. Then $L = K[G_1]$ and as H is locally finite, L is locally finite-dimensional (right or left) over K. It follows that L is a division ring (since finite-dimensional domains over fields are division rings). Consequently $D = L = F[G]$ and G is absolutely irreducible. Since $FG_1 \subseteq D$ is the group ring, $D = K[G_1]$ is a crossed product of K by H, $A \cap G_1 = A_1$ and $G/A \cong H$. □

REMARK. Note that in the above example $A = C_G(A)$. This follows immediately, since by a theorem of Auslander & Lyndon (e.g. Neumann [1], p. 126, especially 43.24), if rank $F \geq 1$ then $C_F(R/R') = R$. We will recall this fact in Chapter 5.

Results such as 1.4.3, 1.4.6 and 1.4.8 enable us to construct various skew linear groups. We now consider conditions under which we can represent the group over a division ring that is locally finite- dimensional over its centre.

1.4.11. *Let* D = E(G) *be a division ring, generated as a division ring by its division subring* E *and the subgroup* G *of* D^* *normalizing* E. *Suppose that:*

a) E *is locally finite-dimensional over its centre* L;

b) G *has an abelian normal subgroup* A *centralizing* E *with* G/A *locally finite.*

c) $L = \Delta_L(G)$, *i.e. each orbit* e^G *for* $e \in L$ *is finite, and* $A \leq \Delta(G)$.

Then D *is locally finite-dimensional over its centre.*

Proof: Consider the division subring B = E(A) of D generated by E and A, and its subfield K = L(A). Then K is central in B by b), and $F = C_K(G)$ is a central subfield of D. Let X be any finite subset of D. It suffices to prove that $F[X] \leq D$ is finite-dimensional over F.

If H is a finitely generated subgroup of G then HA/A is finite by b), B[H] is finite-dimensional (right or left) over B, and so B[H] is a divison ring. Consequently $D = \bigcup_H B[H]$ and $X \subseteq B[H]$ for some such H. If $b \in B$ then the orbit b^H is finite, its order being at most (HA:A) by b). Hence $X \subseteq D_1 = E_1(A)[H]$ for some division subring E_1 of E normalized by H and finitely generated over L. Therefore E_1 is finite-dimensional over L by a). Then $E_1(A) = E_1.L(A)$ is finite-dimensional over L(A) = K, and so D_1 has finite dimension, n say, over K.

By c) the orbit k^G is finite for every $k \in K$. Thus K is algebraic and hence locally finite-dimensional over F. The choice of a left K-basis of D_1 and right multiplication of D_1 on itself determines an embedding Φ of D_1 into $K^{n\times n}$. Clearly the latter is locally finite-dimensional over F. Also since F is central in D we have $F\Phi = F$. Consequently D_1 is locally finite-dimensional over F, and the result follows. □

The first two paragraphs of the above proof show that if we drop condition c) in 1.4.11 then at least D is locally finite-dimensional over its subfield K.

We now need to construct groups G as in 1.4.11.

1.4.12. *Let the group* H *be built out of the finite groups* L_i *for* i = 1, 2,... *by successive split extensions from the top; explicitly let*

$$H = ...(...(L_1 [L_2) [L_3)...) [L_i ... \qquad (*)$$

Then there exists an exact sequence $A \rightarrowtail G \twoheadrightarrow H$ *of groups with* A *abelian,* G *locally poly-*C_∞ *(and hence an* $\underline{O}$*-group) and residually soluble, and* $A \subseteq \Delta(G)$. □

We postpone the proof of 1.4.12 until the end of this section since it is purely group-theoretic. We give three examples of groups H satisfying (*) of 1.4.12.

a) Let q be a prime and let P_j be a Sylow q-subgroup of the symmetric group on q^j symbols and let H be the direct product of the P_j for j = 1, 2,... Since P_j is a wreath power of the cyclic group of order q (Huppert [1], III.15.3) it follows that H satisfies (*) with each L_i a finite elementary abelian q-group. By the same result P_j has derived length j, so H is hyperabelian but not soluble.

b) Let q be a prime and for j = 1, 2,... let C_j be a cyclic group of order q. Set

$$H = \ldots C_i \wr (\ldots C_3 \wr (C_2 \wr C_1) \ldots) \ldots$$

the product being the permutational wreath product with each C_j acting regularly. Again H satisfies (*) with each L_j an elementary abelian q-group. Then H is a locally finite q-group and so is locally nilpotent (it is even a Gruenberg group by Robinson [1], Vol. 2, p. 1). In particular H is radical; that is, $H \in \acute{P}L\,\underline{N}$. However H has no non-trivial finitely generated subnormal subgroups by a result of P. Hall (Robinson [1], Vol. 2, 6.23), so H is not hyperabelian.

c) Let $p_1, p_2,...$ be an infinite sequence of *distinct* primes. For each i let C_i be a cyclic group of order p_i, and let H be the standard wreath product

$$H = \ldots C_i \wr (\ldots C_3 \wr (C_2 \wr C_1) \ldots) \ldots$$

Here H satisfies (*) with each L_i a finite elementary abelian p_i-group. Trivially H is locally soluble, but the Hirsch-Plotkin radical of H is ⟨1⟩, since the Fitting subgroup of $C_i \wr \ldots C_3 \wr (C_2 \wr C_1)\ldots)$ is the base group, which is a p_i-group. In particular H is not a radical group.

Clearly the direct product of a finite number of groups satisfying (*) also satisfies (*). Thus putting together 1.4.3, 1.4.11 with E a prime field, 1.4.12, and the above, we have proved the following:

1.4.13. THEOREM. (Wehrfritz [14], 3.6). *Let* p *be zero or a prime. Then there is a division ring* D *of characteristic* p *that is locally finite-dimensional over its centre and a torsion-free residually soluble subgroup* $G = G_1 \times G_2 \times G_3$ *of* D^* *such that*

a) G_1 *is hyperabelian but not soluble.*

b) G_2 *is radical but not hyperabelian.*

c) G_3 *is locally soluble but not radical.* □

It is immediate from the linear case (Wehrfritz [2], 4.3) that every radical subgroup of GL(n,D), for D locally finite-dimensional over its centre, is locally soluble.

We will give a substantial number of other applications of the above constructions in later chapters. We now turn to the second major technique for constructing division rings. It can frequently be used instead of the Ore construction, although we choose not to make much use of it. However the reader will frequently find it in the literature, especially in the special case of formal power series (which is the case where the ordered group below is infinite cyclic).

1.4.14. *Let* (R, E, G, H) *be a crossed product, where* E *is a division ring and* H *is an ordered group. Then* R *can be embedded in a division ring.*

Proof: Let T be any transversal of $N = E \cap G$ to G , so that $R = \oplus_{t\in T} tE$ by hypothesis. For each $h \in H$ let t_h be the element of T with $t_hN = h$. Let D be the set of all formal infinite series $x = \Sigma_{t\in T} t\varepsilon_t$ where the ε_t are in E and supp $x = \langle tN : \varepsilon_t \neq 0 \rangle \subseteq H$ is well-ordered. There is an obvious embedding of R into D and we regard R as a subset of D. We claim that the ring operations of R can be extended to D in such a way that D becomes a division ring.

Let A and B be well-ordered subsets of H. It is then true that $A \cup B$ and $AB = \{ ab : a \in A \text{ and } b \in B \}$ are well-ordered, and for any $h \in H$ the set $A \cap hB^{-1}$ is finite (Passman [2], 13.2.9). Let $a = \Sigma\, t\alpha_t$ and $b = \Sigma\, t\beta_t$ be elements of D and set A = supp a, B = supp b. Define $a + b = \Sigma\, t(\alpha_t + \beta_t)$. Since $A \cup B$ is well-ordered, $a + b \in D$. Let $t \in T$. Then the set $A \cap (tN)B^{-1}$ is finite and so the element

$$\gamma_t = \Sigma_{u\in A, v\in B, uv=tN} \; f(t_u, t_v)\, \alpha_{t_u}^{t_v}\, \beta_{t_v}$$

is a well-defined element of E, where f is the factor set of the extension $N \rightarrowtail G \twoheadrightarrow H$ with respect to T. Define $ab = \Sigma_{t\in T}\, t\gamma_t$. Since AB is well-ordered, ab is a well-defined element of D. It is now a simple but lengthy exercise to check that D is a ring, containing R as a subring.

Let $a \in D\backslash\langle 0\rangle$ and let h be the first element of the support of a. Then $a = t_h\alpha_h(1-b)$, where $b \in D$ with $B = \text{supp } b \subseteq P$, the positive cone of H. Trivially $t_h\alpha_h$ is a unit of D. It remains to prove that so is 1-b. We claim that $c = 1 + b + \ldots + b^i + \ldots$ is the inverse of 1-b. To see that c is well-defined let $\langle B\rangle$ be the sub-semigroup of P generated by B. Then $\langle B\rangle$ is well-ordered and $\bigcap_{i\geq 0} \langle B\rangle^i = \emptyset$. (Passman [2], 13.2.10). Clearly supp $b^i \subseteq \langle B\rangle^i$. Thus each $h \in H$ lies in only finitely many of the supp b^i and so c can be written as a formal infinite series $\Sigma\, t\gamma_t$ where each $\gamma_t = \Sigma_{i\geq 0}\, \beta_{it} \in E$ and $b^i = \Sigma\, t\beta_{it}$. Finally supp $c \subseteq \langle B\rangle$ is well-ordered. Thus $c \in D$ and it follows easily that c is the inverse of b. □

1.4.15. **REMARKS.** Every torsion-free locally nilpotent group G is orderable. Thus we can apply 1.4.14 to the group ring FG for any field F and obtain the result that for any characteristic p there is a skew linear group of degree 1 and characteristic p isomorphic to G. This we have seen already in 1.4.4.

The construction in 1.4.14 is most unlikely to produce a division algebra that is locally finite-dimensional over its centre. However we can consider the division subring $D_0 = E(G)$ of D generated by E and G. Then 1.4.11 gives usable conditions for D_0 to be locally finite-dimensional over its centre.

Suppose in 1.4.14 that the group H is infinite cyclic. Then $G = \langle g\rangle N$ for some $g \in G$ and $D = \{ \Sigma_{i\geq n}\, g^i\epsilon_i : n \in \mathbb{Z},\ \epsilon_i \in E \}$ is a formal power series ring. Information about this type of ring may be found in Jacobson [3], p. 187.

It can be shown that the construction of 1.4.14 is independent of the choice of the transversal T .

We now return to our discussion of Ore domains and consider generalizations and variations of 1.4.3 and 1.4.8.

1.4.16. *Let* (R, S, G, H) *be a crossed product and let* K *be a subgroup of* G *containing* S ∩ G. *If* R *is an Ore domain then so is* S[K].

Proof: Trivially S[K] is a domain. Let a and b be non-zero elements of S[K]. By hypothesis there are non-zero elements c, d of R with ad = bc. Let T be any right transversal of K to G. Then $R = \oplus_{t\in T} S[K]t$. Suppose $c = \Sigma_{t\in T} c_t t$ and $d = \Sigma_{t\in T} d_t t$ where c_t, d_t are in S[K]. Then $ad_t = bc_t$ for each $t \in T$. Now $d_t \neq 0$ for some $t \in T$ and so S[K] is right Ore. Similarly it is left Ore. □

1.4.17. *Let* (R, E, G, H) *be a crossed product with* E *a division ring and* H *finite. Then* R *is a division ring if and only if* E[P] *is a division ring for every Sylow subgroup* P/(E ∩ G) *of* H.

Proof: Suppose each E[P] is a division ring and let M be a non-zero right ideal of R. Then $(M{:}E)_r = (M{:}E[P])_r(E[P]{:}E)$ divides (R:E). Also (R:E) = |H| while (E[P]:E) = (P:E ∩ G). This is for each P. Thus $(M{:}E)_r = (R{:}E)$ and so M = R. In the same way R has no proper left ideals and hence R is a division ring. The converse is obvious. □

Let $\underline{O}_1$ denote the class of all groups G such that for every field F the group ring FG is an Ore domain. Trivially $\underline{O} \subseteq \underline{O}_1$. Also by 1.4.8, $\underline{O}_1$ contains the class of all torsion-free locally polycyclic-by-finite groups.

1.4.18. *Let* F *be a field and* K *a normal* $\underline{O}_1$*-subgroup of the group* G . *If either*

a) $G/K \in \underline{O}$,

or b) $G/K \in L(\underline{PF})$ *and* FG *is a domain,*

then FG *is an Ore domain.*

Proof: By hypothesis FK is an Ore domain; let E be its division ring of quotients. The action of G on K extends to one of G on E and we can form a crossed product (E[G], E, G, G/K) such that F[G] = FG. If $G/K \in \underline{O}$ then E[G] is an Ore domain by definition of $\underline{O}$. Suppose $G/K \in L(\underline{PF})$ and that FG is a domain. If x, y ∈ E[G] then $x = c^{-1}a$ and $y = bd^{-1}$ for some a, b ∈ FG and c, d ∈ E^*. Thus E[G] is also a domain. Let X be a finitely generated subgroup of G. Then E[X] is Noetherian by P. Hall's theorem and is therefore Ore by Goldie's theorem. Thus in this case too E[G] is an Ore domain. It follows, as in the proof of 1.4.3, that FG is an Ore domain. □

Let $\underline{I}$ be the class of groups G such that every non-trivial finitely generated subgroup of G has an infinite cyclic quotient. This is the class of locally indicable groups considered in 1.4.1. It is quite trivial that $\langle S,L,R,\grave{P}\rangle \underline{I} = \underline{I}$.

1.4.19. a) $\langle L,S\rangle \underline{O}_1 = \underline{O}_1$.

b) $\underline{O}_1\underline{O} = \underline{O}_1$.

c) $\underline{I} \cap \underline{O}_1 L(\underline{PF}) \subseteq \underline{O}_1$.

Proof: Trivially $\underline{O}_1$ is locally closed, and the subgroup closure of $\underline{O}_1$ follows from 1.4.16. Part b) is an immediate consequence of 1.4.18 and Part c) follows from 1.4.1 and 1.4.18. □

1.4.20. **THEOREM.** *Let* G *be a free group and let* $S \subseteq R$ *be normal subgroups of* G *such that* $S \subseteq R'$. *Suppose that* $R/S \in \underline{O}_1$ and $G/R \in \acute{P}L(\underline{S}\ \underline{F})$. Then $G/S \in \underline{O}_1$.

This is a slight generalization of Proposition 1 of Lichtman [3], the real difference being that Lichtman uses the class $\acute{P}(L\underline{S} \cup L\underline{F})$ in place of $\acute{P}L(\underline{SF})$.

Proof: G/R has a series

$$\langle 1\rangle = G_0/R \subseteq G_1/R \ldots\subseteq G_\alpha/R\ldots\subseteq G_\lambda/R = G/R$$

such that each factor is in $L(\underline{S}\ \underline{F})$. We induct on λ. If λ = 0 the conclusion is given. Suppose $G_\alpha/S \in \underline{O}_1$ for all $\alpha < \lambda$. If λ is a limit ordinal then $G/S \in L\underline{O}_1 = \underline{O}_1$ by 1.4.19. If $\mu = \lambda-1$ exists then $G_\mu/S \in \underline{O}_1$ and $S \subseteq R' \subseteq G_\mu'$ Thus we may assume that G/R is locally soluble-by-finite.

Again by the local closure of $\underline{O}_1$ We may assume that G/R is finitely generated and hence soluble-by-finite. Let K/R be *any* soluble subgroup of G/R. We prove by induction on the derived length of K/R that $K/S \in \underline{O}_1$. Thus assume that $K'R/S \in \underline{O}_1$. Then $K' \supseteq R' \supseteq S$, so $K'/S \in \underline{O}_1$ by the subgroup closure of $\underline{O}_1$. Also K/K' is free abelian and hence lies in $\underline{O}$ by 1.4.3. Therefore $K/S \in \underline{O}_1\underline{O} = \underline{O}_1$, by 1.4.19 b).

Let $x \mapsto \bar{x}$ denote the natural map of G onto $G/S = \bar{G}$. Let L/R be a soluble normal subgroup of G/R of finite index and let F be any field. Then $F\bar{L}$ is an Ore domain and so has a division ring E of quotients. Consider the crossed product $(E[\bar{G}], E, \bar{G}, \bar{G}/\bar{L})$ with $F\bar{G}$ the group ring of $\bar{G}$ over F. If P/L is any Sylow subgroup of G/L then P/R is soluble, so $\bar{P} \in \underline{O}_1$. Hence $F\bar{P}$ is an Ore domain and its division ring of quotients is $E[\bar{P}]$. By 1.4.17 the ring $E[\bar{G}]$ is a division ring. Since $E[\bar{G}] = E.F\bar{G} = F\bar{G}.E$ it follows that $F\bar{G}$ is an Ore domain and consequently $\bar{G} \in \underline{O}_1$. □

1.4.21. THEOREM. *Let* G *be a free group,* R *a normal subgroup of* G *and* F *any field. Let* $G/R \in R\acute{P}L(\underline{SF})$ *and assume that* $V(R) \neq R$ *is a verbal subgroup of* R *such that* $R/V(R)$ *is residually torsion-free nilpotent. Then the group algebra of* $G/V(R)$ *over* F *is a subring of a division ring.*

Again this a very slight generalization of a result of Lichtman ([3], Theorem 2). The verbal and ultraproduct techniques used in the proof below do not appear elsewhere in the book and the uninitiated reader can safely skip the proof of 1.4.21.

Proof: There exists a family $\langle N_i : i \in I \rangle$ of normal subgroups of G such that $\bigcap_i N_i = R$ and each $G/N_i \in \acute{P}L(\underline{SF})$. Since $R_o\acute{P}L \leq \acute{P}LS$ we may assume that for all i and j in I there exists $k \in I$ such that $N_i \cap N_j = N_k$. By a theorem of Dunwoody (Robinson [1], 9.43) we have that $V(R) = V(\bigcap N_i) = \bigcap_i V(N_i)$. Thus there exists an ultrafilter Φ on I such that $G/V(R)$ is isomorphic (via the canonical embedding of $G/V(R)$ into $P = \Pi_i G/V(N_i)$) to a subgroup of the corresponding ultraproduct P/Φ (Kegel and Wehrfritz [1], 1.L.8). Suppose that each group ring $FG/V(N_i)$ is a subring of a division ring D_i. Then $FG/V(R)$ is isomorphic to a subring of $\Pi_i FG/V(N_i)/\Phi$, which in turn is a subring of $\Pi_i D_i/\Phi$. This latter is a division ring (ibid., page 65). Thus we are reduced to proving the result for $G/V(N_i)$.

The torsion subgroup of a nilpotent group is fully invariant and a fully invariant subgroup of a free group is verbal (Neumann [1],

12.34). Thus there are verbal subgroups $V_1(R) \supseteq V_2(R) \supseteq \ldots \supseteq V_n(R) \supseteq \ldots$ of R such that $\bigcap_n V_n(R) = V(R)$ and each $R/V_n(R)$ is torsion-free nilpotent. Moreover since $V(R) \neq R$ we can choose $V_1(R) = R'$. Since R is normal in N_i it has rank greater than or equal to the rank of N_i, and so N_i is isomorphic to a free factor R_i of R. Then

$$N_i/V_n(N_i) \cong R_i/V_n(R_i) = R_i/(R_i \cap V_n(R))$$

Hence $V(N_i) = \bigcap_n V_n(N_i)$ and each $N_i/V(N_i)$ is torsion-free nilpotent. Thus $N_i/V_n(N_i) \in \underline{O}_1$ by 1.4.8 and so $G/V_n(N_i) \in \underline{O}_1$ by 1.4.20. Consequently the group ring $FG/V(N_i)$ is a subring of a division ring by the ultraproduct argument again. □

Free groups are residually torsion-free nilpotent. Hence if F is a field and G is a free group then the group ring FG can be embedded into a division ring. This fact also follows from 1.4.14 since free groups are orderable (Passman [2], 13.2.8). Yet a third approach is to use Cohn's theorem on the embeddability of firs (free ideal rings) into division rings (Cohn [2], Theorem 4C.) However apart from the rank 1 case FG is not an Ore domain, so 1.4.14 and 1.4.21 take us beyond the scope of $\underline{O}_1$.

1.4.22. Let $\underline{O}_2$ be the class of all groups H such that every crossed product of a division ring by H, if a domain, is an Ore domain. Trivially $\underline{O} \subseteq \underline{O}_2$, so $C_\infty \in \underline{O}_2$, and the proof of 1.4.3 shows that $\underline{O}_2$ is $\langle P,L \rangle$-closed. Hence it is also $\acute{P}$-closed. Also $\underline{O}_2$ contains all finite groups, for if (R, E, G, H) is a crossed product where E is a division ring, H is finite, and R is a domain, then R is finite-dimensional over its division subring E and is consequently a division ring. Thus, for example, any locally soluble-by-finite group is an $\underline{O}_2$-group, as is any $\acute{P}\underline{A}$-group (that is, any SN^*-group in Kuroš' terminology).

We can now prove the following generalization of 1.4.8.

1.4.23. *Let* G *be locally a soluble-by-finite, residually torsion-free polycyclic-by-finite group (that is, let* $G \in L(\underline{SF} \cap R(\underline{F}^{-S} \cap (\underline{P}\,\underline{F}))$. *If* F *is any field then the group algebra* FG *is an Ore domain; i.e.* $G \in \underline{O}_1$.

Proof: Let $x, y \in FG\setminus\{0\}$ with $xy = 0$, and set $S = \operatorname{supp} x \cup \operatorname{supp} y$. There is, by hypothesis, a torsion-free polycyclic-by-finite group H and a homomorphism ϕ of $\langle S \rangle \subseteq G$ onto H such that

$|S| = |S\Phi|$. Extend Φ by linearity to a homomorphism $\Phi : F\langle S\rangle \longrightarrow FH$. Then $x\Phi$ and $y\Phi$ are both non-zero and yet $x\Phi.y\Phi = 0$. This contradicts the fact that FH is a domain (1.4.8). Therefore FG is a domain. Since G is locally soluble-by-finite we have $G \in \underline{O_2}$ by 1.4.22, and consequently FG is an Ore domain. □

If E is any division ring and G is any torsion-free one-relator group then the group ring EG can be embedded into a division ring; this is a theorem of J. and T. Lewin [1]. Its proof uses powerful results of P. M. Cohn on coproducts of semifirs and the embeddability of semifirs into division rings. Thus every torsion-free one-relator group is isomorphic to a skew linear group. Not all such groups are isomorphic to linear groups since they need not be residually finite. For example the Baumslag-Solitar group $\langle x, y \mid xy^2 = y^3x \rangle$ is not.

Cohn has given necessary and sufficient conditions for a ring to be embeddable into a division ring, see Cohn [3], p. 85, Corollary 2.

1.4.24. *The proofs of* 1.1.5 *and* 1.1.8.

Let Q be a field of characteristic $p \geq 0$, let P be an extension field of Q, and let ω_i for $i = 1, 2,\ldots$ be elements of P such that $P = Q(\omega_i : i \geq 1)$. Suppose that for $i = 1, 2,\ldots$ there is a Q-automorphism Φ_i of P of finite order m_i such that $\omega_j^{\Phi_i} = \omega_j$ if $j \neq i$ and $\omega_i^{\Phi_i} \in \langle\omega_i\rangle \setminus \{\omega_i\}$.

a) *The Division Ring.*

Let A be a free abelian group on the basis $\{ t_i : i \geq 1 \}$. Notice that the above automorphisms Φ_i commute. For each i let t_i act on P as Φ_i, and form the corresponding skew group ring P^*A. Then P^*A has a division ring D of quotients by 1.4.4. Moreover D is locally finite-dimensional over its centre by 1.4.11.

b) *The Reducible Group.*

Let $V = D^2$. For $i = 1, 2,\ldots$ let $v_i = (t_1 + \ldots + t_i, 1) \in V$ and

$$g_i = \begin{bmatrix} \omega_i & 0 \\ \omega_i t_i - t_i \omega_i & \omega_i \end{bmatrix} \in GL(2,D)$$

Clearly $u = (1, 0) \in V$ is an eigenvector of g_i. Also for $1 \leq j \leq i$

$$v_ig_j = (t_1\omega_j + \dots + + t_i\omega_j + \omega_jt_j - t_j\omega_j, \omega_j)$$
$$= (\omega_jt_1 + \dots + \omega_jt_i, \omega_j)$$
$$= \omega_jv_i,$$

since ω_j and t_k commute for $j \neq k$. Set $G_i = \langle g_1,\dots,g_i \rangle$. The above shows that G_i is a conjugate of

$$\langle \omega_j 1_2 : 1 \leq j \leq i \rangle \cong \langle \omega_j : 1 \leq j \leq i \rangle \subseteq P^*.$$

In particular $G = \langle G_i : i \geq 1 \rangle$ is abelian, and in fact

$$G \cong \Omega = \langle \omega_i : i \geq 1 \rangle \subseteq P^*.$$

Also G_i is completely reducible with irreducible constituents of degree 1. A finitely generated subgroup of G is contained in one of the G_i and so every finitely generated subgroup of G is completely reducible.

We prove that G is not completely reducible. Suppose otherwise. Clearly Du is a D-G submodule of V, and so $V = Du \oplus Dv$ for some D-G submodule Dv of V. Then $v = \alpha u + \beta v_i$ for some $\alpha, \beta \in D$ with $\beta \neq 0$. We have $vg_i = \alpha(ug_i) + \beta(v_ig_i) = \alpha.\omega_iu + \beta.\omega_iv_i$, and yet $vg_i = \gamma v$ for some $\gamma \in D$. *Thus* $\gamma\alpha = \alpha.\omega_i$ *and* $\gamma\beta = \beta.\omega_i$. *Now* $\beta \neq 0$, *so either* $\alpha = 0$ *or* $\omega_i^{\beta^{-1}\alpha} = \gamma^\alpha = \omega_i$. Thus in either case $\alpha \in \beta.C_i$, where $C_i = C_D(\omega_i)$. Therefore $v = \beta(\lambda_iu + v_i)$ for some $\lambda_i \in C_i$. Setting $v = (\xi, \eta)$ and inserting the entries of u and v_i we obtain, for each $i \geq 0$, that $\eta = \beta$ and $\eta^{-1}\xi = \lambda_i + t_1 + \dots + t_i$. But $\lambda_i, t_1,\dots, t_{i-1}$ all commute with ω_i while t_i does not, so that $\eta^{-1}\xi \notin \bigcup_{i\geq 1} C_i$. However $C_i \supseteq P(t_j : j \neq i)$ and so $\bigcup C_i = D$. This contradiction shows that G is not completely reducible.

c) *The Irreducible Group.*

Let $x_0 = \begin{bmatrix} 1 & 0 \\ 0 & 1 \end{bmatrix}$ and $x_i = \Pi^1_{j=i} \begin{bmatrix} 0 & 1 \\ 1 & t_j \end{bmatrix} \in GL(2,D)$

for $i = 1, 2,\dots$ Set $H_i = \langle x_i^{-1}(\omega_jx_0)x_i : 1 \leq j \leq i \rangle$ and $H = \langle H_i : i \geq 1 \rangle$.

Since t_i centralizes ω_k for $k < i$ we have $x_i^{-1}(\omega_kx_0)x_i = x_j^{-1}(\omega_kx_0)x_j$ for $k \leq j \leq i$ and so $H_1 \leq H_2 \leq \dots \leq H_i \leq \dots$. In particular H is abelian, and in fact H, like G, is isomorphic to Ω. By construction each H_i is completely reducible with irreducible constituents of degree 1. Thus every finitely generated subgroup of H is reducible. We claim that H itself is irreducible.

Suppose H is reducible. Then there exists $x \in GL(2,D)$ such that D(1, 0) is a D-H^x-submodule of V. In particular

$\begin{bmatrix}\omega_i & 0\\ 0 & \omega_i\end{bmatrix}^{x_i x} = \begin{bmatrix}\varepsilon & 0\\ \varsigma & \eta\end{bmatrix}$ for some ε, ς, η in D. If $x_i x = \begin{bmatrix}a & b\\ c & d\end{bmatrix}$ then

$\begin{bmatrix}\omega_i a & \omega_i b\\ \omega_i c & \omega_i d\end{bmatrix} = \begin{bmatrix}* & b\eta\\ * & d\eta\end{bmatrix}$. Suppose $d \neq 0$. If $b = 0$ then trivially $b \in C_i d$, where again $C_i = C_D(\omega_i)$. If $b \neq 0$ then $\omega_i^b = \eta = \omega_i^d$ and again $b \in C_i d$. If $d = 0$ trivially $d \in C_i b$. Thus with a change of notation

$x_i x = \begin{bmatrix}* & b_i f\\ * & d_i f\end{bmatrix}$ for some b_i, $d_i \in C_i$ and some $f = f_i \in D^*$.

For $i \geq 1$ we have $x_i^{-1} = x_{i-1}^{-1}\begin{bmatrix}-t_i & 1\\ 1 & 0\end{bmatrix}$. Let $x_{i-1}^{-1} = \begin{bmatrix}\alpha & \beta\\ \gamma & \delta\end{bmatrix}$. Note that α, β, γ, δ lie in $Q(t_1,\ldots,t_{i-1}) = C_i \cap C_D(t_i)$.

Then $x = \begin{bmatrix}\alpha & \beta\\ \gamma & \delta\end{bmatrix}\begin{bmatrix}-t_i & 1\\ 1 & 0\end{bmatrix}\begin{bmatrix}* & b_i f\\ * & d_i f\end{bmatrix} = \begin{bmatrix}* & \alpha d_i f - \alpha t_i b_i f + \beta b_i f\\ * & \gamma d_i f - \gamma t_i d_i f + \delta b_i f\end{bmatrix}$.

Thus, since t_i centralizes α and γ, the (1,2) and (2,2) entries of x are

$$x(1,2) = (\alpha d_i + \beta b_i - t_i\alpha b_i)f$$

and

$$x(2,2) = (\gamma d_i + \delta b_i - t_i\gamma b_i)f.$$

We collect some properties of these elements.

i) If $\alpha b_i = \gamma b_i = 0$ then $b_i = 0$:

For since x_{i-1}^{-1} is invertible α and γ are not both zero.

ii) If $\alpha b_i \neq 0$ then $x(1,2) \neq 0$ and $y = x(2,2)x(1,2)^{-1} \notin C_i$:

Since $\lambda = \alpha d_i + \beta b_i$ and $\mu = \alpha b_i$ lie in C_i and $D = \oplus_{j=0}^{m_i-1} t_i C_i$, it is clear that $x(1,2) \neq 0$ and y is defined. Suppose $y \in C_i$. The inverse of $\lambda - t_i\mu = x(1,2)f^{-1}$ has the form $\Sigma_{0\leq j<m_i} t_i^j a_j$ for some $a_j \in C_i$. For $0 < i < m_i$, the coefficient of t_i^j in $(\lambda - t_i\mu)\,\Sigma\, t_i^j a_j$ is zero. Thus $\mu^{t_i^{j-1}} a_{j-1} = \lambda^{t_i^j} a_j$ for $0 < i < m_i$. If $a_{m_i-1} = 0$ then, since by hypothesis $\mu \neq 0$, a trivial induction yields that all the a_j are zero. Therefore $a_{m_i-1} \neq 0$ and

$a_{m_i-2} = \mu^{-t_i^{m_i-2}} \lambda^{t_i^{m_i-1}} a_{m_i-1}$.

By hypothesis $y = (\gamma d_i + \delta b_i - t_i\gamma b_i)(\Sigma\, t_i^j\, a_j) \in C_i$. In this product the coefficient of $t_i^{m_i-1}$ must be zero. Hence

$$(\gamma d_i + \delta b_i)^{t_i^{m_i-1}} a_{m_i-1} = (\gamma b_i)^{t_i^{m_i-2}} a_{m_i-2} = (\gamma\alpha^{-1})^{t_i^{m_i-2}} \lambda^{t_i^{m_i-1}} a_{m_i-1}.$$

Since t_i centralizes α and γ and $a_{m_i-1} \neq 0$ we obtain

$$\gamma d_i + \delta b_i = \gamma\alpha^{-1}(\alpha d_i + \beta b_i).$$

Consequently $(\delta - \gamma\alpha^{-1}\beta)b_i = 0$. Now α, β, γ and δ commute and x_{i-1}^{-1} is invertible, so $\delta \neq \gamma\alpha^{-1}\beta$. Also $b_i \neq 0$ as $\alpha b_i \neq 0$ by hypothesis. This contradiction proves that $y \notin C_i$.

iii) If $\gamma b_i \neq 0$ then $x(2,2) \neq 0$ and $z = x(1,2)x(2,2)^{-1} \notin C_i$: This may be proved exactly as ii) above, with the roles of α and γ, and β and δ interchanged.

Note that y and z do not depend on i. Also since x is invertible at least one of y and z is defined. Further if z is defined and $z \notin C_i$ then $z \neq 0$, y is defined, and $y \notin C_i$, and conversely. From now on assume that y is defined (the argument is essentially the same if z is defined).

Since $y \in D = P(t_i : i \geq 1)$ there exists $h \geq 1$ with $y \in P(t_i : i \leq h)$. Then $y \in C_i$ for every $i > h$. Hence $b_i = 0$ for each $i > h$ by i), ii), and iii) above. However if $i > h + 1$ we have

$$x_i x = \begin{bmatrix} * & 0 \\ * & * \end{bmatrix}, \quad x_i = \begin{bmatrix} 0 & 1 \\ 1 & t_i \end{bmatrix} x_{i-1}, \quad \text{and} \quad x_{i-1}x = \begin{bmatrix} * & 0 \\ * & * \end{bmatrix}. \quad \text{But}$$

$$\begin{bmatrix} * & 0 \\ * & * \end{bmatrix} = \begin{bmatrix} 0 & 1 \\ 1 & t_i \end{bmatrix} \begin{bmatrix} * & 0 \\ * & * \end{bmatrix}$$ is impossible for invertible matrices. This final contradiction shows that H must in fact be irreducible.

We now give specific examples of the above situation.

1) G *and* H *torsion-free*, p *arbitrary.*

Let Q be a prime field of charactersitic $p \geq 0$ and let $\omega_1, \omega_2, ..$ be indeterminates over Q. Set $P = Q(\omega_i : i \geq 1)$. Clearly P has automorphisms ϕ_i such that

$$\omega_j^{\phi_i} = \begin{cases} \omega_j & \text{if } j \neq i \\ \omega_i^{-1} & \text{if } j = i \end{cases} \qquad (*)$$

Here Ω, and hence G and H, are free abelian of countably infinite rank.

2) G *and* H *periodic of rank* 1, p = 0.

Let $Q = \mathbb{Q}$ and let $p_1, p_2, \ldots$ be any infinite sequence of odd primes. For each $i \geq 1$ let ω_i be a primitive p_i-th root of unity in $\mathbb{C}$ and set $P = Q(\omega_i : i \geq 1)$. By Galois theory and elementary properties of the Euler function there exist automorphisms ϕ_i of P satisfying (*) above. Here Ω, G and H are isomorphic to $\times_{i=1}^{\infty} C_{p_i}$.

3) G *and* H *periodic of rank* 1, $p \geq 0$, *and* G *and* H *with no elements of order* p.

Set $Q = GF(p)$. Then there exists an infinite sequence $\{p_i\}$ of distinct primes and a field $P = Q(\omega_i : i \geq 1)$, where ω_i for each i is a primitive p_i-th root of unity, such that P has automorphisms ϕ_i of the required type (but unlike 1 and 2 above ϕ_i does not necessarily invert ω_i). To see this let ω be a primitive $(p-1)^2$-th root of unity over Q and let $\{q_i\}$ be an infinite sequence of distinct primes not dividing $(Q(\omega) : Q)$. Then for some prime p_i not dividing (p-1) the field $GF(p^{q_i})$ contains a primitive p_i-th root of unity. Clearly $Q(\omega_i) = GF(p^{q_i})$ has an automorphism normalizing but not centralizing $\langle\omega_i\rangle$, and the $Q(\omega_i)$ are linearly disjoint over Q. The claim follows. Again Ω, G and H are isomorphic to $\times_{i=1}^{\infty} C_{p_i}$. □

We end this section with a proof of 1.4.12. Suppose we have a finite group $K = J [L$ with L abelian and an exact sequence $A_J \rightarrow G_J \rightarrow J$ with A_J abelian and G_J poly-C_∞. Choose a free *abelian* presentation $B \rightarrow E \rightarrow L$ of L of finite rank. By the Kalužnin-Krasner Theorem we can regard K as a subgroup of $L \wr J = L^{(J)}] J$, where $L^{(J)}$ denotes the direct product of |J| copies of L permuted regularly by J. Now

$$B^{(J)} \times A_J \rightarrowtail E^{(J)}] G_J \twoheadrightarrow L^{(J)}] J$$

with the obvious maps. Let G_K denote the inverse image of $K \subseteq L^{(J)}] J$ in $E^{(J)}] G_J$ and set $A_K = B^{(J)} \times A_J$. Then $A_K \rightarrowtail G_K \twoheadrightarrow K$ is exact and A_K is abelian. Also E and G_J are poly-C_∞, so $E^{(J)}] G_J$ and its subgroup G_K are too.

The embedding of K into $L \wr J$ given by the proof of the

Kalužnin-Krasner Theorem depends upon the choice of a transversal of L to K. Of course we choose J as our transversal. Then the embeddings $J \longrightarrow K$, $K \longrightarrow L^{(J)}] J$ and $J \longrightarrow L^{(J)}] J$ are consistent and L is mapped into $L^{(J)}$. Hence the map

$$E^{(J)}] G_J \longrightarrow L^{(J)}] J$$

maps G_J onto $J \subseteq K$, so $G_J \subseteq G_K$. Also $G_J \cap A_K = A_J$. Since L is embedded into $L^{(J)}$, the subgroup $L \subseteq K$ centralizes A_K. Finally $G_K = (E^{(J)} \cap G_K)] G_J$ so G_J splits off from G_K.

Set $J_i = (\ldots(L_1 [L_2)\ldots)L_i$ and suppose we have constructed an exact sequence $A_i \rightarrowtail G_i \twoheadrightarrow J_i$ with A_i abelian and G_i poly-C_∞. We apply the above construction to $J_{i+1} = J_i [L_{i+1}$ to construct an exact sequence $A_{i+1} \rightarrowtail G_{i+1} \twoheadrightarrow J_{i+1}$ with A_{i+1} abelian, G_{i+1} poly-C_∞, and the following diagram exact.

$$\begin{array}{ccccc} A_i & \longrightarrow & G_i & \longrightarrow & J_i \\ \downarrow & & \downarrow & & \downarrow \\ A_{i+1} & \longrightarrow & G_{i+1} & \longrightarrow & J_{i+1} \end{array}$$

Moreover A_i is centralized by L_{i+1} and G_i is a complement in G_{i+1} of A_{i+1}. Let $G = \varinjlim G_i$ and $A = \bigcup_i A_i \trianglelefteq G$. Then A is an abelian normal subgroup of G, the group G is locally poly-C_∞ and G/A is isomorphic to $\varinjlim J_i \cong H$. Since each G_i is soluble and splits off from G_{i+1}, the group G is residually soluble. If $a \in A$ then $a \in A_i$ for some i and then $C_H(a)$ contains $\langle L_j : j \geq i \rangle$. This latter has finite index $|J_i|$ in H, so $A \subseteq \Delta(G)$ as required. □

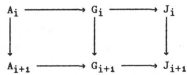

2 FINITE AND LOCALLY FINITE GROUPS

The titles of this chapter and its subsections are in the main self-explanatory. The difficult parts of Chapter 2 concern finite skew linear groups. The generalizations to locally finite groups are usually routine. In the long Section 1 we classify the finite skew linear groups of degree 1, that is, the finite multiplicative subgroups of division rings. Historically this was the foundation of our subject, being now some thirty years old. The positive-characteristic case is almost trivial, whereas the zero-characteristic case is very substantial indeed. The details of the proofs of Section 2.1 will not be required later, but the reader should at least read the first four pages of this section before passing on.

The very brief Section 2.2 collects together a few results on Schur indices required in the later sections. Section 2.3 swiftly deals with finite and locally finite skew linear groups of positive characteristic. The fundamental theorems on finite skew linear groups of characteristic zero are in Section 2.4, while the final Section 2.5 contains our general discussion of finite and locally finite skew linear groups of characterisitic zero.

2.1 FINITE SUBGROUPS OF DIVISION RINGS

It is well-known that the only finite subgroups of the multiplicative group of a field of characteristic $p \geq 0$ are the cyclic p'-groups. The problem of classifying the finite subgroups of the multiplicative groups of division rings was first raised by Herstein in 1953, who gave the solution in the positive characteristic case. The classification in the (much more difficult) zero characteristic case was achieved by Amitsur in 1955 (and independently by J. A. Green at a slightly later date, but not published). We begin with Herstein's result.

2.1.1. (Herstein [1]). *Let G be a finite multiplicative subgroup of a division ring of characteristic* $p > 0$. *Then G is a cyclic* p'*-group.*

Clearly every cyclic p'-group arises as such a group G.

Proof: Let G be a finite subgroup of D^*, where D is a division ring of characteristic p. The subalgebra $\mathbb{F}_p[G]$ of D is a finite domain, and hence a division ring. It is consequently a field by Wedderburn's Theorem (Cohn [2], Vol. 2, p. 196). Thus G is a finite subgroup of a field. The result follows. □

We now commence work on the zero-characteristic case. The approach we take follows Shirvani [1]. For the remainder of this section D denotes a division ring of characteristic zero.

2.1.2. *Let G be a finite subgroup of a* D^*.

a) *If E is a central subfield of D then the subalgebra E[G] of D is a finite-dimensional division algebra.*

b) *G is a Frobenius complement. In particular, the Sylow subgroups of G are cyclic or (generalized) quaternion, and every subgroup of G of order pq , for p and q prime numbers, is cyclic. Abelian subgroups of G are cyclic.*

Recall that a Frobenius complement is a group of fixed-point-free automorphisms of a finite group.

Proof: a) $E[G] \subseteq D$ is finite-dimensional over E and is also a domain. Hence it is a division ring (Jacobson [4], p. 3).

b) The subring $R = \mathbb{Z}[G]$ of D is free abelian of finite rank n, say. The action of G on R by right multiplication is faithful and fixed-point-free, for $rg = r$ implies that $r(g-1) = 0$ in D; whence $r = 0$ or $g = 1$. Regard G as a subgroup of $GL(n,\mathbb{Z})$ via this action. If g is a non-trivial element of G then 1 is not an eigenvalue of g, so $\det(g-1) \neq 0$. If p is a prime then reduction modulo p gives a representation $g \longmapsto \bar{g}$ of G in $GL(n,p)$. For all large p this representation is faithful, and moreover if p does not divide $\det(g-1)$ for all $g \in G \setminus \langle 1 \rangle$ then $\det(\bar{g}-1) \neq 0$. For such a prime p it follows that G acts fixed-point-freely on the finite group $\mathbb{F}_p^{(n)}$.

Thus G is a Frobenius complement. The properties of G listed above are then well-known (e.g. see Passman [1], 18.1). □

The structure of G depends on whether its Sylow 2-subgroups are cyclic or quaternion (we take this term to mean either the quaternion group of order 8 or one of the generalized quaternion groups of order a power of 2.) In the former case all the Sylow subgroups of G are cyclic. Let $\underline{Z}_o$ denote the class of finite groups all of whose Sylow subgroups are cyclic, and let $\underline{Z}$ denote the class of $\underline{Z}_o$-groups that can be embedded into some division ring of characteristic zero. The structure of $\underline{Z}_o$-groups is given in the next result. For the remainder of this section C_n denotes a cyclic group of order n.

2.1.3. *If X is a $\underline{Z}_o$-group then X is a split extension $C_m] C_n$ where the integers m and n are coprime. If in fact $X \in \underline{Z}$ then for all primes $q \mid n$ the C_q subgroup of C_n centralizes C_m.*

Proof: For the first part see M. Hall [1], 9.4.3. If $X \in \underline{Z}$ then X is a Frobenius complement by 2.1.2, so its subgroups of order pq are cyclic. The second part follows. □

The determination of the $\underline{Z}$-groups is the most complicated part of the classification, and we relegate it to the end of this section. But note that the $\underline{Z}_o$-group $C_{p^a}] C_{q^b}$, where p and q are distinct prime numbers, is in $\underline{Z}$ provided the kernel of the action of C_{q^b} on C_{p^a} is sufficiently large. For example, consider the groups $C_7] C_{2^i}$, where i is a positive integer and C_{2^i} acts by inversion. Then $C_7] C_2$ is in $\underline{Z}_o$ but is not a Frobenius complement, $C_7] C_4$ is in $\underline{Z}$, and $C_7] C_8$ is in $\underline{Z}_o$ and is a Frobenius complement but is not in $\underline{Z}$. Finally all of the groups $C_7] C_{2^i}$ with $2^i \geq 16$ are in $\underline{Z}$.

We state the main classification theorem in terms of $\underline{Z}$-groups, followed by the characterisation of $\underline{Z}$.

2.1.4. THEOREM(Amitsur [1]). *Let G be a finite group. Then G is a subgroup of a division ring of characteristic zero if and only if G is isomorphic to one of the following groups:*

a) a $\underline{Z}$ *-group.*

b)i) *the binary octahedral group, of order 48.*

ii) $C_m] Q$, *where* m *is odd,* Q *is quaternion of order* 2^t, *say, an element of* Q *of order* 2^{t-1} *centralizes* C_m, *and an element of* Q *of order 4 inverts* C_m.

iii) $Q \times M$, *where* Q *is quaternion of order 8,* M *is a* $\underline{Z}$*-group of odd order, and 2 has odd (multiplicative) order modulo* $|M|$.

iv) $SL(2,3) \times M$, *where* M *is a* $\underline{Z}$*-group of order coprime to 6, and 2 has odd order modulo* $|M|$.

c) *the binary icosahedral group* $SL(2,5)$, *of order 120.* □

Part b) lists the groups that are soluble and have a quaternion subgroup. The only insoluble finite subgroup of a division ring is $SL(2,5)$. Also note that the conditions on M in b iii) and b iv) are equivalent, since 2 has multiplivative order 2 modulo 3. For brevity write $\gamma(n,s)$, for n and s coprime integers, for the order of n modulo s, that is the least integer γ such that $n^\gamma \equiv 1 \bmod s$.

2.1.5. THEOREM.(Shirvani [1]). *A group* G *is a* $\underline{Z}$*-group if and only if it is of one of the following types:*

a) *cyclic.*

b) $C_m] C_4$, *where* m *is odd and* C_4 *acts by inversion on* C_m.

c) *of the form* $G_o \times G_1 \times \ldots \times G_s$ *where* $s \geq 1$, *the orders* $|G_o|,\ldots,|G_s|$ *are coprime,* G_o *is cyclic, each of* $G_1, \ldots, G_s$ *is non-cyclic of the form*

$$C_{p^a}] (C_{q_1^{b_1}} \times \ldots \times C_{q_r^{b_r}})$$

where $p, q_1,\ldots, q_s$ *are distinct primes with each* $C_{p^a}] C_{q^b}$ *non-cyclic and satisfying the following condition: if* C_{q^α} *is the kernel of the action of* C_{q^b} *on* C_{p^a} *then*

either $q = 2$, $p \equiv -1 \bmod 4$, *and* $\alpha = 1$,

or $q = 2$, $p \equiv -1 \bmod 4$, *and* $2^{\alpha+1} \nmid (p^2-1)$,

or $q = 2$, $p \equiv 1 \bmod 4$, *and* $2^{\alpha+1} \nmid (p-1)$,

or $q > 2$, *and* $q^{\alpha+1} \nmid (p-1)$.

Moreover for each non-cyclic factor $C_{p^a}] C_{q^b}$ *of* G_i,

$$q\gamma(p,q^\alpha) \nmid \gamma(p,|G|/|G_i|).$$ □

Note that the above three types are mutually exclusive, except that the non-cyclic groups $C_{p^a}] C_4$, for $p \equiv -1 \bmod 4$, appear under both

b) and c). Amitsur [1] contains a different characterization of $\underline{Z}$-groups.

We need to recall some definitions and facts from the theory of finite-dimensional central simple algebras.

2.1.6. *Let* B *be a finite-dimensional central simple* F*-algebra contained in an* F*-algebra* A*. Then* $A = B \otimes_F C_A(B)$, *where* $C_A(B)$ *is the centralizer of* B *in* A. □

See Jacobson [5], Vol. 2, Theorem 4.7, p. 218. Note that F is not assumed to be the centre of A.

Let A be a finite-dimensional central simple F-algebra. Thus $A = D^{n\times n}$ where D is a division algebra with centre F. The Schur index m(A) of A is defined to be the square root of $\dim_F D$. The exponent of A is the least positive integer $e = e(A)$ such that $A \otimes_F A \dots \otimes_F A \cong F^{r\times r}$, where there are e factors on the left. In other words, e(A) is the order of A in the Brauer group Br(F). Always $e(A) | m(A)$, with equality if F is an algebraic number field (Reiner [1], Theorem 32.19; see 2.1.8 below).

2.1.7. *Let* D_1 *and* D_2 *be finite-dimensional central division* F*-algebras, where* F *is an algebraic number field. Then* $D_1 \otimes_F D_2$ *is a division algebra if and only if* $m(D_1)$ *and* $m(D_2)$ *are coprime integers.*

Proof: It is well-known (see Jacobson [5], Vol. 2, Corollary 2, p. 219) that $A = D_1 \otimes_F D_2$ is central simple over F. In view of the definition of m(A) and the fact that $e(A) = m(A)$ it follows that A is a division algebra if and only if $m(A) = (\dim_F A)^{1/2} = m(D_1)m(D_2)$, i.e. if and only if $e(A) = e(D_1).e(D_2)$. But $e(D_1)$ and $e(D_2)$ are the orders of D_1 and D_2 in Br(F), while (the class of) A is the product of D_1 and D_2 in the abelian group Br(F). The result follows. □

Let F be an algebraic number field. By a finite prime of F we mean a non-zero prime ideal of the ring of integers O_F of F. An infinite prime of F means an equivalence class of embeddings of F into $\mathbb{C}$. If $\mathfrak{P}$ is a prime (finite or infinite) of F, denote by $F_{\mathfrak{P}}$ the completion of F with respect to the valuation determined by $\mathfrak{P}$. For a proof of the following see

Reiner [1], Theorems 32.17 and 32.19.

2.1.8. *Let* A *be a finite-dimensional central simple* F*-algebra, where* F *is an algebraic number field. Let* $\mathfrak{P}$ *be a prime (finite or infinite) of* F, *and let* $A_\mathfrak{P}$ *denote the central simple* $F_\mathfrak{P}$*-algebra*

$$A_\mathfrak{P} = A \otimes_F F_\mathfrak{P}$$

Then $m(A_\mathfrak{P}) = e(A_\mathfrak{P})$. *Further*

$$m(A) = \text{L.C.M.} \langle\ m(A_\mathfrak{P}) : \mathfrak{P} \text{ a prime of } F\ \rangle$$

$$\& \quad e(A) = \text{L.C.M.} \langle\ e(A_\mathfrak{P}) : \mathfrak{P} \text{ a prime of } F\ \rangle.$$

In particular $e(A) = m(A)$. □

If F is a finite Galois extension of the algebraic number field K and $\mathfrak{P}$ is a prime ideal of K then $\mathfrak{P}\mathcal{O}_F = \Pi\, \mathfrak{p}_i^{e_i}$, where the $\mathfrak{p}_i$ are distinct prime ideals of $\mathcal{O}_F$. Since Gal(F/K) permutes the $\mathfrak{p}_i$ transitively it follows that the e_i are all equal, the common value being the ***ramification index*** $e(\mathfrak{P}, F/K)$ of $\mathfrak{P}$ in F/K. Similarly the numbers f_i defined by $|\mathcal{O}_F/\mathfrak{p}_i| = |\mathcal{O}_K/\mathfrak{P}|^{f_i}$ are all equal, the common value being the ***residue degree*** of $\mathfrak{P}$ in F/K (Lang [1], p. 13). Using the notation for completions introduced above we also have that $|F_{\mathfrak{p}_i} : K_\mathfrak{P}| = e(\mathfrak{P}, F/K).f(\mathfrak{P}, F/K)$ (Lang [1], p.33). As an example consider the cyclotomic field $\mathbb{Q}(\zeta_m)$, where ζ_m is a primitive m-th root of unity. Let p be a prime number, and write $m = p^a n$, where $a \geq 0$ and p does not divide n. Then $e(p\mathbb{Z}, \mathbb{Q}(\zeta_m)/\mathbb{Q}) = \phi(p^a)$, where ϕ is the Euler function, and $f(p\mathbb{Z}, \mathbb{Q}(\zeta_m)/\mathbb{Q}) = \gamma(p,n)$ as defined above (Weiss [1], 7.4.3).

To begin our classification we need to study the quaternion subgroups of division rings, and quaternion algebras. Let

$$\underline{A} = \mathbb{Q} \oplus \mathbb{Q}i \oplus \mathbb{Q}j \oplus \mathbb{Q}ij$$

be the rational quaternion algebra, where $i^2 = j^2 = -1$ and $ij = -ji$. Clearly $\underline{A}$ has centre $\mathbb{Q}$ and Schur index 2. More general quaternion algebras are covered by the next result.

2.1.9. *Let* F *be a finite Galois extension of* $\mathbb{Q}$. *Then* $\underline{A}(F) = \underline{A} \otimes_\mathbb{Q} F$ *is a division algebra if and only if either* F *is a real field or both the ramification index and the residue degree of the rational prime* 2 *in the extension* $F/\mathbb{Q}$ *are odd.*

Proof: First note that $\underline{A}(\mathbb{Q}_p)$ is a division algebra if and only if $p = 2$, see Lam [1], 6.2.5 and 6.2.24. Alternatively this follows from Serre [2], p. 66, Thèoréme 4.6, Corollaire, using the criterion that $\underline{A}(F)$ is a division ring if and only if -1 is not a sum of two squares in F (Albert [1], Theorem 9.27).

If F is real then $\underline{A}(F) \subseteq \underline{A}(\mathbb{R})$ is a division algebra, so we may assume that F has no real embedding. Clearly $\underline{A}(F)$ is a division algebra if and only if $m(\underline{A}(F)) = 2$, which by 2.1.8 happens if and only if $\underline{A}(F_{\mathfrak{P}})$ is a division algebra for some prime $\mathfrak{P}$ of F. Such a $\mathfrak{P}$ is clearly not complex, and is not real by hypothesis. Thus in view of the first paragraph $\mathfrak{P}$ lies over the rational prime 2. Further $\underline{A}(F_{\mathfrak{P}})$ is a division algebra if and only if the index $(F_{\mathfrak{P}}:\mathbb{Q}_2) = e(2,F/\mathbb{Q}).f(2,F/\mathbb{Q})$ is not divisible by $m(\underline{A}(\mathbb{Q}_2)) = 2$ (Reiner [1], 31.10). The result follows. □

The quaternion group of order 2^t, $t \geq 3$, is defined by the presentation

$$Q_{2^t} = \langle x , y \mid x^{2^{t-2}} = y^2 , y^4 = 1 , x^y = x^{-1} \rangle .$$

2.1.10. *Let Q_{2^t} be a quaternion subgroup of some division ring D. Then the subalgebra $\mathbb{Q}[Q_{2^t}]$ of D is isomorphic to $\underline{A}(F)$, where F is the fixed field of the cyclotomic field $\mathbb{Q}(\zeta_{2^{t-1}})$ under the automorphism $\zeta \longmapsto \zeta^{-1}$. In particular $\mathbb{Q}[Q_8] \cong \underline{A}$ and $\mathbb{Q}[Q_{16}] \cong \underline{A}(\mathbb{Q}(2^{1/2}))$. Moreover for any Q_8 subgroup of Q_{2^t} we have $\mathbb{Q}[Q_{2^t}] = \mathbb{Q}[Q_8] \otimes_{\mathbb{Q}} F$.*

We shall see in 2.1.21 that the real quaternion algebra contains copies of all the quaternion groups. However the reader should have no difficulty in proving this elementary fact for himself.

Proof: The subfield $E = \mathbb{Q}[x]$ of $D_1 = \mathbb{Q}[Q_{2^t}]$ is the 2^{t-1}-th cyclotomic field over $\mathbb{Q}$, and y acts by conjugation on E. Since $(D_1:E) \leq 2$ and D_1 is not commutative we have $(D_1:E) = 2$, so the centre of D_1 is the fixed field F of E under conjugation. It is trivial to verify that the elements $i = x^{2^{t-3}}$ and $j = y$ satisfy $i^2 = j^2 = -1$ and $ij = -ji$. Clearly $i \notin F$, so $D_1 = F[i,j]$. Thus D_1 is an image of $\underline{A}(F)$ with $(D_1:F) = 4$; whence $D_1 \cong \underline{A}(F)$.

In the case of Q_8 we have $E = \mathbb{Q}(\sqrt{-1})$, with fixed field $\mathbb{Q}$. For

Q_{16}, let ζ be a primitive 8-th root of unity. Then the fixed field of $\mathbb{Q}(\zeta)$ under conjugation is $\mathbb{Q}(\theta)$, where $\theta = \zeta+\zeta^{-1}$. Since $\zeta^4 = -1$ it follows that $\theta^2 = \zeta^2 + \zeta^{-2} + 2 = \zeta^{-2}(\zeta^4 + 1) + 2 = 2$, so $F = \mathbb{Q}(\theta) \cong \mathbb{Q}(2^{1/2})$, as claimed.

Finally consider any Q_8 subgroup of Q_{2^t}. Then $\mathbb{Q}[Q_8] = \underline{A}$, so by 2.1.6 we can write $D = \underline{A} \otimes_{\mathbb{Q}} \Gamma$, where Γ is the centralizer of $\underline{A}$ in D. Since $m(D) = 2 = m(\underline{A})$ it follows that Γ is a field. Therefore $\Gamma = F$. □

2.1.11. THEOREM. *The only finite insoluble subgroup of a division ring* D *is* SL(2,5).

Proof: By a theorem of Zassenhaus (Passman [1], Theorem 18.6), if X is an insoluble Frobenius complement then X has a normal subgroup Y of index at most 2 such that $Y \cong SL(2,5) \times M$, where M is a Frobenius complement belonging to $\underline{Z}_0$, of order coprime to 30.

Assume first that $G \cong SL(2,5)$ is a subgroup of a division ring. We prove that $D = \mathbb{Q}[G] \cong \underline{A}(\mathbb{Q}(5^{1/2}))$. Now SL(2,5) is generated by

$$x = \begin{bmatrix} -1 & 1 \\ -1 & 0 \end{bmatrix} \qquad y = \begin{bmatrix} 1 & 0 \\ 1 & 1 \end{bmatrix}$$

and in fact these generators yield the presentation

$$\langle x,y,z \mid x^3 = y^5 = z^2 = 1,\ z = (xy)^2,\ [x,z] = [y,z] = 1 \rangle$$

of SL(2,5), see Passman [1], 13.7. (To see that they do generate the whole group note that $\langle x, y, z\rangle$ clearly has order at least $3\times4\times5$, while SL(2,5) has no subgroup of order 60.

Let $S = \langle 1,y,y^2,y^3,x,xy,xy^2,xy^3\rangle$, and let T be the $\mathbb{Q}$-subspace of D spanned by S. We claim that $D = T$. For $x \neq 1 = x^3$ implies that $x^2 = -x-1 \in T$, and $y \neq 1 = y^5$ implies that $y^4 = -y^3-y^2-y-1 \in T$. From these it follows that $xT \subseteq T$ and $Ty \subseteq T$. Also $z \neq 1 = z^2$ implies that $z = -1$, so $yx = x^{-1}zy^{-1} = -x^2y^4 = (x+1)(-y^3-y^2-y-1) \in T$. Consequently $yT \subseteq \Sigma Ty^i = T$. We have now proved that $SS \subseteq T$. Therefore $D = T$, as claimed.

The above shows that $(D:\mathbb{Q}) \leq 8$. Set $E = \mathbb{Q}(y)$. Then $(E:\mathbb{Q}) = \phi(5) = 4$, since y is a primitive 5-th root of unity. Also D is not commutative, so $(D:\mathbb{Q}) = 8$, E is a maximal subfield of D, $m(D) = 2$, and the centre F of D has degree 2 over $\mathbb{Q}$. But E has a unique quadratic subfield isomorphic to $\mathbb{Q}(5^{1/2})$, so $F \cong \mathbb{Q}(5^{1/2})$ (explicitly if $y_1 = 2(y+y^4)+1$ then $y_1^2 = 5$ and $F = \mathbb{Q}(y_1)$). Let Q be any Sylow 2-subgroup of G. Then Q is quaternion of order 8. Now $\mathbb{Q}[Q] \cong \underline{A}$ by 2.1.10, and we can write $D = \underline{A} \otimes_{\mathbb{Q}} C_D(Q)$ by

2.1.6. It follows that $C_D(Q)$ has dimension 2 over $\mathbb{Q}$. Thus $C_D(Q)$ is a field and therefore is the centre of D. In other words $D = \underline{A} \otimes_{\mathbb{Q}} F \cong \underline{A}(\mathbb{Q}(5^{1/2}))$.

We now show that SL(2,5) is the only insoluble finite subgroup of a division ring. First, no group of the form $G = SL(2,5) \times C$, where C is cyclic of prime order p, embeds into a division ring. For assume that $G \subseteq D^*$ for some D. By 2.1.2 we have $p \geq 5$. Consider the algebra

$$B = \underline{A}(\mathbb{Q}(5^{1/2})) \otimes_{\mathbb{Q}} \mathbb{Q}(\zeta) \cong \underline{A} \otimes_{\mathbb{Q}} \mathbb{Q}(5^{1/2}) \otimes_{\mathbb{Q}} \mathbb{Q}(\zeta)$$

where ζ is a primitive p-th root of unity. Then $\mathbb{Q}(5^{1/2}) \otimes \mathbb{Q}(\zeta)$ is a field since $p \geq 5$ and the unique quadratic subfield of $\mathbb{Q}(\zeta_p)$ is one of $\mathbb{Q}(\sqrt{p})$ or $\mathbb{Q}(\sqrt{-p})$. Thus B is simple. Since B maps homomorphically onto $\mathbb{Q}[G] = \mathbb{Q}[SL(2,5) \times C]$ it embeds into D. But $\mathbb{Q}(\zeta)$ has no real embedding, and 2 has residue degree 2 in the extension $\mathbb{Q}(5^{1/2})$ of $\mathbb{Q}$ (Weiss [1], 6.2.1). This contradicts 2.1.9.

In view of Zassenhaus' theorem quoted above, it remains only to show that if G has a normal SL(2,5) subgroup of index 2 then G cannot be embedded into a division ring. Let S be the normal SL(2,5) subgroup of G, and let Q be a fixed Sylow 2-subgroup of G. Then Q is quaternion of order 16. The subalgebras $\mathbb{Q}[S] \cong \underline{A}(\mathbb{Q}(5^{1/2}))$ and $\mathbb{Q}[Q] \cong \underline{A}(\mathbb{Q}(2^{1/2}))$ of $D = \mathbb{Q}[G]$ are non-isomorphic and of dimension 8 over the field $\mathbb{Q}$. Also $(D:\mathbb{Q})$ is at most $(\mathbb{Q}[S]:\mathbb{Q})(G:S) = 16$. Therefore $(D:\mathbb{Q}) = 16$. Further $S \cap Q$ is quaternion of order 8. By 2.1.6 we have $D = \underline{A} \otimes_{\mathbb{Q}} \Gamma$, where $\underline{A} = \mathbb{Q}[S \cap Q]$ and Γ is the centralizer of $\underline{A}$ in D. Clearly $(\Gamma:\mathbb{Q}) = 4$. If Γ were not a field it would be central simple over $\mathbb{Q}$, so $m(\Gamma) = 2 = m(\underline{A})$, contradicting 2.1.7. Hence Γ is a field. Trivially Γ contains the centres of $\mathbb{Q}[S]$ and $\mathbb{Q}[Q]$. Thus Γ contains subfields isomorphic to $\mathbb{Q}(5^{1/2})$ and $\mathbb{Q}(2^{1/2})$; whence $\Gamma = \mathbb{Q}(2^{1/2},5^{1/2})$

Now $X = \langle x \rangle$ is a Sylow 3-subgroup of S. It follows from the Frattini argument that $G = SP$ for P a Sylow 2-subgroup of $N_G(X)$. The normalizer of a 3-cycle in $Alt5 \cong S/\langle z \rangle$ has order 6 and inverts the 3-cycle. Thus $P \cap S$ is cyclic of order 4 and inverts X. Therefore $|P| = 8$, and since X has no automorphism of order 4, the group P is not cyclic, and hence is quaternion of order 8. By 2.1.10 and 2.1.6 again

$$D = \mathbb{Q}[P] \otimes_{\mathbb{Q}} C_D(P) \supseteq \mathbb{Q}[P] \otimes_{\mathbb{Q}} \mathbb{Q}(2^{1/2},5^{1/2}) \cong D,$$

so $D = \mathbb{Q}[P] \otimes_{\mathbb{Q}} \mathbb{Q}(2^{1/2},5^{1/2})$. Let $P = \langle i,j \rangle$, where i centralizes x and j inverts x. Then $x \in C_D(i) = \mathbb{Q}(2^{1/2},5^{1/2},i)$. But $x^3 = 1$, so in the complex numbers $-(1 \pm i3^{1/2})/2 \in \mathbb{Q}(2^{1/2},5^{1/2},i)$. Thus $3^{1/2} \in \mathbb{Q}(2^{1/2},5^{1/2})$. A direct calculation shows that this is false. The result follows.

For an explicit embedding of SL(2,5) into $\underline{A}(\mathbb{R})$ see 2.1.21. □

We have now proved Part c) of 2.1.4, and we set about the proof of Part b). Unless otherwise stated, G denotes a finite subgroup of a division ring of characteristic zero, and $D = \mathbb{Q}[G]$ is the $\mathbb{Q}$-subalgebra generated by G. We need a number of preliminary results.

2.1.12. a) *If* n, r *and* s *are mutually coprime then*

$$\gamma(n,rs) = \text{L.C.M.}(\ \gamma(n,r)\ ,\ \gamma(n,s)\ )$$

b) *Let* q *be a prime dividing* n−1 *and let* i *be a positive integer.*

i) *If* $q = 2$ *and* $n \equiv -1 \bmod 4$ *let* $n^2-1 = 2^d t$ *where* $2 \nmid t$. *Then*

$\gamma = \gamma(n,2) = 1$ *and* $n^\gamma - 1$ *is exactly divisible by* 2.

$\gamma = \gamma(n,2^2) = \ldots = \gamma(n,2^d) = 2$ *and* $n^\gamma - 1$ *is exactly divisible by* 2^d.

$\gamma = \gamma(n,2^i) = 2^{i-d+1}$ *if* $i \geq d$, *and* $n^\gamma - 1$ *is exactly divisible by* 2^i.

ii) *In all cases not covered by* i) *let* $n-1 = q^d t$ *where* $q \nmid t$. *Then*

$\gamma = \gamma(n,q) = \ldots = \gamma(n,q^d) = 1$, *and* $n^\gamma - 1$ *is exactly divisible by* q^d

$\gamma = \gamma(n,q^i) = q^{i-d}$ *if* $i \geq d$, *and* $n^\gamma - 1$ *is exactly divisible by* q^i.

c) *In general if* q *is a prime and* m *and* i *are positive integers with* $q \nmid m$ *then* $\gamma(m,q^i) = \gamma(m,q)q^a$ *where* $a = a(i) \geq 0$.

Proof: a) This is clear if one considers the multiplicative order of n in the ring $(\mathbb{Z}/r\mathbb{Z}) \times (\mathbb{Z}/s\mathbb{Z})$.

b) Note first the following: assume, for some k and s, that n^k-1 is exactly divisible by q^s, where either $q \geq 2$ or $s \geq 2$. Thus $n^k = 1+mq^s$, where $q \nmid m$. Then $n^{kq} = (1+q^s m)^q \equiv (1+q^{s+1}m) \bmod q^{s+2}$, so $n^{qk}-1$ is exactly divisible by q^{s+1}.

Suppose we also have $k = \gamma(n,q^s)$. Then k divides $\gamma(n,q^{s+1})$, which in turn divides kq. Since q^{s+1} does not divide n^k-1 we have $\gamma(n,q^s) \neq \gamma(n,q^{s+1})$, whence $\gamma(n,q^{s+1}) = q\gamma(n,q^s)$. This clearly establishes b ii). In the remaining case $\gamma(n,2) = 1$ and n−1 is exactly divisible by 2, say $n = 4m-1$. The argument of the first paragraph applies to $n^2-1 = 8(2m^2-m)$, thereby proving b i).

c) If $q \mid (m-1)$ then this is Part b). Suppose $q \nmid (m-1)$, so $q \neq 2$, and let k be the least integer for which q divides m^k-1 , i.e. $k = \gamma(m, q)$. If q^d is the highest power of q dividing m^k-1 then the first paragraph of the proof shows that $\gamma(m,q) = \ldots = \gamma(m,q^d) = k$ and $\gamma(m,q^i) = q^{i-d}k$ for $i \geq d$, as claimed. □

2.1.13. *Suppose that* G *is soluble, and let* Q *be any quaternion* *subgroup of G of order 2^t.*

a) *If* $t = 3$ *then* $C_G(Q) \cong C_2 \times M$ *where* $M \in \underline{Z}$ *has odd order* m *and* $\gamma(2,m)$ *is odd.*

b) *If* $t \geq 3$ *then* $C_G(Q) \cong C_2$.

Proof: a) The centralizer of Q in any quaternion group containing it is the centre of that group. Thus $\langle -1 \rangle$ is the Sylow 2-subgroup of $C_G(Q)$. Therefore $C_G(Q) = \langle -1 \rangle \times M$ for some subgroup M of odd order. Now $\mathbb{Q}[Q] \cong \underline{A}$ by 2.1.10. Let $g \in M$ have prime order p. Then by 2.1.6, in D

$$\mathbb{Q}[\underline{A}, g] \cong \underline{A} \otimes_{\mathbb{Q}} \mathbb{Q}(g)$$

is a division ring. Since g is a primitive p-th root of unity and p is odd, the field $\mathbb{Q}(g)$ has no real embedding. The residue degree of 2 in $\mathbb{Q}(g)$ is $\gamma(2,p)$ (Weiss [1], 7.4.3). By 2.1.9 this must be odd. It follows from 2.1.12 that $\gamma(2,m)$ is odd.

b) It is enough to prove the result in the case where Q has order 16. Let $g \in C_G(Q)$ have odd prime order p. Then, using 2.1.10, in D we have

$$\mathbb{Q}[Q,g] \cong \mathbb{Q}[Q] \otimes_{\mathbb{Q}(2^{1/2})} \mathbb{Q}(2^{1/2},g) \cong \underline{A} \otimes_{\mathbb{Q}} \mathbb{Q}(2^{1/2},g).$$

But $\mathbb{Q}(g)$ has no real embedding, and 2 is a square in $\mathbb{Q}(2^{1/2})$, so the ramification index of 2 in $\mathbb{Q}(2^{1/2},g) \supseteq \mathbb{Q}(2^{1/2})$ is even. This, however, contradicts 2.1.9. It follows that $C_G(Q)$ is a 2-group, as required. □

Recall that $O_2(G)$ denotes the maximal normal 2-subgroup of G.

2.1.14. *Suppose* $Q = O_2(G)$ *is quaternion of order 8, and that* G *is soluble. Then* G *is isomorphic to one of the following groups:*

a) $Q \times M$ *where* M *is a* $\underline{Z}$*-group of odd order* m *with* $\gamma(2,m)$ *odd.*

b) $SL(2,3) \times M$ *where* M *is a* $\underline{Z}$*-group of order* m *coprime to* 6 *with* $\gamma(2,m)$ *odd.*

c) *the binary octahedral group, of order 48.*

Proof: Consider the normal subgroup $K = QC_G(Q)$ of G. Since $G/C_G(Q)$ embeds into $\mathrm{Aut}(Q) \cong \mathrm{Sym}(4)$ and $|Q \cap C_G(Q)| = 2$, it follows that G/K embeds into Sym(3). By 2.1.13 we have $C_G(Q) = \langle -1 \rangle \times M$ where $m = |M|$ is odd and $\gamma(2,m)$ is also odd.

If $(G:K) = 1$ then (a) holds. Suppose $(G:K) = 3$. Since $\gamma(2,3) = 2$ (that is, $2^2 = 1+3$) we have $3 \nmid m$. Hence the Sylow 3-subgroups of G have order 3 and do not centralize Q. Let C be such a Sylow subgroup. Then $L = \langle Q,C\rangle = Q \rtimes C \cong SL(2,3)$, and $G = LM$. We claim that $\mathbb{Q}[L] = \mathbb{Q}[Q]$. We can choose c, a generator of C, and $Q = \langle\, x,y \mid x^2 = y^2,\ y^4 = 1,\ x^y = x^{-1} \,\rangle$, such that c acts on Q by cyclically permuting y, x, and xy. But conjugation by the element $t = -(1+x+y+xy)/2 \in \mathbb{Q}[Q]$ also has the same effect. Thus ct^{-1} centralizes $\mathbb{Q}[Q]$, and hence t. Therefore t and c commute. But now t and c are both cube roots of unity in the field $\mathbb{Q}[t,c]$. It follows that $c \in \langle t\rangle$. Hence $\mathbb{Q}[L] = \mathbb{Q}[Q]$, as claimed. But M centralizes Q, so it centralizes $\mathbb{Q}[L]$, and hence L. Therefore in this case $G \cong SL(2,3) \times M$ and we have (b).

Now assume that (G:K) is even. Then a Sylow 2-subgroup Q_1 of G is quaternion of order 16. By 2.1.6 we can write $\mathbb{Q}[G] = \underline{A} \otimes_{\mathbb{Q}} \Gamma$ where $\underline{A} = \mathbb{Q}[Q]$ and Γ is the centralizer of Q in $\mathbb{Q}[G]$. In particular $M \subseteq \Gamma$. Let Z be the centre of Γ. Then $\underline{A} \otimes_{\mathbb{Q}} \Gamma \cong \underline{A}(Z) \otimes_Z \Gamma$. It follows from 2.1.7 that $m(\Gamma)$ is odd. By 2.1.10 there exists a central element g of $\mathbb{Q}[Q_1]$ with $g^2 = 2$ and $\mathbb{Q}[Q_1] = \underline{A} \otimes_{\mathbb{Q}} \mathbb{Q}(g)$. Clearly $g \in \Gamma$. Let E be a maximal subfield of Γ containing g. Then $(Z(g):Z)$ divides $(E:Z) = m(\Gamma)$, which is odd. Since $g^2 = 2$ we must have $g \in Z$. But then M centralizes g and Q, whence it centralizes $\mathbb{Q}[Q_1]$. This implies that $M = \langle 1\rangle$, by 2.1.13. Thus $K = Q$, $G/K \cong Sym(3)$, and $G/\langle -1\rangle \cong Sym(4)$, the automorphism group of Q. Consequently we have an extension

$$1 \longrightarrow C_2 \longrightarrow G \longrightarrow Sym(4) \longrightarrow 1$$

such that the Sylow 2-subgroup of G is quaternion. Thus G is a group of order 48 that is uniquely determined up to isomorphism. It is in fact the binary octahedral group.

The uniqueness of G can be seen as follows. Let $C = \langle c\rangle$ be a Sylow 3-subgroup of G, P a Sylow 2-subgroup of $N_G(C)$, and choose $Q_1 \supseteq P$. Then $G = CQP$ by the Frattini argument and $Q_1 = QP$. Clearly $|P \cap Q| = 2$, so P is cyclic of order 4. Let $x \in Q_1$ have order 8 and let y generate P. Then $Q_1 = \langle x, y\rangle$, $y \in Q_1 \backslash Q = xQ$, and $Q = \langle x^2, x^{-1}y\rangle$.

Either c or c^{-1} conjugates $\langle x^2\rangle$ to $\langle x^{-1}y\rangle$, so we can pick c such that x^{2c} is either $x^{-1}y$ or $y^{-1}x$. But $y^{-1}x = x^{-1}y^{-1}$. Thus replacing y by y^{-1} if necessary we may assume that $x^{2c} = x^{-1}y$. Suppose $(x^{-1}y)^c = xy$. Then $cx = x^{-1}ycy^{-1} = x^{-1}c^2 = (cx)^{-1}$. Hence $cx \in \langle x^4\rangle$, and so $c \in \langle x\rangle$. This is impossible and therefore $(x^{-1}y)^c = xy^{-1}$. Consequently G is given by the presentation

$$\langle\, x, y, c \mid x^4=y^2,\ y^4=1,\ x^y=x^{-1},\ x^{2c}=x^{-1}y,\ (x^{-1}y)^c=xy^{-1},\ c^y=c^2,\ c^3=1 \,\rangle.$$

A shorter but less explicit presentation of G is given by

$$\langle r, s \mid r^2 = s^3 = (rs)^4,\ r^4 = 1 \rangle$$

where we set $r = x^2y$ and $s = -c$ in the above notation. □

2.1.15. *Let $G = L \rtimes Q$ be the split extension of a cyclic group L of odd order by a quaternion group Q of order 2^t. Then either $t = 3$ and $G = L \times Q$, or an element of order 2^{t-1} of Q centralizes L and an element of order 4 of Q inverts L.*

Proof: Suppose P is a quaternion subgroup of order 8 of D^*. It follows from 2.1.6 and 2.1.10 that $D = \underline{A}(F) \otimes_F \Gamma$, where F is the centre of D and $\Gamma = C_D(P)$. Since D is a division ring and $m(\underline{A}(F)) = 2$, the Schur index $m(\Gamma)$ must be odd, by 2.1.7. Thus $m(D) = m(\underline{A}(F)).m(\Gamma)$ is not divisible by 4. Therefore neither is $(E:E \cap F) = (EF:F)$ for any subfield E of D for which E is a Galois extension of $E \cap F$.

We use the presentation

$$Q = \langle x,y \mid x^{2^{t-2}} = y^2,\ y^4 = 1,\ x^y = x^{-1} \rangle$$

for Q. Consider first the case $t \geq 3$. Suppose x does not centralize L. Then x acts as an automorphism of $E = \mathbb{Q}[L]$ of order a power of 2. By the above x must act as an involution on E, so x^2 centralizes E. But then x and y act as distinct automorphisms of order 2 on $E(x^2)$. This contradicts the first paragraph. Thus x centralizes L. By 2.1.13 no Sylow subgroup of L is centralized by y, and inversion is the only automorphism of order 2 of a cyclic group of odd prime-power order. Hence y inverts L.

Now consider the case $t = 3$. If neither x nor y centralizes L then they must determine the same automorphism of order 2 of $E = \mathbb{Q}[L]$ (by the first paragraph again); whence xy centralizes L. Therefore, by a change of notation we may assume that x centralizes L. If y does too, then $G = L \times Q$.

Suppose y does not centralize L, so y inverts at least one Sylow subgroup P of L. We have to show that y inverts every Sylow subgroup of L. Let P have order p^a, say, and let $D_1 = \mathbb{Q}[PQ]$. Clearly $E_1 = \mathbb{Q}[P,x] \cong \mathbb{Q}(\zeta_{p^a}, \zeta_4)$ is a maximal subfield of D_1. Let Z_1 be the centre of D_1. We claim that the rational prime 2 is unramified in the extension E_1/Z_1. For if not, since $(E_1:Z_1) = 2$ we have $e(2,E_1/Z_1) = 2$, and so in the ring of integers $\mathcal{O}$ of E_1, the prime 2 generates the square of a prime ideal $\mathfrak{P}$. Then y normalizes $\mathfrak{P}$ and acts on $\mathcal{O}/\mathfrak{P}$. Also y inverts $P \subseteq \mathcal{O}$ and $P \cap (1 + \mathfrak{P}) = \{1\}$

since p is odd and $\mathrm{char}(O/\mathfrak{P}) = 2$. Consequently y acts non-trivially on $O/\mathfrak{P}$ and $f(2,E_1/Z_1) \geq 1$. This contradiction proves that $e(2,E_1/Z_1) = 1$.

Now $e(2, E_1/\mathbb{Q}) = \Phi(4) = 2$, so the multiplicativity of e implies that $e(2, Z_1/\mathbb{Q}) = 2$. Let Z be the centre of $D = \mathbb{Q}[G]$. Since $Z_1 \subseteq C_{E(x)}(y) \subseteq Z$ we have that $e(2, Z/\mathbb{Q})$ is even. Also D contains the subalgebra $\mathbb{Q}[Q].Z \cong \underline{A}(Z)$, by the simplicity of the latter. Hence the field Z is real by 2.1.9. This means that no Sylow subgroup of L is centralized by y, for otherwise Z would contain an odd order root of unity. Thus y inverts L. □

2.1.16. *Suppose G is soluble with $O_2(G)$ cyclic. Then G is a split extension of a 2′-group by a 2-group.*

Proof: The Fitting subgroup N of G is cyclic. So $G' \subseteq N$ since $C_G(N) \subseteq N$, and therefore $G = O_{2,2',2}(G)$. Also $\mathrm{Aut}(O_2(G))$ is a 2-group. The conclusion now follows from the Schur-Zassenhaus Theorem. □

2.1.17. *Let G be a soluble subgroup of a division ring of characteristic zero, and assume that the Sylow 2-subgroups of G are quaternion. Then G is isomorphic to one of the groups listed in* 2.1.4 b).

Proof: Assume first that $O_2(G)$ is quaternion. If $|O_2(G)| \geq 16$ then its automorphism group is a 2-group. $C_G(O_2(G))$ is also a 2-group by 2.1.13. Therefore G is a 2-group and b ii) holds (with $m = 1$). If $O_2(G)$ is quaternion of order 8 then the conclusion follows from 2.1.14.

Now suppose that $O_2(G)$ is cyclic. We have $G = M \rtimes Q$ for some 2′-subgroup M of G, by 2.1.16. It remains to show that M is cyclic, for then 2.1.15 is applicable and we have b ii).

M is a $\underline{Z}$-group, say $M = L_1 \rtimes L_2$, where L_1 and L_2 are cyclic of coprime order. Also Q normalizes L_1, and by the Frattini argument we can choose L_2 also to be normalized by Q. Let c_i be a generator of L_i, and suppose we can find an element z of Q which inverts both L_1 and L_2. Then $[c_1,c_2] \in L_1$ and

$$[c_1,c_2]^{-1} = [c_1,c_2]^z = [c_1^z,c_2^z] = [c_1^{-1},c_2^{-1}] = [c_1,c_2]^{c_2^{-1}c_1^{-1}}$$

But $c_2^{-1}c_1^{-1}$ and $[c_1,c_2]$ are both of odd order. Hence $[c_1,c_2] = 1$; whence M is abelian, and consequently cyclic by 2.1.2.

We have to show the existence of an element z with the above property. If $|Q| \geq 16$ then, in the notation of 2.1.15, the element y inverts both L_1 and L_2. Now suppose $|Q| = 8$. Let $L_3 \neq \langle 1 \rangle$ be the kernel of the action of L_2 on L_1. The action of Q on L_2 is either trivial or by inversion. Also L_3 is normalized by Q. By 2.1.15 applied to $L \rtimes Q$, where $L = L_1L_3$, either Q centralizes L or y (say) inverts L. In the first case it follows from 2.1.3 that Q centralizes L_2 and hence M, contrary to the assumption that $O_2(G)$ is cyclic. In the second case y inverts L_3 and hence L_2, so we have the element z, as required. □

We now come to the proof of 2.1.5. In what follows ζ_i always denotes a primitive i-th root of unity. Assume that $G = C_m \rtimes C_n$ is non-cyclic. We may assume that every Sylow subgroup of C_n acts non-trivially on C_m. Let $n = ab$ where C_a is the kernel of the action of C_n on C_m, and let $E = \mathbb{Q}(\zeta_m, \zeta_a)$. Then E has an automorphism of order b fixing ζ_a, and we may form the crossed product ($E[G]$, E, G, C_b), where G acts on E in the obvious fashion. Then $E[G]$ is simple (Reiner [1], 29.6, or Cohn [2], Vol. 2, p. 377). If G is a subgroup of some division ring D then $\mathbb{Q}[G] \subseteq D$ is a homomorphic image of $E[G]$, and hence $E[G] \cong \mathbb{Q}[G]$ is a division ring. Trivially if $E[G]$ is a division ring then G is a subgroup of a division ring.

Suppose $n = q_1^{b_1} \times \ldots \times q_r^{b_r}$ where the q_i are distinct primes. Write $b_i = \alpha_i+\beta_i$, where $C_{q_i^{\alpha_i}}$ is the kernel of the action of $C_{q_i^{b_i}}$ on C_m. By 1.4.17, the crossed product $E[G]$ is a division ring if and only if for each $1 \leq i \leq r$ the ring $E[C_m](C_{q_i^{b_i}} \times \Pi_{j \neq i} C_{q_j^{\alpha_j}})]$ is a division ring, and the latter is clearly the crossed product associated with the group $C_m \rtimes (C_{q_i^{b_i}} \times \Pi_{j \neq i} C_{q_i^{\alpha_i}})$. Therefore $C_m \rtimes C_n \in \underline{Z}$ if and only if

$$C_m \rtimes C_{q_i^{b_i}} \times \Pi_{j \neq i} C_{q_i^{\alpha_i}} \in \underline{Z} \quad \text{for each } i.$$

The case $G = C_{p^a} \rtimes C_{q^b}$ is of fundamental importance. Let $b = \alpha + \beta$, where q^α is the order of the kernel of the action of C_{q^b} on C_{p^a}. Then q^β divides $p-1$. To avoid trivial cases assume that $\alpha, \beta \geq 1$. This implies, in particular, that p is odd. If $q = 2$ and $p \equiv -1 \bmod 4$ write $p^2-1 = 2^d t$ where $2 \nmid t$; otherwise let $p-1 = q^d t$, where $q \nmid t$. Let r be any positive integer prime to pq, and consider the group $G = (C_{p^a} \rtimes C_{q^b}) \times C_r$ and the crossed product ($E[G]$, E, G, C_{q^β}) as above, where $E = \mathbb{Q}(\zeta_{p^a}, \zeta)$ and $\zeta = \zeta_{rq^\alpha}$. At one point we require the following result.

2.1.18. *Let* $k = \mathbb{Q}_p$ *and* $K = \mathbb{Q}_p(\zeta_{p^a})$ *where* p *is an odd prime and* $a \geq 1$. *Suppose* α *is a root of unity in* k *such that* $\alpha = N_{K/k}(\beta)$, *the norm of* β, *for* *some* $\beta \in K$. *Then* $\alpha = 1$.

Proof: We may clearly assume that $a = 1$. Also we may choose β to be a unit in K (Serre [1], p. 88). Now K/k is totally ramified and $(K:k) = p-1$. If $\mathfrak{P}$ is the prime of K and if σ is a k-automorphism of K then $\beta^\sigma \equiv \beta \bmod \mathfrak{P}$. Thus

$$\alpha = N_{K/k}(\beta) = \Pi_\sigma \beta^\sigma \equiv \beta^{p-1} \equiv 1 \bmod \mathfrak{P}$$

since the residue field of K has p elements. Thus $\alpha \equiv 1 \bmod (k \cap \mathfrak{P}) = (p)$. But α is a root of unity and $p \neq 2$. Hence $\alpha = 1$ (Serre [2], § 3.1). (For this argument we are indebted to C. Bushnell.) □

2.1.19. LEMMA. *The crossed product* $R = E[G]$ *is a division ring if and only if*

either $q = 2$, $\alpha = \beta = 1$, *and* $r = 1$

or $q = 2$, $p \equiv -1 \bmod 4$, $\alpha = 1$, *and* $2 \nmid \delta = \gamma(p,rq^\alpha)/\gamma(p,q^\alpha)$

or $\alpha \geq d$ *and* $q \nmid \delta$.

Proof: The centre Z of R is a number field, so $e(R) = m(R)$ by 2.1.8. Thus R is a division ring if and only if $e(R) = q^\beta$. Now R is a cyclic algebra by definition, so by Theorem 7.17 of Albert [1], $e(R)$ is the least positive integer e for which $\zeta^e \in N_{E/Z}$, the group of norms from E to Z. By the Hasse Norm Theorem (Cassels & Fröhlich [1], p. 185), $\zeta^e \in N_{E/Z}$ if and only if $\zeta^e \in N_{E_\mathfrak{p}/Z_\mathfrak{P}}$ for all primes (finite and infinite) $\mathfrak{P}$ of Z and primes $\mathfrak{p}$ of E extending $\mathfrak{P}$. There are several cases to consider.

a) If $\mathfrak{P}$ is a complex prime then $Z_\mathfrak{P} = \mathbb{C} = E_\mathfrak{p}$ and ζ is trivially a norm.

b) $\mathfrak{P}$ lies over a rational prime $k \neq p$. Consider the following diagram of field extensions:

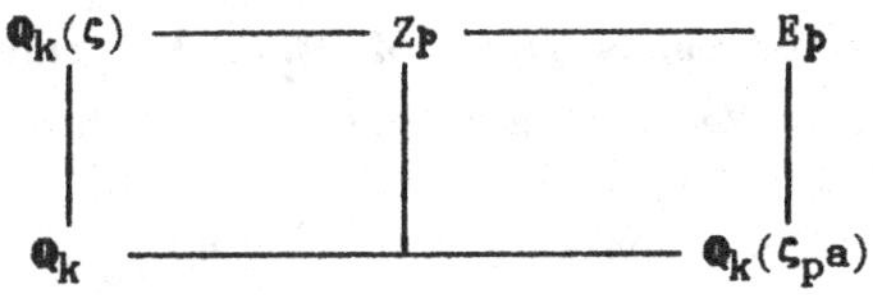

The ramification index of the rational prime k in $\mathbb{Q}(\zeta_{p^a})$ is 1 since $k \nmid p^a$ (Weiss [1], 7.2.4), so the extension $\mathbb{Q}_k(\zeta_{p^a})$ of the k-adic field $\mathbb{Q}_k$ is unramified. From the above diagram it follows that $E_\mathfrak{p}$ is an unramified extension of $Z_\mathfrak{P}$. But then ζ, being a unit, is automatically a norm (Cassels & Frölich [1], p. 29, Corollary).

c) $\mathfrak{P}$ is real, so $Z_\mathfrak{P} = \mathbb{R}$. Since $q \mid (p-1)$ we have $p^a \geq 2$ and therefore $\mathfrak{p}$ is not real. Thus $E_\mathfrak{p} = \mathbb{C}$. But $(E_\mathfrak{p}:Z_\mathfrak{P})$ always divides $(E:Z) = q^\beta$, so $q = 2$. Also $\zeta \in Z \subseteq \mathbb{R}$ and $\zeta \neq 1$. Thus $\zeta = -1$; whence $\alpha = 1 = r$. Consequently in this case we need to consider when $(-1)^e$ is a norm from $\mathbb{C}$ to $\mathbb{R}$. The norm map is simply the square of the modulus. Therefore e must be even. Conversely note that if $rq^\alpha = 2$ and $\beta = 1$ then some element of G acts on E by complex conjugation, Z is real, and E[G] is a division ring, so this case does occur.

d) $\mathfrak{P}$ lies over the rational prime p. Consider the following diagram of field extensions:

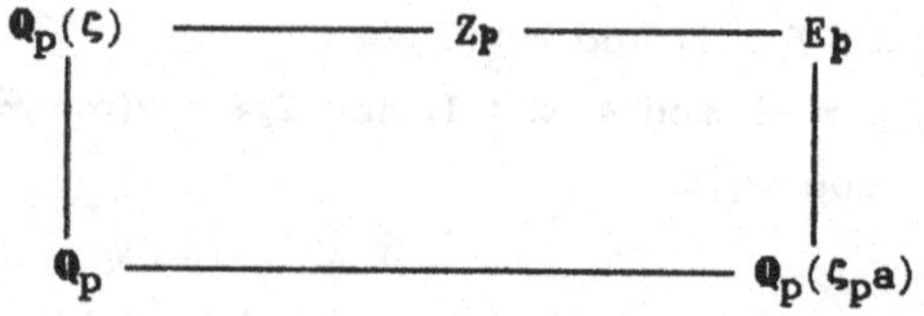

The bottom extension is totally ramified and therefore has degree $\Phi(p^a) = p^{a-1}(p-1)$. Also $(E_\mathfrak{p}:Z_\mathfrak{P}) = (E:Z) = q^\beta$ since $\mathfrak{P}$ does not split on adjoining ζ_{p^a}. Thus $h = (Z_\mathfrak{P}:\mathbb{Q}_p(\zeta)) = (p-1)p^{a-1}q^{-\beta}$. Using the general properties of the norm map we have:

$$\zeta^e \in N_{E_\mathfrak{p}/Z_\mathfrak{P}} \iff N_{Z_\mathfrak{P}/\mathbb{Q}_p(\zeta)}(\zeta^e) \in N_{E_\mathfrak{p}/\mathbb{Q}_p(\zeta)} \quad \text{(Lang [1], p. 207)}$$

$$\iff \zeta^{eh} \in N_{E_\mathfrak{p}/\mathbb{Q}_p(\zeta)}$$

$$\iff N_{\mathbb{Q}_p(\zeta)/\mathbb{Q}_p}(\zeta)^{eh} = N_{\mathbb{Q}_p(\zeta)/\mathbb{Q}_p}(\zeta^{eh}) \in N_{E_\mathfrak{p}/\mathbb{Q}_p}$$

Now $(\mathbb{Q}_p(\zeta):\mathbb{Q}_p) = \gamma(p, rq^\alpha) = \gamma$, say, and the extension $\mathbb{Q}_p(\zeta)/\mathbb{Q}_p$ is unramified. Thus some generator of the Galois group of $\mathbb{Q}_p(\zeta)/\mathbb{Q}_p$ raises ζ to its p-th power (Reiner [1], 5.11). Hence if $s = 1+p+\ldots+p^{\gamma-1}$ then $N_{\mathbb{Q}_p(\zeta)/\mathbb{Q}_p}(\zeta) = \zeta^s$ and so $\zeta^e \in N_{E_\mathfrak{p}/Z_\mathfrak{P}}$ if and only if $\zeta^{ehs} \in N_{E_\mathfrak{p}/\mathbb{Q}_p}$, which by 2.1.18 happens if and only if $\zeta^{ehs} = 1$. Thus $\zeta^e \in N_{E_\mathfrak{p}/Z_\mathfrak{P}}$ if and only if rq^α divides *ehs*. Now

$$ehs = ep^{a-1}(p-1)q^{-\beta}(1+p+\ldots+p^{\gamma-1}) = eq^{-\beta}(p^\gamma - 1)p^{a-1}.$$

By definition r divides $p^{\gamma(p,r)} - 1$, which in turn divides $p^{\gamma}-1$ by 2.1.12 a). If $m = \gamma(p,q^{\alpha})$ then

$$p^{\gamma}-1 = p^{m\delta}-1 = (p^m-1)\Sigma_{0\leq i<\delta}\, p^{im} \equiv \delta(p^m-1) \bmod q(p^m-1).$$

since $p^m \equiv 1 \bmod q$. Thus the least e, such that $q^{\alpha}|ehs = eq^{-\beta}(p^{\gamma}-1)p^{a-1}$, is q^{β}, if and only if q^{α} is the exact power of q dividing $p^{\gamma}-1$, which in view of the above calculation is the case if and only if $q\nmid\delta$ and q^{α} is the exact power of q dividing p^m-1. By 2.1.12 the latter condition holds if and only if

either $\quad q = 2$, $p \equiv -1 \bmod 4$, and $\alpha = 1$,

or $\quad \alpha \geq d$.

Collecting the above together we see that the least e for which $\zeta^e \in N_{E/Z}$ is q^{β}, if and only if

either $\quad q = 2$, $\alpha = \beta = 1$, and $r = 1$,

or $\quad q = 2$, $p \equiv -1 \bmod 4$, $\alpha = 1$, and $2\nmid\delta$,

or $\quad \alpha \geq d$ and $q\nmid\delta$.

The result follows. □

2.1.20. *Let p, q, r and s be distinct primes. Then the following groups G are not embeddable into division rings:*

a) $G = (C_p \times C_r) \rtimes C_{q^b}$, *where* $q^b \neq 4$ *and* C_{q^b} *acts non-trivially on both* C_p *and* C_r.

b) $G = (C_p \times C_r \times C_s) \rtimes C_4$, *where* C_4 *inverts* C_p *and* C_r *and centralizes* C_s.

Proof: a) Suppose G is embeddable into a division ring $D = \mathbb{Q}[G]$ with centre F. Then $m(D) = q^{\delta}$ for some δ. Consider the subgroup $H = C_p \rtimes C_{q^b}$ of G and the subalgebra $A = \mathbb{Q}[H]$ of D. Then A is a cyclic algebra with centre Z, say, and $m(A) = q^{\beta}$ in our usual notation.

By 2.1.8 there exists a prime $\mathfrak{p}$ of F such that for $D_{\mathfrak{p}} = D \otimes_F F_{\mathfrak{p}}$ we have $m(D_{\mathfrak{p}}) = m(D) = q^{\delta}$, so $D_{\mathfrak{p}}$ is a division ring. The subfield of $D_{\mathfrak{p}}$ generated by Z and the completion of $\mathbb{Q}$ is a completion, Z_P say, of Z. Thus $D_{\mathfrak{p}}$ contains the subalgebra $A.Z_P$, which is a homomorphic image of the simple algebra $A_P = A \otimes_Z Z_P$. Thus A_P is also a division algebra, and in particular $m(A_P) = q^{\beta}$. The proof of 2.1.19 shows that this is only possible if either P is real and $q^b = 4$, or $p \in P$. Thus here $p \in P \subseteq \mathfrak{p}$. In the same way $r \in P$. This contradiction proves a).

b) Repeat the above proof with $H = (C_p \rtimes C_4) \times C_s$. Since $s \neq 1$

the proof of 2.1.19 shows that $\mathbb{P}$ is not real and so $p \in \mathbb{P} \subseteq \mathfrak{P}$. Similarly $r \in \mathfrak{P}$. The proof is complete. □

Proof of 2.1.5: Let $G = C_m \,]\, C_n$ be a $\underline{Z}$-group, and assume that G is not cyclic, and if $n = 4$ that C_4 does not invert C_m. We may suppose that every Sylow subgroup of C_n acts non-trivially on C_m. If G_o is the centralizer of C_n in C_m then G_o is a Hall subgroup of G which is also a direct factor. Clearly G_o is cyclic.

Let $n = q_1^{b_1} \times \ldots \times q_r^{b_r}$ and $m/|G_o| = p_1^{a_1} \times \ldots \times p_s^{a_s}$ where the p_i, q_j are distinct primes and the $a_i, b_j \geq 1$. From 2.1.20 (and our assumption that G is not of type b)) it follows that each Sylow subgroup of C_n acts non-trivially on a unique Sylow subgroup of C_m. Let G_i be generated by the Sylow p_i-subgroup of C_m and the Sylow subgroups of C_n acting non-trivially on it. Then $G = G_o \times G_1 \times \ldots \times G_s$, where for $i = 1, 2, \ldots$ we have

$$G_i = C_{p_i^{a_i}} \,]\, (C_{q_1^{b_1}} \times \ldots \times C_{q_t^{b_t}}).$$

Let $G_i = C_{p^a} \,]\, (C_{q_1^{b_1}} \times \ldots \times C_{q_t^{b_t}})$ be as above, and consider some prime power $q^b = q_i^{b_i}$. Then in the notation of 2.1.19, either $q^b = 4$ and $p \equiv -1 \bmod 4$, or $\alpha \geq d$. By definition of d we have $\alpha \geq d$ if and only if $q^{\alpha+1} \nmid (p-1)$ or (p^2-1), as appropriate. Finally, by 2.1.19 again we have that $q \nmid \gamma(p,rq^{\alpha})/\gamma(p,q^{\alpha})$ for all prime powers r dividing $|G|/|G_i|$. By 2.1.12 a) this is equivalent to $q \nmid \gamma(p,q^{\alpha}|G|/|G_i|)/\gamma(p,q^{\alpha})$, which is in turn equivalent to $q\gamma(p,q^{\alpha}) \nmid \gamma(p,|G|/|G_i|)$. Thus G satisfies the conditions specified in 2.1.5.

Conversely suppose that G satisfies the conditions of 2.1.5. If G is cyclic it is embeddable in $\mathbb{C}$. If $G = C_m \,]\, C_4$ is as in b) then G is isomorphic to the subgroup $\langle \zeta_m, j \rangle$ of the quaternion algebra $\underline{A}(\mathbb{R})$, where $\zeta_m \in \mathbb{C} = \mathbb{R} + \mathbb{R}i$. Finally assume that G is as in 2.1.5 c). Write $n = \prod q_i^{b_i}$ where the q_i are distinct primes, and suppose the kernel of the action of $C_{q_i^{b_i}}$ on C_m has order $q_i^{\alpha_i} < q_i^{b_i}$. Set $H_i = C_m \,]\, C_{q_i^{b_i}} \times \prod_{j \neq i} C_{q_j^{\alpha_j}}$. Then $H_i = C_{p^a} \,]\, C_{q_i^{b_i}} \times C_r$ for a suitable prime power p^a and integer r co-prime to pq_i. The conditions of 2.1.5 c), with 2.1.12, are enough to ensure that H_i is a $\underline{Z}$-group, by 2.1.19. Consequently G is also a $\underline{Z}$-group, by 1.4.17. The proof is complete. □

Finally we come to the sufficiency of 2.1.4.

2.1.21. *The groups listed in 2.1.4 are all embeddable into division rings of characteristic zero.*

Proof: $\underline{Z}$-groups are embeddable into division rings by definition. Consider the groups in Part b). The groups $L \rtimes Q$ of b ii) are subgroups of $\underline{A}(\mathbb{R})$. The algebra $\underline{A}$ contains copies of Q_8 (generated by i and j) and SL(2,3) (generated by i, j, and $-(1+i+j+ij)/2$). Let M be a $\underline{Z}$-group as in b iii) or b iv). Then M embeds into a division ring D with centre $F \subseteq \mathbb{Q}(\zeta_m)$ and odd Schur index dividing $m = |M|$. This follows from the construction of the crossed product division algebra $\mathbb{Q}[G]$ associated with G (cf. remarks before 2.1.18). The prime 2 is unramified in $\mathbb{Q}(\zeta_m)$, and by assumption $\gamma(2,m)$ is odd. Thus $\underline{A}(F) \subseteq \underline{A}(\mathbb{Q}(\zeta_m))$ is a division ring by 2.1.9. Since $m(\underline{A}(F)) = 2$ and m(D) is odd, $\underline{A}(F) \otimes_F D$ is also a division algebra. This algebra clearly contains copies of $Q_8 \times M$ and $SL(2,3) \times M$.

The binary octahedral group and SL(2,5) are both subgroups of $\underline{A}(\mathbb{R})$. The former is generated by

$$j, \quad (1+i)/\sqrt{2}, \quad -(1+i+j+ij)/2,$$

and the latter by

$$y, \quad x = (y^2-y + (y^2-1)j)/\sqrt{5},$$

where y is a primitive 5-th root of unity in $\mathbb{C}$ and x and y correspond to the generators in the presentation of SL(2,5) used in the proof of 2.1.11. □

In fact the group of quaternions of norm 1 maps homomorphically onto the 3-dimensional rotation group $O(3,\mathbb{R})$ (Lam [1], Theorem 3.1), the kernel consisting of $\{\pm 1\}$. The classification of the finite subgroups of the orthogonal group (due to F. Klein) is well-known. In particular since $O(3,\mathbb{R})$ contains finite subgroups isomorphic to Sym(4) and A_5, the quaternions contain subgroups isomorphic to the binary octahedral group and SL(2,5).

2.2 THE SCHUR INDEX

For the remainder of this chapter we study finite and locally finite skew linear groups. The entire positive-characteristic case and much of the zero-characteristic case depend on Schur index arguments. Suppose that F is a field, G is a group, and μ is a homomorphism of the group algebra FG *onto* the simple Artinian ring $D^{n\times n}$, where D is a division ring. The *Schur Index* of μ is then m(D), the Schur index of D. If G is a finite group then an irreducible FG-module V, or equivalently an irreducible representation Φ of G over F, determines such a μ, and then we speak of the Schur index of V or of Φ. The following reduces us to a consideration of finite groups.

2.2.1. *Let* F *be a field and* G *a locally finite group such that for some integer* m *the Schur index of every irreducible* F*-representation of every finite subgroup of* G *is at most* m. *Let* μ *be a homomorphism of* FG *onto a simple Artinian ring. Then* μ *has Schur index at most* m.

Proof: Let $(FG)\mu = ED \cong D^{n\times n}$ where E is a set of matrix units and D is a division ring. Let H be any finite subgroup of G such that $E \subseteq (FH)\mu$. Then $(FH)\mu = EC \cong C^{n\times n}$, where $C = D \cap (FH)\mu$ is the centralizer of E in $(FH)\mu$ (Passman [2], 6.1.5). Clearly C is a finite-dimensional F-algebra, and also a domain. Hence C is a division ring.

Let Z be the centre of C, so $\dim_Z C \leq m^2$. Choose H such that (C:Z) is maximal. If H_1 is any finite subgroup of G containing H and Z_1 is the centre of $C_1 = D \cap (FH_1)\mu$ then $(C:Z) \leq (C_1:Z_1)$ by Jacobson [1], Theorem 1. The maximal choice of (C:Z) implies that $(C:Z) = (C_1:Z_1)$. But the same theorem now implies that $CZ_1 = C_1$. Thus ZZ_1 is contained in the centre Z_1 of C_1, i.e. $Z \subseteq Z_1$. It follows that Z is central in D and that D has dimension at most m^2 over its centre. □

The following is an alternative proof of the finite-dimensionality of D: if K is a maximal subfield of C then $(C:K) = k \leq m$ (Cohn [2], Volume 2, p. 366) and C is isomorphic to a subring of $K^{k\times k}$. It follows that C satisfies the standard polynomial identity s_{2m} of degree 2m (Passman [2], 5.1.9). This is for all such C. Thus D also satisfies s_{2m}, and the result follows from Kaplansky's Theorem (ibid. 5.3.4).

2.2.2. *Let* G *be a finite group and* F *a field that contains the algebraic closure* $\mathbb{A}$ *of* $\mathbb{Q}$ *if* charF = 0. *Then every irreducible* FG*-module has Schur index* 1.

Proof: Let $F_o = \mathbb{A}$ if charF = 0 and $F_o = GF(p)$ if charF = p.

Let μ be a homomorphism of FG onto the simple ring R. Set $R_o = (F_oG)\mu$. Clearly $R = F[R_o]$. If $\underline{n}$ is a nilpotent ideal of R_o then $F\underline{n}$ is a nilpotent ideal of R. Thus R_o is semisimple. Any central idempotent of R_o is central in R. Hence R_o is simple, say $R_o \cong E^{n\times n}$, with E a finite-dimensional division F_o-algebra. The definition of F_o and Wedderburn's theorem on finite division rings imply that E is in fact a field.

Now $R = F[R_o] \cong F[E^{n\times n}] = (FE)^{n\times n}$, and FE is commutative. Any ideal of FE gives rise to an ideal of the simple ring R. Hence FE is a field and so the Schur index is 1. □

For a proof of the following result see Passman [2], 12.4.7.

2.2.3. (P. Roquette, 1958). *Let* G *be a finite nilpotent group and* F *a field of characteristic zero. Then any irreducible* FG*-module has Schur index at most* 2, *and has Schur index* 1 *if the Sylow* 2*-subgroup of* G *is abelian.* □.

At one point only we need the following result.

2.2.4. (Janusz [1]). *Let* F *be a field of characteristic zero and let* q *be a prime power. Every irreducible representation of* SL(2,q) *(resp.* PSL(2,q)) *over* F *has Schur index at most* 2 *(resp.* 1). □

A number of other bounds for Schur indices of finite groups are known. For example if G is a finite soluble group with all its Sylow subgroups elementary abelian then its Schur indices are all 1 (Passman [1], 12.4.8). On the other hand, by 2.1.5 there exist metacyclic groups with arbitrarily large Schur indices. However by a result of Solomon ([1], Theorem 6), if G is a finite group of order $\Pi_i p_i^{a_i}$, where the p_i are distinct primes, then every Schur index divides $2\Pi_i(p_i-1)$.

2.3 LOCALLY FINITE SKEW LINEAR GROUPS OF POSITIVE CHARACTERISTIC

The following theorem reduces this case almost entirely to linear groups and leaves little more to be said. Throughout this section D denotes a division ring of characteristic $p > 0$.

2.3.1. THEOREM. (Zalesskiĭ [1]). *Let G be a locally finite subgroup of GL(n,D). Then $O_p(G) = s(G) = u(G)$ and $G/O_p(G)$ is isomorphic to a subgroup of $GL(n,\bar{P})$, where $\bar{P}$ is the algebraic closure of the field P of p elements.*

Thus the study of these groups is reduced to the linear case, the study of stability subgroups, and an extension problem. 2.3.3 gives an example of a group G as in 2.3.1 that is not linear, since it does not satisfy the minimal condition on centralizers and does not satisfy Sylow's Theorem (as periodic linear groups do; see Wehrfritz [2], 9.10).

Proof: That $O_p(G) = s(G) = u(G)$ follows from 1.3.2 and 1.3.11. Also by 1.3.11 we may assume that $u(G) = \langle 1 \rangle$ and that G is completely reducible. Then P[G] is semisimple Artinian by 1.1.14c). Let $P[G] = \oplus E_i^{n_i \times n_i}$ where the E_i are division rings. As P[G] is a locally finite ring Wedderburn's theorem on finite division rings implies that each E_i is a field. Hence each E_i is isomorphic to a subfield of $\bar{P}$. Also $\Sigma\, n_i \leq n$ by 1.1.9. Consequently

$$G \hookrightarrow \times_i\, GL(n_i,E_i) \hookrightarrow \times_i\, GL(n_i,\bar{P}) \hookrightarrow GL(n,\bar{P})$$

and the result follows. □

Alternative Proof: Suppose first that G is finite. Then R = P[G] is a finite ring with radical J, say. Also $R/J = \times E_i^{n_i \times n_i}$ where the E_i are now finite fields. Idempotents of R/J lift over J to R, so again $\Sigma\, n_i \leq n$. Also $O_p(G) = G \cap (1+J)$, so we have

$$G/O_p(G) \hookrightarrow \times_i\, GL(n_i,\bar{P}) \hookrightarrow GL(n,\bar{P})$$

as before.

If $X \leq Y$ are finite subgroups of G then $X \cap O_p(G) = \bigcap_{Y \geq X} X \cap O_p(Y)$ since O_p is a radical rule in the sense of Kegel & Wehrfritz [1], § 1.B. There exists Y with $X \cap O_p(G) = X \cap O_p(Y)$, so by the finite case $X/(X \cap O_p(G))$ is isomorphic to a subgroup of $GL(n,\bar{P})$. By Mal'cev's Local Theorem (Wehrfritz [2], 2.7) the group $G/O_p(G)$ is isomorphic to a subgroup of GL(n,F) for some field F of characteristic p and by the proof of Winter's Theorem (Wehrfritz [2], 9.5) we can choose $F = \bar{P}$.

2.3.2. *Let G be a locally finite subgroup of GL(n,D).*

a) *The maximal p-subgroups of G are all conjugate.*

b) *If* u(G) = ⟨1⟩ *then for every prime* q *the maximal* q-*subgroups of* G *are all conjugate.*

Proof: This follows at once from 2.3.1 and the corresponding theorem for linear groups (Wehrfritz [2], 9.10). □

2.3.3. (Zalesskiĭ [1]). *Let* p *and* q *be distinct primes. Then there exists a division ring* D *of characteristic* p *and a metabelian* pq-*subgroup* G *of* GL(2,D) *such that the maximal* q-*subgroups of* G *are not all conjugate.*

Proof: Let Q be the Prüfer q^{∞}-subgroup of the multiplicative group of the algebraic closure of P, the field of p elements, and let K = P[Q]. The field K certainly has an automorphism x of infinite order. The skew group ring K⟨x⟩ is a Noetherian domain and so by Goldie's theorem it has a classical ring D of quotients, which is a division ring. Set

$$G = \left\{ \begin{bmatrix} a & 0 \\ b & a \end{bmatrix} : a \in Q,\ b \in D \right\} \leq GL(2,D)$$

Then G is an extension of an elementary abelian p-group by Q and so is certainly a metabelian pq-group. Let Q_i be the subgroup of Q of order q^i. Then the index $(\langle x\rangle : C_{\langle x\rangle}(Q_i))$ is finite, ⟨x⟩ is infinite, and $\bigcap_i C_{\langle x\rangle}(Q_i) = C_{\langle x\rangle}(K) = \langle 1\rangle$. It follows that $C_D(Q) \neq C_D(X)$ for any finite subset X of Q. By the lemma on page 343 of Hartley [1] the maximal q-subgroups of G are not all conjugate. □

QUESTION: Can the division ring D in 2.3.3 be chosen to be locally finite-dimensional over its centre?

2.4 THE THEOREMS OF ZALESSKIĬ AND HARTLEY & SHAHABI

In this section we consider finite skew linear groups over division rings of characteristic zero. The main theorem, due to Hartley & Shahabi, is that such groups have a large metabelian normal subgroup, see 2.4.8. The soluble case was solved earlier by Zalesskiĭ, who also conjectured the general case. The latter is proved using a case by case analysis and the classification of the finite simple groups.

Throughout D denotes a division ring of characteristic zero.

2.4.1. (Zalesskiĭ [1]). *Let* G *be a finite subgroup of* GL(n,D) *such that every irreducible* $\mathbb{Q}$G-*module has Schur index at most* m. *Then* G *is isomorphic to a subgroup of* GL(mn,$\mathbb{C}$).

Proof: $R = \mathbb{Q}[G] \leq D^{n\times n}$ is semisimple, say $R = \oplus E_i^{n_i \times n_i}$, where the E_i are finite-dimensional division $\mathbb{Q}$-algebras. By 1.1.9 we have $\Sigma n_i \leq n$. Let F_i be a maximal subfield of E_i, so by assumption $(E_i:F_i) \leq m$ (Cohn [2], Vol. 2, p. 366). Then F_i is isomorphic to a subfield of $\mathbb{C}$, and we have

$$E_i \hookrightarrow F_i^{m\times m} \hookrightarrow \mathbb{C}^{m\times m};$$

whence $$G \hookrightarrow \times GL(n_i,E_i) \hookrightarrow \times GL(\Sigma m n_i,\mathbb{C}) \hookrightarrow GL(mn,\mathbb{C}).$$ □

2.4.2. *Let* G *be a finite subgroup of* GL(n,D) *that is either nilpotent or isomorphic to* SL(2,q) *or* PSL(2,q) *for some prime power* q. *Then* G *is isomorphic to a subgroup of* GL(2n,$\mathbb{C}$).

Proof: Apply 2.4.1, 2.2.3, and 2.2.4. □

2.4.3. *Let* G *be a finite subgroup of* GL(n,D).

a) *Every 2-subgroup of* G *can be generated by* 2n *elements.*

b) *Every* p-*subgroup of* G, *for* p *an odd prime, can be generated by* n *elements.*

c) *Every abelian subgroup of* G *can be generated by* n *elements.*

d) *For odd primes* $p \geq n$, *every* p-*subgroup of* G *is abelian.*

e) *If* $G \cong$ SL(2,q) *or* PSL(2,q) *then* q *is bounded by a function of* n *only.*

Proof: Part c) follows from 2.4.1 with m = 1. Let p be a prime and let P be a p-subgroup of G. If p is odd then P is isomorphic to a subgroup of GL(n,$\mathbb{C}$) by 2.4.1 and 2.2.3, and the image of P is monomial. Part d) follows. Certainly P is generated by n+n! elements, which is sufficient for our purposes, but in fact P can be generated by n elements by a theorem of Wehrfritz [9]. If p = 2 then P is isomorphic to a subgroup of GL(2n,$\mathbb{C}$). Thus P can be generated by 2n+(2n)! elements, but in fact 2n will suffice. For the image of P in GL(2n,$\mathbb{C}$) is monomial, so P has an abelian normal subgroup A such that G/A is isomorphic to a 2-subgroup of Sym(2n). As such G/A can be generated by n elements by the proposition of Wehrfritz [9], as can A by Part c). Consequently G can be generated by 2n elements. If G is as in e) then G is also isomorphic to a subgroup of $GL(2n,\mathbb{C})$ and the result follows from Jordan's Theorem (Curtis & Reiner [1], 36.13, or Dixon [1], 5.6). □

The bounds in parts a) and b) are the best possible; for if D is the real quaternion algebra then GL(n,D) diagonally contains the direct product of n quaternion groups, and for each odd prime p, an elementary abelian p-group of rank n.

2.4.4. THEOREM. *There is an integer-valued function* $f_s(n)$ *of the positive integer* n *only such that for any division ring* D *of characteristic zero and any finite subgroup* G *of* GL(n,D) *with* $C_G(N) \leq N$, *for* N *the Fitting subgroup of* G, *the group* G *contains a metabelian normal subgroup of index at most* $f_s(n)$.

If G is a finite soluble group then always $C_G(N) \leq N$. This case of 2.4.4 is due to Zalesskiĭ. Amitsur's Theorem shows that in 2.4.4 "metabelian" cannot be replaced by "nilpotent", even if G is soluble and $n = 1$ (cf. 2.1.5). We also know, by 2.1.4, that the best value for $f_s(1)$ is 6.

Proof: By 2.4.2 the group N is isomorphic to a monomial subgroup of $GL(2n,\mathbb{C})$. Hence $A = N^{(2n)!}$ is an abelian normal subgroup of G. Also any abelian subgroup of G has rank at most n by 2.4.3, and so (N:A) divides $((2n)!)^{n+1}$ for example. Set $J = \mathbb{Q}[A] \leq D^{n\times n}$. Then J is a direct sum of $r \leq n$ (by 1.1.9) fields Q_i. Let $Y = \bigcap_i N_G(Q_i)$. Then (G:Y) divides n!.

The image of A in Q_i is cyclic, and the automorphism group of a cyclic group is abelian of rank at most 2. Also $C_G(A) = \bigcap_i C_G(Q_i)$, so $Y/C_G(A)$ is abelian of rank at most 2n. Set $K = C_G(N/A)$, so (G:K) divides $((2n)!)^{n+1}!$. Then $C_K(A)/C_G(N)$ is abelian by stability theory. Since $C_G(N) \subseteq N$ by hypothesis, $NC_K(A)$ is nilpotent, so $C_K(A) \subseteq N$. Hence $C_K(A)/A$ has exponent dividing (2n)! and is central in $(Y \cap K)/A$. Further $(Y \cap K)/C_K(A)$ is abelian of rank at most 2n. Let B/A be the centre of $(Y \cap K)/A$. Then B is a metabelian normal subgroup of G, and $(Y \cap K :B)$ divides $(2n)!^{2n}$. The result follows. □

The proof of the result for insoluble groups involves a number of extra complications. The strategy of the proof is, however, quite simple.

We find a normal subgroup L of G such that 2.4.4 applies to L, and with the property that the index (G:L) is bounded by a function of n only. This clearly implies the result. We begin our construction by dealing with simple groups. This uses the classification of the finite simple groups in the weak form that there are only finitely many that are not alternating and not of Lie type.

2.4.5. *Let* r *be a positive integer. Then apart from the* PSL(2,q) *for certain* q *there are only finitely many finite simple groups* G *such that for every prime* p *and every* p*-subgroup* P *of* G *we have that* P *can be generated by at most* r *elements, and is abelian whenever* $p \geq r$.

Proof: If G is alternating of degree n then G contains an elementary abelian 3-subgroup of rank [n/3], and so here $n \leq 3(r+1)$. (In fact $n \leq 2r+4$.)

Let $G = {}^{*}X_t(q)$ be of Lie type, where $q = p^s$ with p prime, X is one of the letters A, B,..., G, and * is blank or 2 or 3, as appropriate. The order of a Sylow p-subgroup of G is q^x for some integer $x \geq 0$ which can be read off from the order of G given in Carter [1], 8.6.1 and 14.3.1. Also G is the image of a linear group of degree y+1 and characteristic p, where y = y(t) can be found from ibid. 11.3.2. (If $^{*}X$ is such that t is bounded, e.g. E_t, then this is trivial anyway.) Thus a Sylow p-subgroup of G has a series of length y with elementary abelian factors, and consequently G has an elementary abelian p-section of rank at least sx/y. Thus $sx/y \leq r$. Certain values of x/y are given in the table below. In each case $s(t-1)/2 \leq sx/y \leq r$, so both s and t are bounded. In the remaining cases t is bounded anyway and hence so is y, and then $sx/y \leq r$ bounds s directly. Finally the Sylow p-subgroup of G is abelian only if $^{*}X_t = A_1$ (this follows from Chevalley's commutator formula, see Carter [1], 5.2.2 and § 13.6). Thus provided G is not of type A_1, i.e. provided $G \not\cong PSL(2,q)$ for some q, we also have $p \leq r$. The result now follows from the existence of only finitely many sporadic simple groups.

$^{*}X_t$	A_t , ^2A_t	B_t	C_t	D_t , ^2D_t
x/y	(t+1)/2	t/2	$t^2/(2t-1)$	$t(t-1)/(2t-1)$

□

A *quasisimple* group is a non-trivial perfect group that is simple modulo its centre.

2.4.6. (Hartley & Shahabi [1]). *Given* n *there are, up to isomorphism, only finitely many finite quasisimple groups* G *such that, for some division ring* D *of characteristic zero,* G *is isomorphic to a subgroup of* GL(n,D).

Proof: Let Z be the centre of G, and set $H = G/Z$. Now $|Z| \leq |H_2(H,\mathbb{Z})|$, the Schur multiplicator of H, which is finite. Apply 2.4.3 to G and then 2.4.5 to H. Thus omitting a finite number of possibilities for H, and hence for G, we have $H \cong PSL(2,q)$ for some q. Now the Schur covering group of PSL(2,q) is SL(2,q) apart from small q (Huppert [1], Chapter V, 25.7). Thus apart from small q (i.e. $q \leq 9$) we have $G \cong SL(2,q)$ or $G \cong PSL(2,q)$. The theorem now follows from 2.4.3 e). □

It would be nice to have a proof of 2.4.6 that did not rely on the classification of finite simple groups.

We need a suitable analogue of the Fitting subgroup. Let G be a finite group. The ***extended Fitting subgroup*** $F^*(G)$ of G is, by definition, the subgroup F(G)E(G), where F(G) denotes the Fitting subgroup of G (this is a more convenient notation here than our usual $\eta_1(G)$) and E(G) is the subgroup generated by all the quasisimple subnormal subgroups of G.

2.4.7. *Let* G *be a finite group.*

a) $C_G(F^*(G)) \subseteq F(G)$.

b) $E(G) = E_1E_2 \ldots E_t$, *the product of quasisimple subgroups* E_i *of* G *where* $[E_i,E_j] = \langle 1 \rangle$ *for all* $i \neq j$. □

For a proof see Huppert & Blackburn [1], Chapter X. By 13.10, 13.14, and 13.18 a) of that work their definition of $F^*(G)$ agrees with ours, and then 2.4.7 follows from their 13.12 and 13.18.

2.4.8. THEOREM. (Hartley & Shahabi [1]). *There exists an integer-valued function* f(n) *of* n *only such that if* G *is a finite subgroup of* GL(n,D), *where* D *is any division ring of characteristic zero, then* G *has a metabelian normal subgroup of index at most* f(n).

By Amitsur's Theorem, the best value for f(1) is 60, attained by SL(2,5). Thus by taking monomial representations of the permutational wreath product of SL(2,5) by Sym(n) one sees that $f(n) \geq 60^n n!$. The bound obtained below is very much larger than this.

Proof: Let $L = C_G(E(G))$. Then L is normal in G and therefore $E(L) \subseteq L \cap E(G) \subseteq \zeta_1(L)$. Consequently $E(L) = \langle 1 \rangle$, so $F^*(L) = F(L)$ and $C_L(F(L)) \subseteq F(L)$ by 2.4.7. It follows from 2.4.4 that L has a metabelian normal subgroup M of index at most $f_s(n)$. Set $M_G = \bigcap_{x \in G} M^x$. Then M_G is a metabelian normal subgroup of G with

$$(G{:}M_G) \leq f_s(n)^{(G:L)}(G{:}L).$$

It remains to bound (G:L).

If $E(G) = \langle 1 \rangle$ there is nothing to prove. Assume the notation of 2.4.7. By the Feit-Thompson Theorem each $E_i/\zeta_1(E_i)$ contains an involution. Thus E(G) has an elementary abelian 2-section of order 2^t. By 2.4.3 we have $t \leq 2n$, and by 2.4.6 there are only finitely many possibilities (up to isomorphism) for each E_i. This bounds the order of E(G) and hence the index (G:L). □

More recently, Hartley & Shahabi [2] have determined all the finite quasisimple subgroups of GL(2,D), for an arbitrary division ring D of charactersitic zero. Their result is as follows.

2.4.9. *A finite quasisimple group* G *is isomorphic to a subgroup of* GL(2,D), *for some division ring* D *of characteristic zero, if and only if* G *is isomorphic to either* SL(2,5) *or* SL(2,9). □

Their proof is based on the classification of finite groups of sectional 2-rank at most 4, due to Gorenstein & Harada. The classification of all finite subgroups of GL(2,D) has been carried out by Banieqbal [1], but has not (at the time of writing) appeared in print. Finally a result of Tits [1] shows that the double cover of the Hall-Janko group J_2 is embeddable in GL(3,D), where D is a finite-dimensional division algebra of characteristic zero. The classification of finite skew linear groups of degree $n \geq 3$ is therefore likely to be difficult.

2.5. LOCALLY FINITE SKEW LINEAR GROUPS OF CHARACTERISTIC ZERO

Here we present a number of routine generalizations and extensions of the results of the previous section. We also include a classification of the locally finite subgroups of division rings. Throughout D is a division ring of characteristic zero, n is a positive integer, and G is a locally finite subgroup of GL(n,D).

2.5.1. *If* G *is abelian then* G *is isomorphic to a subgroup of* GL(n,**C**). *In particular* G *has rank at most* n.

Proof: By 2.4.1 and Mal'cev's local theorem (Kegel & Wehrfritz [1], pp 65-6) G is isomorphic to a subgroup of GL(n,F), for some field F of characteristic zero. Thus G has rank at most n. In particular G is countable, and hence we may choose F countable. Consequently F is isomorphic to a subfield of **C**. □

2.5.2. *If* G *is locally nilpotent then* G *is isomorphic to a subgroup of* GL(2n,**C**). *In particular* G *has an abelian normal subgroup of rank at most* n *and index dividing* (2n)!.

Proof: By 2.4.2 and Mal'cev's local theorem again G is isomorphic to a subgroup of GL(2n,F) for some field F of characteristic zero. Therefore G has an abelian normal subgroup A with (G:A) dividing (2n)!. By 2.5.1 above A has rank at most n. Thus G is countable and we can choose F to be a subfield of **C**. □

The quaternion group of order 8 shows that in general we cannot replace 2n by n in 2.5.2.

2.5.3. *If* G *is locally nilpotent with its Sylow 2-subgroup abelian then* G *is isomorphic to a subgroup of* GL(n,**C**). *In particular* G *has an abelian normal subgroup of rank at most* n *and index dividing* n!.

Proof: This follows from 2.2.3 and 2.4.1; cf. the previous proof. □

2.5.4. (Zalesskiĭ [1]). *If* G *is locally soluble then* G *has a metabelian normal subgroup of finite index at most* $f_s(n)$, *the function of* 2.4.4. *In particular* G *is soluble with bounded derived length.*

Proof: This follows at once from 2.4.4 and a standard inverse limit argument (Kegel & Wehrfritz [1], §1.K). □

2.5.5. (Hartley & Shahabi [1]). G *has a metabelian normal subgroup of finite index at most* f(n), *the function of* 2.4.8.

Proof: This follows at once from 2.4.8 and the same inverse limit argument. □

2.5.6. COROLLARY. *There exists a function* g(n) *of* n *only such that for some abelian normal subgroup* A *of* G *the factor* G/A *is isomorphic to a subgroup of* $GL(g(n),\mathbb{C})$.

Note that by the Jordan-Schur Theorem (Wehrfritz [2], 9.4), 2.5.5 is a consequence of 2.5.6, with a possibly different value for f(n).

Proof: Let H be a metabelian normal subgroup of G with $(G:H) \le f(n)$. By 2.4.3 the abelian group H/H' has rank at most 2n. Thus H/H' is isomorphic to a subgroup of $GL(2n,\mathbb{C})$, and consequently G/H' is isomorphic to a subgroup of $GL(f(n)+2n,\mathbb{C})$. □

As an immediate consequence of 2.5.2 we have the following result.

2.5.7. *For each prime* p *the group* G *satisfies* Min-p, *the minimal condition on* p-*subgroups.* □

2.5.8. *The group* G *contains an abelian normal subgroup* A *such that* G/A *is abelian-by-finite and for every prime* p *has all its* p-*subgroups finite. If* π(G) *is finite then* G *is abelian-by-finite.*

Actually every soluble-by-finite periodic group G with Min-p for all p has an abelian normal subgroup A such that all p-subgroups of

G/A are finite (Kegel & Wehrfritz [1], 3.31).

Proof: Let H be a metabelian normal subgroup of G of finite index, and let $A \supseteq H'$ be an abelian normal subgroup of G, maximal subject to $A \subseteq H$. Since A has finite rank, A is a union of finite normal subgroups of G. Thus $G/C_H(A)$ is residually finite. Any countable p-subgroup of $G/C_H(A)$ is the image of a p-subgroup of G (Kegel & Wehrfritz [1], 1.D.4), and $C_H(A)$ contains every Prüfer p^∞-subgroup of G. Thus every p-subgroup of $G/C_H(A)$ is finite. By the maximal choice of A the centre of $C_H(A)$ is A. Also $C_H(A)$ is nilpotent, so by 2.5.2 and the linear case (Wehrfritz [2], 3.13), it is centre-by-finite. In other words, $C_H(A)/A$ is finite. The result follows. □

We can illustrate the above results for locally finite subgroups of division rings (the case n = 1). These were first considered by Faudree [1]. We need a slight generalization of the number $\gamma(p,s)$ of 2.1.12. Let $\{p_i : i = 1,2,...\}$ be the set of all primes, and let p be any prime. If G is a locally finite group with no elements of order p, then $\gamma(p,|G|)$ is the formal product $\gamma = \Pi\, p_i^{a_i}$, where $0 \leq a_i \leq \infty$, defined by

$$\gamma(p,|G|) = \text{L.C.M. } \{\, \gamma(p,|H|) : H \text{ a finite subgroup of } G \,\}$$

The classification is as follows (see Shirvani [1]). (As before C_n denotes the cyclic group of order n.)

2.5.9. *Let G be a locally finite subgroup of a division ring of characteristic zero. Then G is isomorphic to one of the following groups.*

a) i) *A periodic locally cyclic group.*

ii) $L \rtimes C_4$ *where L is a periodic locally cyclic 2′-group and* C_4 *inverts L.*

iii) *Of the form* $G_0 \times \mathsf{X}_{i\in I} G_i$ *where the sets* $\pi(G_0)$, $\pi(G_i)$, $i \in I$, *are disjoint,* G_0 *is periodic and locally cyclic, each* G_i *is of the form* $C_{p^a} \rtimes (C_{q_1^{b_1}} \times \ldots \times C_{q_r^{b_r}})$ *with* $0 \leq a \leq \infty$ *and* $b_1, \ldots, b_r$ *finite such that the arithmetic conditions of 2.1.5 c) are satisfied.*

b) i) *The binary octahedral group.*

ii) *A group* $L \rtimes Q$ *where L is a periodic locally cyclic 2′-group and Q is a locally quaternion group, such that a subgroup of index 2 in Q centralizes L, and a subgroup of order 4 in Q inverts L.*

iii) *A group* $Q \times M$ *where Q is quaternion of order 8 and M is a*

$2'$*-group as in* Part a) *with* $\gamma(2,|M|)$ *odd.*

iv) *A group* $SL(2,3) \times M$ *where* M *is a* $\{2,3\}'$*-group as in* a) *with* $\gamma(2,|M|)$ *odd.*

c) $SL(2,5)$.

The above groups are all countable. Note also that they can all be embedded into division rings, since each of their finitely generated subgroups can be so embedded, and Mal'cev's local theorem (Kegel & Wehrfritz [1], pp 65–6) then applies.

Proof: Suppose first that G contains no quaternion subgroup. If G is abelian, then G is locally cyclic by 2.5.1. If G contains a subgroup $E = C_m \rtimes C_4$, **where** C_4 **inverts** C_m **and m is divisible by at least two primes, then by 2.1.5 every finite subgroup of G containing E has the same form. Thus** $G = O_{2'}(G).C_4$, **where** $O_{2'}(G)$ **is locally cyclic and is inverted by** C_4. **Therefore a ii) holds. Now assume that G has no subgroup of the form** $C_m \rtimes C_4$ **as above. If H is any finite subgroup of G there exists a finite subgroup** $K \supseteq H$ **of G such that** $H \cap G' = H \cap K'$. **Thus by 2.1.2 the groups** G' **and** G/G' **are locally cyclic and** $\pi(G') \cap \pi(G/G') = \emptyset$. **Since** G/G' **is countable,** G' **has a complement R in G; cf. Kegel & Wehrfritz [1], 1.D.4. The automorphism group of a periodic locally cyclic group is residually finite. Consequently, any Sylow subgroup of R that acts non-trivially on** G' **is finite (cyclic) and (by 2.1.5) acts non-trivially on exactly one Sylow subgroup of** G'. **Write** $G' = \times_{i\in I} P_i$, **where the** P_i **are the non-trivial Sylow subgroups of** G'. **Let Q denote a Sylow subgroup of R. Set**

$$G_o = \langle\, Q : [G',Q] = \langle 1\rangle \,\rangle$$

$$G_i = \langle P_i, \text{ all } Q : [P_i,Q] \neq \langle 1\rangle \,\rangle \text{ for each } i \in I.$$

Then a iii) holds.

Now assume that G contains a quaternion subgroup Q of order 8. If G has a subgroup isomorphic to either the binary octahedral group or to SL(2,5), then G is equal to this subgroup, since these groups are maximal among those listed in 2.1.4. We may therefore assume that locally G satisfies b ii), b iii), or b iv) of 2.1.4. If an element x of G of order 3 normalizes Q then $S = \langle x\rangle Q \cong SL(2,3)$ **(note that by 2.1.4 the element x cannot centralize Q), and every finite subgroup of G containing S satisfies b iv) of 2.1.4. Thus here** $G = S \times O_{\{2,3\}'}(G)$ **and b iv) holds. Suppose no such x exists but that Q centralizes an element** $y \neq 1$ **of G of odd order.**

Hence here $G = Q \times O_{2'}(G)$ and b iii) holds. We are left with the case where every finite subgroup of G containing Q satisfies b ii) of 2.1.4. Then G is locally an extension of a cyclic 2′-group by a quaternion group. Consequently $L = O_{2'}(G)$ is locally cyclic and G/L is locally quaternion (and so countable). The extension splits (Kegel & Wehrfritz, loc. cit.), that is, $G = Q_1L$, where Q_1 is locally quaternion. By b ii) of 2.1.4 the action of Q_1 on L is as specified in b ii) above. □

2.5.10. EXAMPLE. (Zalesskiĭ [1]). *There exists a locally finite-dimensional division algebra* D *of characteristic zero and a metabelian locally finite subgroup* G *of* D* *such that* G *is not isomorphic to any linear group (of any degree over any field). Also* G *does not have a locally nilpotent subgroup of finite index.*

2.5.10 should be compared with 2.5.1, 2.5.2, 2.5.3, and 2.5.6.

Proof: Let $H = C_m \rtimes C_n$ be a $\underline{Z}$-group of odd order. Suppose that $p_1,\ldots,p_r$ are the prime divisors of m, and $q_1,\ldots,q_s$ those of n. By Dirichlet's Theorem on primes in an arithmetic progression (Serre [2], p. 102) there is a prime $q \geq mn$ with $q \equiv 2 \bmod(q_1\ldots q_s)$. Then q is a unit modulo $q_1\ldots q_s$, so $tq \equiv 1 \bmod q_1\ldots q_s$ for some integer t. Hence $1+tq \equiv 2 \bmod q_1\ldots q_s$ and therefore by the same theorem there exists a prime $p \geq mn$ with $p \equiv (1+tq) \bmod (qq_1\ldots q_s)$. By the choice of q, no q_i divides q-1, and by the choice of p and t, no q_i divides p-1, while q divides p-1. Since $q \geq mn \geq \phi(mn) \geq \gamma(p,mn)$ it follows that q does not divide $\gamma(p,mn)$. Also for any integer α and for all i and j, no q_i divides $\gamma(p_j,pq^{\alpha+1})$ since the latter divides $\phi(pq^{\alpha+1}) = (p-1)(q-1)q^{\alpha}$. By 2.1.5 there exists a $\underline{Z}$-group $H \times K$ where K is non-cyclic of the form $C_p \rtimes C_{q^{\alpha+1}}$.

A simple induction and Mal'cev's local theorem leads to the existence of a metabelian group $G = \times_{i=1}^{\infty} G_i$ of a division ring of characteristic zero, where each G_i is a non-cyclic $\underline{Z}$-group. Clearly G is locally finite and the division subring $\mathbb{Q}[G]$ is locally finite-dimensional over $\mathbb{Q}$. Also G contains an infinite descending chain of centralizers, so G is not isomorphic to any linear group. Finally the Hirsch-Plotkin radical $\eta(G)$ of G is the direct product of the maximal cyclic subgroups of the G_i, and $G/\eta(G)$ is infinite. □

We now return to our general hypothesis that G is an arbitrary locally finite subgroup of GL(n,D).

2.5.11. *For each prime* p, *the maximal* p*-subgroups of* G *are all conjugate.*

In fact a result of Šunkov [1] shows that every locally finite group, with Min-p for all p, has the above conjugacy property.

Proof: Pick A as in 2.5.8. Certainly the maximal p-subgroups of G/A are conjugate, so we may assume that G/A is a finite p-group. Since rank A $\leq$ n, the group G is isomorphic to a subgroup of GL(n+(G:A), $\mathbb{C}$), and the result follows from the linear case (Wehrfritz [2], 9.10). □

2.5.12. *If* G *is soluble and* π *is either a finite or a cofinite set of primes, then the maximal* π*-subgroups of* G *are all conjugate.*

Proof: If π is finite a modification of the proof of 2.5.11 gives the result. Assume that π' is finite. Let A be as in 2.5.6. The maximal π-subgroups of G/A are all conjugate, so we may assume that G/A is a π-group. Then $B = O_{\pi'}(G) \leq A$ is abelian and satisfies the minimal condition. Thus if $C = C_G(B)$, then G/C is finite (Kegel & Wehrfritz [1], 1.F.3). By the Schur-Zassenhaus Theorem applied locally $C = B \times O_\pi(C)$. The maximal π-subgroups of $G/O_\pi(C)$ are conjugate since they are finite. The result follows. □

2.5.13. EXAMPLE. *Let* G *be the metabelian group of* 2.5.10 *and let* $\pi = \{q_i : i = 1,2,\ldots\}$. *Then the maximal* π*-subgroups of* G *are not all conjugate.*

Proof: We have $G = \bigtimes_i G_i$, where each $G_i = \langle x_i \rangle] \langle y_i \rangle$ is non-cyclic. Then $\langle y_i : i = 1,2,\ldots \rangle$ and $\langle y_i^{x_i} : i = 1,2,\ldots \rangle$ are two non-conjugate maximal π-subgroups of G. □

We conclude this section with the following generalization of 2.5.5.

2.5.14. *There exists a function* h(n) *of* n *only such that* G *has a characteristic metabelian subgroup* H *of finite index at most* h(n) *with the following properties :*

a) *there exists an abelian characteristic subgroup* A *of* G *contained in* H *such that* H/A *is abelian with all its primary components finite;*

b) *for every prime* p *the maximal* p*-subgroups of* H *are abelian.*

Proof: By 2.5.5 we may choose a metabelian normal subgroup M of G with $(G:M) \leq f(n)$. Let p be a prime. By Min-p the maximal p-subgroup of M/M' is covered by a maximal p-subgroup P of M (Kegel & Wehrfritz [1], 3.13). Let Q be a maximal abelian subgroup of P containing $M' \cap P \supseteq P'$. Then Q is normal in P, the index (P:Q) divides (2n)! (by 2.5.2 and the linear case, Wehrfritz [2], 1.12), and $Q = C_P(Q)$.

Pick one such Q for each prime p, and let N be the subgroup generated by all the Q's. Then $M' \subseteq N$, so N is normal in M and of finite index dividing (2n)!. Also N/M' is the direct product of the QM'/M', so QM'/M' is the maximal p-subgroup of N/M'. Further $M' \cap Q = M' \cap P$ is a maximal p-subgroup of M'. It follows from the modular law that Q is a maximal p-subgroup of N. Since Q is abelian it follows from 2.5.11 that every p-subgroup of N is abelian. This is for every prime p. By 2.5.1 the group Q has rank at most n, and hence so do M' and N/M'. Set $H = G^t$, where $t = f(n)!\,(2n)!$. Then H is a characteristic subgroup of G and $(G:H) \leq h(n)$ for $h(n) = t^{2n+1}$. Clearly $H \subseteq N$.

Let A be the Hirsch-Plotkin radical of H. Then A is characteristic in G, and A is abelian since the primary subgroups of H are all abelian. Also by 2.5.8 there exists an abelian normal subgroup B of G with every primary subgroup of G/B finite. Then $H \cap B \subseteq A$, so every primary subgroup of H/A is finite. Clearly $H' \subseteq A$ and H/A is abelian. □

EXERCISE. (Zalesskiĭ [1]). Let G be a locally finite subgroup of GL(n,D). Prove that the subring $\mathbb{Q}[G]$ of $D^{n\times n}$ is semisimple Artinian. Hence show that G is a subdirect product of a finite number of absolutely irreducible skew linear groups over locally finite-dimensional division $\mathbb{Q}$-algebras. (Hint. Use 1.1.10).

3 LOCALLY FINITE-DIMENSIONAL DIVISION ALGEBRAS

Throughout this chapter F is a field of characteristic $p \geq 0$ and D is a locally finite-dimensional division F-algebra. (That is $\dim_F F[X]$ is finite for every finite subset X of D.) We study the subgroups of GL(n,D). If X is any finitely generated subgroup of GL(n,D) then $m = (F[X]:F)$ is finite and X is isomorphic to a subgroup of GL(m,F). Thus GL(n,D) is locally linear over F in the obvious sense. Clearly then the properties of skew linear groups over D are closely related to the corresponding properties of linear groups over F.

The results of this chapter both depend and extend the results of Chapter 2 via the exercise after 2.5.14, and 2.3.1. However, apart from a brief first section on general techniques, we concentrate on solubility, nilpotence and related concepts. The foundations of this theory were laid by Zalesskiĭ [1] in 1967. In spite of much subsequent work (Wehrfritz [10], [11], [12], [14], [15]) one important problem raised by Zalesskiĭ remains open, namely, is every locally nilpotent subgroup of GL(n,D) hypercentral. (We know from 1.4.4 that this is false for arbitrary division rings.)

As usual if p is a prime a p'-group is a periodic group with no elements of order p. A $0'$-group simply is a periodic group.

3.1. GENERAL TECHNIQUES

From 1.3.11 we have:

3.1.1. *Every unipotent subgroup of* GL(n,D) *is a stability subgroup of* GL(n,D) *and as such is conjugate to a subgroup of* $Tr_1(n,D)$. *Every subgroup* G *of* GL(n,D) *has a unique maximal unipotent normal subgroup* u(G). □

1.3.11, 1.1.2 and its proof show that:

3.1.2. *Let* G *be a subgroup of* GL(n,D). *Then there is an* F*-algebra homomorphism* Φ *of* F[G] $\leq$ $D^{n\times n}$ *into* $D^{n\times n}$ *such that* $G\Phi$ *is a completely reducible subgroup of* GL(n,D) *and* $G \cap (1 + \ker\Phi) = u(G)$. □

Let $R = D^{n\times n}$. Each finite-dimensional F-subspace of R carries its Zariski topology and these are compatible. Thus R carries a 'Zariski' topology by defining $C \subseteq R$ to be closed if and only if $C \cap V$ is closed in every finite-dimensional F-subspace V of R. (For details of the Zariski topology, see Chapter 5 of Wehrfritz [2].)

Suppose X is a subgroup of the finitely generated subgroup Y of GL(n,D). The topology on R induces topologies on X and Y that make X and Y into CZ-groups (again see Wehrfritz [2], Chapter 5.) These topologies are, of course, the same as the topologies induced by the Zariski topology on F[Y]. Let X^o denote the connected component of X containing the identity; it is a normal subgroup of finite index in X (ibid. 5.2). Also $X^o \subseteq X_1^o \subseteq Y^o$ whenever $X \subseteq X_1 \subseteq Y$. Let G be any subgroup of GL(n,D) and set $G^+ = \bigcup_Y (G\cap Y)^o$, where Y ranges over all the finitely generated subgroups of GL(n,D). The following is immediate.

3.1.3. G^+ *is a connected normal subgroup of* G *with* G/G^+ *locally finite.*□

G^+ is the best substitute we have found for the connected component of the identity of a linear group. Note that it depends on the choice of the central subfield F as well as on D, a fact that the notation suppresses. Also care is required because for a linear group G, in general $G^+ \neq G^o$ since, for example, if G is the torsion subgroup of $\mathbb{C}^* = GL(1,\mathbb{C})$, where $F = D = \mathbb{C}$, then $G^+ = \langle 1\rangle$ while $G^o = G$. The equality $G^+ = G^o$ holds, however, if G is finitely generated.

REMARK. The above definition of G^+ is not the same as that used in Wehrfritz [10], where G^+ is taken to be $\bigcup_X X^o$, with X ranging over the finitely generated subgroups of G. If we call this latter subgroup G^- then 3.1.3 remains true with G^- in place of G^+, and $G^- \subseteq G^+$. Suppose $F = D$ is the field of rational functions in one variable x over the field $\mathbb{F}_q$ of $q \geq 0$ elements and set $G = Tr_1(2,\mathbb{F}_q[x])$. The reader can check that

$G^- = \langle 1 \rangle$ while $G^+ = G$. Probably G^- can now be assigned to history.

Before we put our 'Zariski' topology to work we require some definitions. Let $\underline{S}_g$ denote the class of all finitely generated groups X such that every finite image of X is soluble, and let $\overline{\underline{S}} = L\underline{S}_g$ be the class of locally $\underline{S}_g$-groups. In practise $\overline{\underline{S}}$ is about the largest class of generalized soluble groups that one can expect to say much about. In particular it contains two of the most important classes of generalized soluble groups, as Part a) below shows.

3.1.4. a) $L\underline{S} \subseteq \overline{\underline{S}}$ *and* $\acute{P}L\underline{N} \subseteq \overline{\underline{S}}$.

b) $\overline{\underline{S}} \cap \underline{F} \subseteq \underline{S} \subseteq \overline{\underline{S}} = L(\underline{G} \cap \overline{\underline{S}}) = Q\,\overline{\underline{S}}$.

c) $S_f\,\overline{\underline{S}} = \overline{\underline{S}}$, *where* S_f *is the subgroup-of-finite-index operator.*

d) *If* $\underline{X}$ *is any class of groups satisfying* $\underline{X} \cap \underline{F} \subseteq \underline{S}$ *and* $Q\,\underline{X} = \underline{X} = L(\underline{G} \cap \underline{X})$ then $\underline{X} \subseteq \overline{\underline{S}}$.

e) *Linear* $\overline{\underline{S}}$*-groups are soluble.*

We have already seen that skew linear $\overline{\underline{S}}$-groups need not be soluble, see 1.4.4 and 1.4.13.

Proof: Part a) is obvious, as is Part b) once one notes that $\underline{G} \cap \overline{\underline{S}} = \underline{S}_g$. To prove Part c), let H be a subgroup of finite index of the $\overline{\underline{S}}$-group G, let X be an $\underline{S}_g$-subgroup of G and let N be normal of finite index in $H \cap X$. Then N has finite index in X and hence so does $N_X = \bigcap_{x \in X} N^x$. Thus X/N_X is soluble by hypothesis and consequently so is $(H \cap X)/N$. It follows that $H \cap X \in \underline{S}_g$, whence $H \in \overline{\underline{S}}$, as required. Part d) is an immediate consequence of the definitions. Part e) follows at once from b) and a result of Gruenberg (Wehrfritz [2],4.3). □

EXERCISE. Prove that $\overline{\underline{S}}$ contains every group generated by its Engel elements.

3.1.5. COROLLARY. *Let* G *be an* $\overline{\underline{S}}$*-subgroup of* GL(n,D). *Then* G *is locally soluble and* $(G^+)'$ *is unipotent. In particular* G *is nilpotent-by-abelian-by-locally finite.*

This is effectively taken from Wehrfritz [10]. Zalesskiĭ [1] proves the special cases $G \in \acute{P}L\underline{N}$ and $G \in L\underline{S}$, $p = 0$, by very different techniques.

Proof: G is locally linear and hence is locally soluble by 3.1.4e). If Y is any finitely generated subgroup of GL(n,D) and $X = (G \cap Y)^o$ then X is locally soluble and connected linear. By the Lie-Kolchin theorem (Wehrfritz [2], 5.8) X is triangularizable and X' is unipotent. But $G^+ = \bigcup_Y X$ by definition, and therefore $(G^+)' = \bigcup_Y X'$ is unipotent. □

We recall some facts about Jordan decomposition. Let $\bar{F}$ be an algebraic closure of the field F. Then an element g of $GL(n,\bar{F})$ can be written uniquely in the form $g = g_u g_d = g_d g_u$, where g_u is a unipotent and g_d a diagonalizable element of $GL(n,\bar{F})$. This is called the (multiplicative) Jordan decomposition of g. For linear groups this normally suffices. In the skew linear case, however, ground field extension presents problems. Thus it is useful to be more precise about where the Jordan decomposition takes place. The following is due to C. Chevalley.

3.1.6. *Let $g \in GL(n,F)$ and assume that the characteristic polynomial $q(X)$ of g is separable over F. Then the Jordan factors g_u, g_d of g lie in $GL(n,F)$.*

Proof: Let $q = \Pi_i q_i^{e_i}$ where the q_i are distinct irreducible polynomials and set $f = \Pi_i q_i$ and $e = \max\{e_i\}$. Then $f(g)^e = 0$ and $(f',f) = 1$ since q is separable. Thus $1 = f'h + fk$ for some elements h, k of the polynomial ring F[X].

Let $\Phi : F[X] \longrightarrow F[X]$ be the F-algebra homomorphism defined by $X\Phi = X - fh$. Replacing X by $X - fh$ and applying Taylor's Theorem we obtain $f\Phi \equiv (f - fhf') \bmod (fh)^2$, so $f\Phi = f - (1-fk)f + (fh)^2 d = f^2 k_1$, for some $k_1 \in F[X]$. By induction $f\Phi^i = f^{2^i}.k_i$ for some $k_i \in F[X]$.

Choose m with $2^m \geq e$ and set $s = X\Phi^m$. Let $\Psi : F[X] \longrightarrow F^{n\times n}$ be the F-algebra homomorphism given by $X\Psi = g$. Then

$$f(s(g)) = f\Phi^m \Psi = (f^{2^m}.k_m)\Psi = f(g)^{2^m} k_m(g) = 0.$$

Since $(f',f) = 1$ it follows that $s(g)$ is diagonalizable (Wehrfritz [2], p. 91). Now f divides $-fh = X\Phi - X$. If $X\Phi^i - X = fh_i$ for some h_i then

$$X\phi^{i+1}-X = (X+fh_i)\phi - X = (f\phi)(h_i\phi) - fh = (f^2k_1)(h_i\phi) - fh$$

is also divisible by f. By induction f divides $X - X\phi^m = X - s = t$, say. Thus $t(g)^e = 0$ and so t(g) is nilpotent. Trivially s and t commute, whence s(g) and t(g) commute, and $g = s(g) + t(g)$.

Since the eigenvalues of s(g) and g are the same s(g) is invertible. Set $g_d = s(g) \in GL(n,F)$ and $g_u = 1 + s(g)^{-1}t(g)$. Since t(g) is nilpotent and s and t commute, g_u is unipotent. Clearly $g_u \cdot g_d = g = g_d \cdot g_u$. □

For the reader familiar with the usual proof of the existence of the Jordan constituents of g (e.g. that given in Borel [1]) we offer the following alternative proof: the usual proof shows that we can write $g = g_d + g_n$, where g_d is diagonalizable in GL(n,P), P the splitting field of the characteristic polynomial q(X) of g over F, and g_n is nilpotent. In particular $g_n \in P^{n\times n}$. By assumption q is separable, so the extension P/F is Galois. Let $\sigma \in Gal(P/F)$. Then $g = g^\sigma = g_d^\sigma + g_n^\sigma$. Now $x^{-1}g_d x = diag(\lambda_1, \ldots, \lambda_n)$ for some $x \in GL(n,P)$. Hence g_d^σ is also diagonalizable. Clearly g_n^σ is nilpotent. The uniqueness of the additive decomposition implies that $g_d^\sigma = g_d$ and $g_n^\sigma = g_n$. This is for each $\sigma \in Gal(P/F)$, so $g_d, g_n \in F^{n\times n}$. The multiplicative decomposition follows from this by putting $g_u = 1 + g_d^{-1}g_n$ as before.

For the moment let R be any locally finite-dimensional F-algebra, but of course the example we have in mind is $R = D^{n\times n}$. As before an element g of R is *unipotent* if $g - 1$ is nilpotent. Say g is a ***semisimple*** element of R if the subalgebra F[g] of R is semisimple Artinian. This is equivalent to demanding that R is completely reducible as right (or left) F[g]-module. Call g a ***d-element*** of R if g is a unit and $\bar{F} \otimes_F F[g]$ is semisimple Artinian for $\bar{F}$ an algebraic closure of F, that is, if $1 \otimes g$ is a semisimple element of $\bar{F} \otimes_F R$. Then g is a d-element of R if and only if g acts as a d-element in the usual sense (i.e. g acts invertibly and diagonalizably, see Wehrfritz [2], p. 90) on every finitely generated right F[g]-submodule of R. Note that $g^{-1} \in F[g]$ as $\dim_F F[g]$ is finite.

Suppose first that F is perfect. Then a unit g of R has a Jordan decomposition in R, meaning that R contains a unique unipotent element g_u and a unique d-element g_d satisfying $g_u g_d = g = g_d g_u$. For if S is any finite-dimensional subalgebra of R containing ⟨g⟩ then g has a

Jordan decomposition in S by 3.1.6. This applies to any such S, so the uniqueness in S yields the uniqueness in R.

Suppose now that F is not perfect. Let $\tilde{F}$ be a perfect closure of F and regard R as a subring of $\tilde{R} = \tilde{F} \otimes_F R$. Then a unit g of R again has a Jordan decomposition, but now g_u and g_d are elements of $\tilde{R}$ rather than of R. If G is any group of units of R set $G_u = \{g_u : g \in G\}$ and $G_d = \{g_d : g \in G\}$. The following is an immediate consequence of the linear case (Wehrfritz [2], 7.14 and 7.11).

3.1.7. *Let G be a locally nilpotent subgroup of the group of units of R. Then $g \longmapsto g_u$ and $g \longmapsto g_d$ are homomorphisms of G onto G_u and G_d respectively. In particular G_u and G_d are locally nilpotent subgroups of the group of units of $\tilde{R}$. Also $[G_u,G_d] = \langle 1\rangle$ and $G.G_u = G.G_d = G_u \times G_d$.* □

We conclude this section with the following rather special result, which we need only occasionally in subsequent proofs. For any group G we denote the hypercentre of G by $\zeta(G)$ and the α-th term of the upper central series of G by $\zeta_\alpha(G)$.

3.1.8. *Let G be any subgroup of GL(n,D) and set $K = \zeta(G)$ and $\tilde{G} = K_uG$. Then $\zeta(\tilde{G}) = K_u \times K_d$ and $\zeta_i(G) = G \cap \zeta_i(\tilde{G})$ for all $i \leq \omega$.*

Note that $\tilde{G}$ is a subgroup of the group of units of $\tilde{R}$ introduced above, not of $R = D^{n\times n}$.

Proof: The map $g \mapsto g_u$ is a G-map, meaning that $(g^x)_u = (g_u)^x$ for all $g, x \in G$, and similarly so is $g \mapsto g_d$. This is a consequence of the uniqueness of the Jordan decomposition. By 3.1.7 the sets K_u and K_d are hypercentral groups and

$$K_u \cong_G K_uK_d/K_d \cong_G KK_d/K_d \cong K/(K \cap K_d)$$

and $K_d \cong_G K/(K \cap K_u)$. Thus $K_uK_d \leq \zeta(\tilde{G})$. Then the modular law yields that

$$\zeta(\tilde{G}) = K_uG \cap \zeta(\tilde{G}) = K_u(G \cap \zeta(\tilde{G})) \subseteq K_uK = K_uK_d$$

Therefore $\zeta(\tilde{G}) = K_uK_d$.

Let $0 \leq i < \omega$. Trivially $G \cap \zeta_i(\tilde{G}) \subseteq \zeta_i(G)$. Let $z \in \zeta_i(G)$ and choose $g_1,\ldots, g_i \in \tilde{G}$. There exists a finitely generated subgroup X of G with

$z \in X$ and $g_1,\dots, g_i \in \zeta(X)_u.X = Y$, say. By the linear case (Wehrfritz [2], 7.17) we have $\zeta_i(X) \subseteq \zeta_i(Y)$. Thus $[z,g_1,\dots,g_i] = 1$. Consequently $\zeta_i(G) = G \cap \zeta_i(\tilde{G})$ for all $i < \omega$. Finally

$$\zeta_\omega(G) = \cup_{i<\omega} \zeta_i(G) = \cup_{i<\omega} (G \cap \zeta_i(\tilde{G})) = G \cap \cup_{i<\omega} \zeta_i(\tilde{G}) = G \cap \zeta_\omega(\tilde{G}). \square$$

3.2 LOCALLY NILPOTENT GROUPS

The basic structure of locally nilpotent skew linear groups over locally finite-dimensional division algebras was elucidated by Zalesskiĭ [1] and further refined in Wehrfritz [10], [11], [14]. The fundamental concept of a primary group we take from Zalesskiĭ. There is a relativized version in Wehrfritz [11] but we shall avoid using it here.

Let q be any prime. A group G is *q-primary* if G modulo its centre $\zeta_1(G)$ is a locally finite q-group. In the long term it is not clear whether the hypothesis that $G/\zeta_1(G)$ be locally finite should be included or not, but it is certainly convenient to include it here.

3.2.1. *For some prime* q *let* Q *be a* q*-primary group. Then* Q *is locally nilpotent and* Q' *is a locally finite* q*-group.*

Proof: Q is locally nilpotent since $Q/\zeta_1(Q)$ is locally finite. If X is any finitely generated subgroup of Q then $(X:\zeta_1(X))$ is a power of q. Thus by Schur's theorem (e.g. Robinson [1], 4.12) X' is a finite q-group. Therefore Q' is a locally finite q-group. □

3.2.2. *Let* G *be a locally nilpotent group with centre* Z *and suppose that* G/Z *is locally finite. For each prime* q *let* G_q/Z *be the maximal* q*-subgroup of* G/Z. *Then*

a) G_q *is* q*-primary for each prime* q;

b) $G = \langle G_q : q \text{ prime} \rangle$;

c) $[G_q,G_r] = \langle 1 \rangle$ *for all primes* $q \neq r$;

d) $G' = \langle G_q' : q \text{ prime} \rangle$ *is locally finite.*

Proof: Parts a) and b) are immediate. For Part c) note that the group G_r stabilizes the series $\langle 1 \rangle \leq Z \leq G_q$, so $G_r/C_{G_r}(G_q)$ is isomorphic to a subgroup of the torsion subgroup of $\mathrm{Hom}(G_q/Z,Z)$ by elementary stability theory (e.g. Kegel & Wehrfritz [1], 1.C.3). The latter group is a q-group, so G_r centralizes G_q. Part d) follows from a), 3.2.1, b) and c). □

The following lemma is crucial for our later development.

3.2.3. *Let* G *be a locally nilpotent subgroup of* GL(n,D).

a) $[G^+,G] \subseteq u(G)$.

b) *If* $H = G/u(G)$ *then* $H/\zeta_1(H)$ *is a locally finite* p'*-group.*

c) *Let* C *be the set of* d*-elements of* G^+. *Then* C *is a central subgroup of* G. *If* $p \geq 0$ *then* G/C *is locally finite.*

Proof: a) Let Y be any finitely generated subgroup of GL(n,D) and set $X = G \cap Y$. By the linear case (Wehrfritz [2], 4.23 and 3.13) we have $[X^o,X] \subseteq u(X)$. But $u(X) = X \cap u(G)$ by 3.1.7. Thus $[G^+,G] \subseteq u(G)$.

b) G/G^+ is always locally finite. Thus $H/\zeta_1(H)$ is locally finite by Part a). Let $p \geq 0$. Clearly $O_p(H) = \langle 1\rangle$. If $H_p/\zeta_1(H)$ is the maximal p-subgroup of $H/\zeta_1(H)$, then $H_p' = \langle 1\rangle$ and $H_p = \zeta_1(H)$ by 3.2.2. Thus $H/\zeta_1(H)$ is a p'-group. (This latter conclusion can also be derived from the non-modularity of nilpotent linear d-groups, Wehrfritz [2], 7.6, 7.7, cf. the proof of 3.4.5 below.)

c) By 3.1.7 certainly C is a subgroup of G and clearly G normalizes C. Also $[C,G] \subseteq C \cap u(G) = \langle 1\rangle$ by a), so C is central in G. Let $p > 0$. Then $(G^+)_u$ is a locally finite p-group and $G^+/C \cong (G^+)_u$, by 3.1.7. Since G/G^+ is always locally finite the result follows. □

3.2.4. COROLLARY. (Zalesskiĭ [1]). *Let* G *be a locally nilpotent subgroup of* GL(n,D) *such that either* $u(G) = \langle 1\rangle$ *or* $p > 0$. *Then* G *is centre-by-periodic. In particular* G *is the central product of primary groups, one for each prime* q.

3.2.4 follows immediately from 3.2.2 and 3.2.3. If $p = 0$ and $u(G) \neq \langle 1\rangle$ then the situation is not really more complicated than that described in 3.2.4, the following being an immediate consequence of 3.1.7.

3.2.5. *Let* G *be a locally nilpotent subgroup of* GL(n,D) *where* $p = 0$. *Then*

$$G \subseteq G.G_u = G.G_d = G_u \times G_d \subseteq GL(n,D)$$

where G_u *is torsion-free nilpotent of class less than* n *and* G_d *is locally nilpotent with* $u(G_d) = \langle 1 \rangle$. *In particular* 3.2.4 *applies to* G_d. □

We will see below that most of the really knotty problems involve p-primary groups (for $p \geq 0$) where the ground field is imperfect. Our immediate goal is to describe the locally nilpotent subgroups of D^* (the case $n = 1$). We start with the primary ones.

3.2.6. *Let* G *be a q-primary subgroup of* D^* *for some prime* q.

a) G' *is locally cyclic.*

b) *If* G' *is finite then* G *is nilpotent.*

c) *If* G' *is infinite*

either i) G *is nilpotent of class* 2

or ii) $p = 0$, $q = 2$, G *is hypercyclic with central height exactly* $\omega + 1$ *and* G *has a subgroup* H *of index* 2 *with* $|H'| \leq 2$, *so* H *is normal in* G *and* H *is nilpotent of class* 1 *or* 2.

Proof: a) Certainly G' is a q-group (3.2.1). It is sufficient to assume that G is finitely generated and prove that G' is cyclic. Now G' is finite. A finite q-subgroup of D^* must be cyclic unless $q = 2$ and charD = 0, in which case it can be quaternion (2.1.2). Thus assume that G' is quaternion of order 2^n. If $2^n \geq 16$ then G' has a characteristic cyclic subgroup A of order 2^{n-1}. Then $G/C_G(A)$ embeds in Aut(A), which is abelian. In other words $G' \subseteq C_G(A)$, which is false. Now assume that G' is quaternion of order 8. Then $G/C_G(G')$ embeds into $\mathrm{Aut}(G') \cong \mathrm{Sym}(4)$ (Passman [1], 9.9). Since $G' \cap C_G(G') = \zeta_1(G')$ is cyclic of order 2, the image of G' in $G/C_G(G')$ is isomorphic to the Klein 4-group. However $G/\zeta_1(G)$ is a 2-group by assumption, so $G/C_G(G')$ is contained in a Sylow 2-subgroup S of Sym(4). But S is dihedral of order 8, so S' is cyclic of order 2. This is a contradiction. Hence G' is cyclic, as required.

b) If G' is finite then clearly G is nilpotent, of class at most $|G'|$ for example.

c) If $q \geq 2$ then $\mathrm{End}(G')$ is isomorphic to $\mathbb{Z}_q$, the ring of q-adic integers, which contains no elements of order q. Thus here G' is central and G is nilpotent of class 2. If $p > 0$ and $q = 2$ then the subfield of D generated by G' is isomorphic to $\varinjlim_{i\to\infty} GF(p^{2^i})$. This field has no automorphism of order 2 since it has no subfield of index 2. Thus again G' is central in G and G is nilpotent of class 2.

Finally let $p = 0$, $q = 2$, and assume that G is not nilpotent of class 2. Then $H = C_G(G') \neq G$. Now ± 1 are the only 2-elements of $\mathbb{Z}_2$. Hence $(G:H) = 2$ and if $g \in G\backslash H$ then g inverts G'. In particular G is hypercentral of central height $\omega + 1$. Also $[G,H] \subseteq G'$ is central in H, so $[[G,H],H] = \langle 1\rangle$. By P. Hall's Three Subgroup Lemma, $[G,H'] = \langle 1\rangle$, i.e. H' is central in G. But $H' \subseteq G'$, so g inverts H'. Thus $|H'| \leq 2$. □

3.2.7. COROLLARY. *Let G be a locally nilpotent subgroup of* D^*.

a) G *is metabelian.*

b) G *has a subgroup* H *of index at most* 2 *such that* H *is a central product of nilpotent groups (i.e.* $H \in Q D \underline{N}$*). In particular* H *is a Fitting group.*

Proof: By 3.2.4 the group G is a central product of primary groups G_q, one for each prime q. By 3.2.6 a) each G_q' is abelian. It follows that G is metabelian. Also G_q is nilpotent for all $q > 2$ by 3.2.6 b) and c), and G_2 has a nilpotent normal subgroup H_2 of index at most 2. Set $H = \langle H_2, G_q : q \geq 3 \rangle$. Then $(G : H) \leq 2$ and $H \in Q D \underline{N}$. Trivially H is a Fitting group. □

3.2.8. EXAMPLES.

a) *In* 3.2.6 b) *there is no bound on the nilpotency class of* G.

b) *There are groups satisfying* 3.2.6 c), Part i).

c) *There exist* 2-*primary skew-linear groups of central height* $\omega + 1$. *They can even be chosen not to be abelian-by-finite.*

d) *A locally nilpotent subgroup of* D^* *need not be nilpotent-by-finite.*

Proof: a) Let $\langle z\rangle$ be a cyclic group of order q^r, where q is a prime number, and let $G = \langle z\rangle \rtimes \langle x\rangle$, where x has infinite order and acts on z according to the automorphism $z \mapsto z^{1+q^t}$, where $t \geq 2$ is a fixed number. It is easy to verify that G is q-primary and nilpotent of class $-[-r/t]$. Also $G' \subseteq \langle z\rangle$ is finite. It remains to embed G into some division ring D^*. To this end we need a field E containing z, regarded as a primitive q^r-th root of unity, and a field automorphism $\theta : z \mapsto z^{1+q^t}$. For then the crossed product of E by $\langle x \rangle$ embeds into a division ring D, which can be chosen to be finite-dimensional over its centre (1.4.3 and 1.4.11).

In characteristic zero take $E = \mathbb{Q}(\zeta)$ and $t = 2$, say, where ζ is a primitive q^r-th root of unity. In characteristic p let F be the smallest field containing a q^2-th root of unity, let q^t be the highest power of q dividing $|F^*|$, and let $E = F(\zeta)$ for ζ a primitive q^r-th root of unity with $r \geq t$. It is easy to check the existence of the automorphism θ (use 2.1.12).

b) Let q be a prime and let F be a field containing, for all $r \geq 1$, a primitive q^r-th root ζ_r of unity (so char$F \neq q$). Let $E = F(y_1, y_2 \ldots)$ be the field of rational functions in the indeterminates $y_1, y_2, \ldots$. Let X be a free abelian group on the basis $x_1, x_2, \ldots$ and let X act as a group of F-automorphisms of E by:

$$y_j^{x_i} = \begin{cases} \zeta_i y_i & \text{if } i = j \\ y_j & \text{if } i \neq j \end{cases}$$

Form the split extension G of $\langle \zeta_i, y_i : i = 1,2,\ldots \rangle$ by X. By 1.4.3 and 1.4.11 there exists a locally finite-dimensional algebra D containing E such that $G \subseteq D^*$. Since the centre of G contains $\langle x_i^{q^i}, y_i^{q^i}, \zeta_i : i = 1,2,\ldots \rangle$ the group G is q-primary, and $G' = \langle \zeta_i : i = 1, 2,\ldots \rangle$. In particular G is nilpotent of class 2. Note that the construction works for all primes q and all characteristics different from q.

c) Let $p = 0$ and $q = 2$. We need to construct 2-primary groups with central height $\omega + 1$. The obvious example is the infinite locally quaternion group, which is a 2-subgroup of the real quaternion algebra. A variant is the following. Let A be a Prüfer 2^∞-group and let G be the split extension of A by an infinite cyclic group whose generator inverts A. Then by 1.4.3 and 1.4.11, G embeds into a locally finite-dimensional division algebra D. Note that both of these examples have an abelian subgroup of index 2.

We now construct an example with *no abelian normal subgroup of finite index.* Let C be a Prüfer 2^∞-group with involution c. For i = 1, 2, ... let $\langle x_i, y_i \rangle$ be the quotient of a free nilpotent group of class 2 and rank 2 by the square of its centre. In terms of presentations:

$$\langle x_i, y_i \rangle = \langle x_i, y_i \mid [x_i, y_i, x_i] = [y_i, x_i, y_i] = [x_i, y_i]^2 = 1 \rangle$$

Let B be the central product of C and the $\langle x_i,y_i\rangle$ for i = 1, 2, ... amalgamating c and $[x_i,y_i]$ for all i. Let G be the split extension of B by the infinite cyclic group $\langle g\rangle$, where g inverts C and centralizes each $\langle x_i,y_i\rangle$; this is well-defined. Then G is a 2-primary group with $G' = C$ and central height $\omega + 1$ (note that for each i, x_i^2 and y_i^2 are central in G). G does not have an abelian normal subgroup of finite index since the group $\langle x_i, y_i : i = 1, 2,\ldots\rangle$ has none. Since G/C is free abelian on the images of g, the x_i and the y_i, the crossed product $(\mathbb{C}[G], \mathbb{C}, G, G/C)$, where C is the Prüfer 2^∞-subgroup of $\mathbb{C}^*$ and g acts on $\mathbb{C}$ by complex conjugation, can be embedded in a locally finite-dimensional division algebra, by 1.4.3 and 1.4.11.

d) Let π be an infinite set of odd primes. For each $q \in \pi$ let r_q be a positive integer and let ζ_q be a primitive q^{r_q}-th root of unity. Set $E = \mathbb{Q}(\zeta_q : q \in \pi)$. By elementary Galois theory and the properties of the Euler function, for each q there is an automorphism θ_q of E that fixes ζ_r for $r \neq q$ and maps ζ_q to ζ_q^{1+q}. Let X be free abelian on the basis $\{x_q : q \in \pi\}$ and let X act on E by each x_q acting as θ_q. Let G be split extension of $\langle \zeta_q : q \in \pi \rangle$ by X. By 1.4.3 and 1.4.11 G can be embedded in a locally finite-dimensional division algebra D of characteristic zero. Each $G_q = \langle \zeta_q, x_q \rangle$ is nilpotent of class r_q (cf. Part a) above). If N is a nilpotent normal subgroup of G of finite index then $G = N.\Pi_{q\in\omega} G_q$ for some *finite* subset ω of π, whence G is nilpotent. Thus if we choose the r_q unbounded then G is not nilpotent-by-finite.

Similar examples exist in positive characteristics, but here the set π of primes must be chosen with care (cf. the proof of 3 on p. 42). □

We now consider the general case where n > 1. We start with the characteristic prime. Recall that F denotes a central subfield of D and p is the characteristic of F.

3.2.9. *Let G be a p-primary subgroup of* GL(n,D). *Then* G/u(G) *is abelian. If also F is perfect then G is nilpotent of class at most* max(1,n−1).

Clearly the bound is attained for all n and p.

Proof: G' is a p-group by 3.2.1, so $G' \leq u(G)$, which proves the first part. Now assume that F is perfect. Then $G \leq G_u \times G_d \leq GL(n,D)$ as in

3.1.7, and G_u is nilpotent of class at most n-1 by 3.1.1. Also G_d is p-primary since it is a homomorphic image of G, and $u(G_d) = 1$. Thus G_d is abelian by the first part. □

3.2.10. *Let G be a q-primary subgroup of GL(n,D) where $q \neq p$. Then G is hypercentral and has a normal subgroup H of finite index that is nilpotent of class at most 2.*

Proof: G' is a q-group by 3.2.1, so G' is isomorphic to a linear group of degree 2n and characteristic p (by 2.3.1 for $p > 0$ and 2.5.2 for $p = 0$). In particular G' is a Černikov group; let A be its minimal subgroup of finite index. Then A is normal in G and A is a union of finite characteristic subgroups. Thus $A \leq \zeta_\omega(G)$. Clearly the derived group of G/A is finite, so G/A is nilpotent and G is hypercentral.

Let $K = C_G(A)$. Any periodic group of automorphisms of A is finite (Kegel & Wehrfritz [1], 1.F.3) and trivially $\zeta_1(G) \leq K$. Hence (G:K) is finite. Set $H = C_K(K')$. Clearly $H' \leq K' \cap H \leq \zeta_1(H)$, so H is nilpotent of class at most 2, and is also normal in G. Since $(AK':A) \leq (G':A)$ is finite, $K/C_K(AK'/A)$ is also finite. By elementary stabiltiy theory (Kegel & Wehrfritz [1], 1.C), $C_K(AK'/A)/H$ is isomorphic to a subgroup of Hom(AK'/A,A). But AK'/A is finite and A has finite rank, so this homomorphism group is also finite. Therefore H has finite index in G. □

3.2.11. THEOREM. (Zalesskiĭ [1]). *Let G be a locally nilpotent subgroup of GL(n,D) and suppose that either $u(G) = \langle 1 \rangle$ or F is perfect. Then G is hypercentral.*

Proof: By 3.2.4 the group G is centre-by-periodic. Thus G is a central product of primary groups by 3.2.2 and the theorem follows from 3.2.9 and 3.2.10. □

A bound of the form (ω + finite) for the central height of the group in 3.2.11 can be easily extracted from the above proofs. Using somewhat different techniques the precise bounds can be computed; see Section 3.3.

As mentioned above, a major outstanding question is whether every locally nilpotent subgroup G of GL(n,D) is hypercentral. A little can be said in general. Specifically G can be shown to be a Gruenberg group, see 3.5.5 below.

3.2.12. EXAMPLES. a) We have already seen (3.2.8 d) that the group G in 3.2.11 need not be nilpotent-by-finite in either case. Example 3.2.8 b) shows that the q-primary group in 3.2.10 need not be abelian-by-finite. Moreover example 3.2.8 a) shows that in 3.2.10 one cannot choose H so that (G:H) is bounded in terms of p, q and n only. However we have yet to see examples of non-trivial p-primary groups.

b) *In general* p-*primary subgroups of* GL(n,D) *where* charD = p, *need not be nilpotent.* Let C be a cyclic group of order p, let E_i be an elementary abelian p-group of order p^i and let $H = \mathsf{X}_{i=1}^{\infty} C \wr E_i$. Clearly $H = A] E$ is a split extension of the product A of the base groups and $E = \mathsf{X} E_i$. Let X be a free abelian group mapping onto E, and form the split extension $G = A] X$. Let F be the field of rational functions, over the field of p-elements, in a basis of A, so now we regard A as an additive subgroup of F. The action of X on A extends to one of X on E. By 1.4.3, 1.4.11, and 1.4.6 b) the group G is isomorphic to a subgroup of GL(2,D) for some locally finite-dimensional division algebra D of characteristic p.

Certainly G is *locally nilpotent*, p-*primary*, but *not nilpotent.* It is, of course, hypercentral. We note a number of other properties of G whose significance will only become apparent in Section 3.4 below, see 3.4.7. *For every integer* $m \geq 1$ *the groups* $G/\zeta_m(G)$ *and* $[G,_m G]$ *are infinite* p-*groups and in particular are not* p′-*groups. Also* $\zeta_{i+1}(G)/\zeta_i(G)$ *is infinite for every* i *with* $0 \leq i < \omega$. We leave the verification to the reader.

We have already seen that locally soluble subgroups of GL(n,D) need not be soluble (or even hyperabelian), see 1.4.13. However locally nilpotent subgroups of GL(n,D), although not necessarily nilpotent, are always soluble.

3.2.13. THEOREM. *Let* G *be a locally nilpotent subgroup of* GL(n,D). *Then* G/u(G) *contains a metabelian normal subgroup* N/u(G) *such that* G/N *is isomorphic to a* p′-*subgroup of* Sym(n). *In particular* (G:N) *divides* n!.

This is Theorem 2 of Wehrfritz [14] with an improved bound.

Since u(G) is nilpotent of class less than n it follows from 3.2.13 that G is soluble, with derived length at most n+n! for example. The best bounds for the derived length of G are known and are given in the next section.

Proof: By 1.1.2 we may assume that u(G) = ⟨1⟩ and that G is completely reducible. By 3.2.11 the group G is hypercentral. Let A be a maximal abelian periodic normal subgroup of G and let Q be the maximal periodic subgroup of G. We claim that $C_Q(A) = A$. For if not $C_Q(A)/A$ is a non-trivial normal subgroup of the hypercentral group G/A and hence contains a non-trivial element aA of its centre (e.g. Kegel and Wehrfritz [1], 1.B.1). Then ⟨a,A⟩ is abelian since $a \in C_Q(A)$, is periodic since it is contained in Q, and is normal in G since $[a,G] \leq A$. This contradiction completes the proof that $C_Q(A) = A$.

By Clifford's theorem (1.1.7) A is completely reducible, so by 1.1.12 the algebra F[A] is semisimple Artinian. Let E be its set of primitive idempotents. It is easy to see that G normalizes E. Set $N = C_G(E)$. Then G/N is isomorphic to a subgroup of Sym(n), since $|E| \leq n$ by 1.1.9. Also $G/\zeta_1(G)$ is a p′-group by 3.2.3 b), so clearly G/N is a p′-group. It remains to prove that N is metabelian.

Let $e \in E$. Then eA is a periodic subgroup of the field eF[A], so eA is locally finite-cyclic with its automorphism group abelian. Hence N′ centralizes each eA and consequently centralizes A. But $N' \leq Q$ by 3.2.2 d) and thus the first paragraph of the proof yields that $N' \leq A$. It follows that N is metabelian. □

One can be a little more precise.

3.2.14. *Let* G *be a locally nilpotent subgroup of* GL(n,D). *Then* G *has normal subgroups* $u(G) \leq A \leq N \leq G$ *such that*

a) G/N *is isomporphic to a* p′*-subgroup of* Sym(n).

b) N/A *is a periodic abelian* p′*-group.*

and c) A/u(G) *is abelian.*

The first result of the next section yields the further information that $A/u(G) \leq \Delta(G/u(G))$, the set of elements of G/u(G) with finite conjugacy classes.

Proof: Again we may assume that $u(G) = \langle 1 \rangle$. By 3.2.13 there exists a metabelian normal subgroup N of G with G/N isomorphic to a p'-subgroup of Sym(n). If Z denotes the centre of G then by 3.2.3 b) we know that G/Z is a locally finite p'-group. Set $A = (N \cap Z)N'$. Then N/A is a periodic abelian p'-group. Also since N' is abelian and Z is central, A is abelian. Clearly A is normal in G. □

3.3 LOCALLY SOLUBLE GROUPS

We have already seen (3.1.5) that in the context of locally finite-dimensional division algebras, the most general class of generalized soluble groups that it is sufficient and worthwhile to consider is the class of locally soluble gorups. Also if G is a locally soluble subgroup of GL(n,D) then G/u(G) is abelian-by-locally finite. Further u(G) is nilpotent (3.1.1) and by 1.4.13 we know that G need not be soluble (or even hyperabelian).

In spite of this, we show in this section that quite strong statements can be made about the length of the derived series of such a group G. However, virtually nothing can be said about the lengths in such groups of ascending normal series with abelian factors, even if the group is hyperabelian.

We start with a lemma that is useful in a number of contexts.

3.3.1. LEMMA. (Wehrfritz [15]). *Let* N *be a nilpotent normal subgroup of the subgroup* G *of* GL(n,D) *with* $u(N) = \langle 1 \rangle$. *Then*

$$N \leq \Delta(G) = \{g \in G : |g^G| < \infty\}$$

Proof: Assume first that N is abelian. Since $u(N) = N \cap u(G)$, we can pass to G/u(G) via 1.1.2 and assume that G is completely reducible. Then so is N by Clifford's theorem (1.1.7) and so $R = F[N]$ is commutative and (by 1.1.12) semisimple Artinian. Thus R is a direct sum of a finite number of fields, say $R = \oplus_{i=1}^{r} L_i$.

Let $a \in N$ and $g \in G$. Now a is a root of some polynomial f(x) over F, as is every element of $D^{n \times n}$. Clearly a^g is also an element of $N \subseteq R$ and a root of f(x). But each L_i contains at most deg(f) roots of f(x) and therefore R contains at most $(\deg f)^r$ roots of f(x). Consequently $|a^G| < \infty$.

In the general case let A denote the centre of N. Then N/A is periodic by 3.2.4. Also $u(A) \subseteq u(N) = \langle 1\rangle$, and so $A \subseteq \Delta(G)$ by the first part of the proof. Let $a \in N$ and denote the order of a modulo A by r. Then $K = C_G(a^r)$ has finite index in G. Let $k \in K$ and set $H = \langle a,a^k\rangle A$. Clearly $(a^k)^r = a^r$, and so $(a^{-1}a^k)^r \in H'$. Also $H/H'A$ has exponent dividing r and for all $i \geq 1$ there is a homomorphism of $\otimes^i(H/H'A)$ onto $\gamma^iH/\gamma^{i+1}H$, where γ^iH is the i-th term of the lower central series of H (cf. Robinson [1], Vol 1, pp. 54–5). Thus H' has finite exponent dividing r^{c-1}, where c is the nil-potency class of N. Consequently $(a^{-1}a^k)^{r^c} = 1$.

Set $T = \langle b \in N : b^{r^c} = 1 \rangle$. Then T/T' has exponent dividing r^c and Robinson [1] again yields that T has exponent dividing r^{c^2}. But $u(T) = \langle 1\rangle$, so T is isomorphic to a linear group of degree 2n and characteristic p, by 2.3.1 and 2.5.3. Also $u(T) = \langle 1\rangle$ implies that T is a p'-group. By a theorem of Burnside (Wehrfritz [2], 1.23), T is finite. Finally $a^k \in aT$ for all $k \in K$ and consequently $|a^K| \leq |T|$. Therefore $|a^G| \leq |T|(G:K) < \infty$. □

For any ordinal α the α-th term of the derived series of a group G is denoted by $G^{(\alpha)}$.

3.3.2. LEMMA. *Let* G *be a locally soluble subgroup of* GL(n,D) *such that* $u(G) = \langle 1\rangle$, *and let* R *be the finite residual of* G, *that is,* R *is the intersection of the normal subgroups of* G *of finite index. Then* $G^{(\omega)} \leq R$ *and* R *is abelian. In particular* $G^{(\omega+1)} = \langle 1\rangle$.

Proof: Any finite image of G is soluble, so trivially $G^{(\omega)} \leq R$. If R is nilpotent then $R \leq \Delta(G)$ by 3.3.1. In this case R centralizes R and the result follows. It remains to show that R is nilpotent.

By 3.1.5 there exists an abelian normal subgroup A of G with G/A periodic. By 3.3.1 we have $A \leq \Delta(G)$, so R centralizes A. Thus R is centre-by-(locally finite) and Schur's theorem (Robinson [1], 4.12) yields that R' is locally finite. We can now apply 2.3.1 and 2.5.4 to R'. Certainly R' satisfies the minimal condition on q-subgroups for all primes $q \neq p$, since linear groups of characteristic p have this property. If $p > 0$ then $O_p(R') = u(R') = \langle 1\rangle$, and by 2.3.1 and the linear case, R' is abelian-by-finite. Thus R' also satisfies the minimal condition on p-subgroups.

Clearly R' is soluble. Let Q be the Sylow q-subgroup of some factor of the derived series of R'. Then A centralizes Q and so $G/C_G(Q)$ is locally finite. But Q is a Černikov group, since it is a homomorphic image of a q-subgroup of R' (Kegel & Wehrfritz [1], 3.13). Consequently $G/C_G(Q)$ is finite (ibid, 1.F.3). Thus R centralizes the factors of the derived series of R', and therefore R is nilpotent, as required. □

A unipotent subgroup of GL(n,D) is nilpotent of class at most n-1, and therefore has derived length at most $-[-\log_2 n]$. Thus the above two lemmas, together with 3.1.5, yield the following result.

3.3.3. THEOREM (Wehrfritz [14, 15]). *Let* G *be a locally soluble subgroup of* GL(n,D).

a) $\bar{G} = G/u(G)$ *has an abelian normal subgroup* $\bar{A}$ *with* $\bar{G}/\bar{A}$ *locally finite and* $\bar{A} \leq \Delta(\bar{G})$.

b) $G^{(\omega+1)} \leq u(G)$ *and* G *has derived length at most* $\omega + (1-[-\log_2 n])$. □

Thus as in the linear case the derived length is bounded, but now the bound is an infinite one. In fact the bound of 3.3.3 b) is the best possible, as the following example shows.

3.3.4. EXAMPLE. (Wehrfritz [14]). *Let* p *be zero or a prime, and let* n *be any positive integer. Then there exists a locally finite-dimensional division algebra* D *of characteristic* p *and a hyperabelian subgroup of* GL(n,D) *of derived length exactly* $\omega + (1-[-\log_2 n])$.

Proof: Suppose we have completed the case n = 1; that is, suppose we have constructed a suitable division algebra D and a hyperabelian subgroup G of D* such that G has derived length $\omega + 1$. Let $\{e_{ij}\}$ be the set of standard matrix units in $D^{n\times n}$. Direct calculation shows that for any $a \in D$ and $g \in D^*$ we have

$$[1_n + (g-1)e_{jj}, 1_n + ae_{ij}] = 1_n + a(1-g)e_{ij} \qquad (*)$$

Set $G_n = \{ (g_{ij}) \in Tr(n,D) : g_{ii} \in G \text{ for } i = 1, 2,\dots, n \}$. Then (*) and the trivial fact that $D(1-g) = D$ if $g \neq 1$, show that $Tr_1(n,D) \leq G_n'$. This argument has only used the fact that $G \neq \langle 1\rangle$. Thus an elementary induction shows that $Tr_1(n,D) \leq G_n^{(\omega+1)}$. Therefore G_n has derived length exactly $\omega + 1-[-\log_2 n]$.

We are left with the case n = 1. If one can construct an

abstract group G with the following properties:

a) G has an abelian normal subgroup A such that $A \leq \Delta(G)$ and G/A is locally finite,

b) G is torsion-free and locally polycyclic,

c) G is hyperabelian and $G^{(\omega)} \neq G^{(\omega+1)} = \langle 1 \rangle$:

then G can be embedded into a suitable division algebra, by 1.4.8 and 1.4.11.

Suppose $K = J \wr L$ is finite, where $L = \langle x \rangle \times L_1$ is abelian and x is central in K, and assume that we have constructed an exact sequence $A_J \rightarrowtail G_J \twoheadrightarrow J$ with A_J abelian and G_J poly-C_∞. Then J, as a transversal of L to K determines, via the Kalužnin-Krasner Theorem, an embedding η of K into $L \wr J = L^{(J)}] J$ such that $\eta|_J$ is the identity and $x\eta = (x,x,\ldots,x)$; the latter since x is central in K.

Pick a free abelian presentation $B \rightarrowtail E \twoheadrightarrow L$ of L of finite rank such that $E = \langle e \rangle \times E_1$, where $e \mapsto x$ and $E_1 \twoheadrightarrow L_1$. Then

$$B^{(J)} \times A_J \to E^{(J)}] G_J \to L^{(J)}] J$$

is exact. Let G_K be the full inverse image of $K\eta$ in $E^{(J)}] G_J$, and set $A_K = B^{(J)} \times A_J$. Then $A_K \rightarrowtail G_K \twoheadrightarrow K$ is exact with A_K abelian. Let $g = (e,e,\ldots,e) \in E^{(J)}$. Then g maps to $x\eta$, so $g \in G_K$. Clearly g is central in $E^{(J)}] G_J$, and so g is central in G_K. Also $\langle g \rangle$ is a direct factor of $E^{(J)}$, so $G_K/\langle g \rangle$ and G_K are poly-C_∞.

Let q be any prime. For i = 1, 2, ... let $K_i = J_i \wr L_i$ be a finite q-group such that L_i is abelian of exponent q and contains a non-trivial central element x_i of K_i lying in $K_i^{(i-1)}$. For example K_i could be a Sylow q-subgroup of $\mathrm{Sym}(q^i)$. Set

$$H = \left[\times_{i=1}^{\infty} K_i \right] / \langle x_i x_{i+1}^{-1} : i = 1, 2, \ldots \rangle$$

and denote the image of x_1 in H by h. It is easy to see that H is a hypercentral q-group with $H^{(\omega)} = \langle h \rangle$ cyclic of order q. By the above, for each i there exists an exact sequence $A_i \rightarrowtail G_i \twoheadrightarrow K_i$ with A_i abelian, and a central element g_i of G_i of infinite order mapping onto x_i such that $G_i/\langle g_i \rangle$ is poly-C_∞. Set

$$G = \left[\times_{i=1}^{\infty} G_i \right] / \langle g_i g_{i+1}^{-1} : i = 1, 2, \ldots \rangle$$

and let A, g denote the images of $\times_i A_i$, g_1 respectively in G. Since

$G/\langle g\rangle \cong \times_{i=1}^{\infty} G_i/\langle g_i\rangle$, the group G is locally poly-C_∞. The construction is complete. □

Obviously in the above construction $(G_n/u(G_n))^{(\omega)}$ is torsion-free, since G is. It is possible to modify the construction so as to produce, for any integer $r > 1$ and prime to p, a group G_n satisfying the requirements of 3.3.4 with $(G_n/u(G_n))^{(\omega)}$ a direct product of n cyclic groups of order r, see Proposition 2 of Wehrfritz [15].

3.3.3 should be regarded as the analogue of the Mal'cev and Zassenhaus theorems for linear groups. Curiously, ascending series in this context behave badly in both hyperabelian and radical groups. If G is any group let $\{\eta^{(\alpha)}(G)\}$ be the upper locally nilpotent series of G and let hph(G) (= Hirsch-Plotkin height of G) denote the least ordinal γ for which $\eta^{(\gamma)}(G) = \eta^{(\gamma+1)}(G)$. Similarly let hah(G) (= hyperabelian height of G) denote the least ordinal γ for which there is an ascending series

$$\langle 1\rangle = G_0 \le G_1 \le \ldots \le G_\alpha \le \ldots \le G_\gamma$$

of normal subgroups of G with abelian factors, such that G/G_γ has no non-trivial abelian normal subgroup. Trivially $\mathrm{hph}(G) \le \mathrm{hah}(G)$ for any hyperabelian group G.

3.3.5. EXAMPLE. (Wehrfritz [15]). *Let* p *be* 0 *or a prime and let* $1 \le \beta \le \gamma$ *be ordinals. Then there exists a locally finite-dimensional division algebra* D *of characteristic* p *and a torsion-free hyperabelian subgroup* G *of* D^* *such that* $\mathrm{hph}(G) = \beta$ *and* $\mathrm{hah}(G) = \gamma$. □

We refer the reader to Wehrfritz [15] for the details of the construction. Again the approach is to construct a suitable abstract group G to which 1.4.8 and 1.4.11 can be applied.

If G is any hyperabelian group then hph(G) = 0 if and only if hah(G) = 0. It remains only to discuss the case where $\beta = 1$, i.e. where the group is also locally nilpotent. Here the possibilities for hah(G) are very limited. For if G is a locally nilpotent subgroup of GL(n,D) then G is soluble by 3.2.13, and hah(G) (which now is just the derived length of G) is bounded by a function of n alone, for example n+n! In fact one can give the precise bound for the derived length of G, though this, not surprisingly, involves the characteristic.

3.3.6. *Let* G *be a p-primary subgroup of* GL(n,D).

a) *If* F *is perfect then* G *has derived length at most* $\max(1, -[-\log_2 n])$ *and this bound is attained for all* n *and* $p \geq 0$.

b) *In general* G *has derived length at most* $1 - [-\log_2 n]$ *and this bound too is attained for all* n *and* $p \geq 0$.

Proof: a) By 3.1.7 we have $G \leq G_u \times G_d \leq GL(n,D)$. Being unitriangularizable G_u is soluble of derived length at most $-[-\log_2 n]$. Also G_d is p-primary since it is an image of G and so G_d is abelian by 3.2.9. The bound therefore holds, and it is attained, since it is attained in the linear case.

b) Let $p \geq 0$. By 3.2.9 we have $G' \leq u(G)$, so G has derived length at most $1 - [-\log_2 n]$. Trivially this is attained if $n = 1$. Suppose $n \geq 1$. Let $E = \mathbb{F}_p(x)$ where x is an indeterminate and let $\langle y \rangle$ be an infinite cyclic group. Let y act on E by the field automorphism determined by $x \mapsto x-1$. The ring D of quotients of the skew group ring $E^{\cdot}\langle y \rangle$ is a division ring of characteristic p by 1.4.3. Since y^p centralizes E the algebra D is finite- dimensional over its centre. Let $\{ e_{ij} : 1 \leq i,j \leq n \}$ be the set of standard matrix units of $D^{n \times n}$ and set

$$G = \langle 1_n + xe_{i+1,i},\ y1_n : 1 \leq i < n \rangle \leq GL(n,D).$$

Then $$[y1_n, 1_n + xe_{i+1,i}] = 1_n + e_{i+1,i},$$

so $Tr_1(n,p) \leq G'$ and the former group has derived length $-[-\log_2 n]$. Thus G has derived length at least $1 - [-\log_2 n]$. Finally $y^p 1_n$ is central in G, and $G/\langle y^p 1_n \rangle$ is a finite p-group, so G is p-primary. □

3.3.7. *Let* G *be a q-primary subgroup of* GL(n,D), *where* $q \neq p$. *Then* G *has derived length at most*

$$\begin{cases} 2 & \text{if} \quad n < q \\ 1 + \lceil \log_q n \rceil & \text{if} \quad n \geq q \quad \text{and} \quad p+q \geq 2 \\ 2 + \lceil \log_2 n \rceil & \text{if} \quad p+q = 2 \end{cases}$$

and this bound is attained for all n, p, *and* $q \neq p$, *even if* F *is required to be perfect.*

Proof: We induct on n. By hypothesis $G/\zeta_1(G)$ and G' are q-groups and G is hypercentral by 3.2.10. Since $p \neq q$ we have $G' \cap u(G) = \langle 1 \rangle$. Consequently if $G^{(d)} \subseteq u(G)$ for some $d \geq 1$ then $G^{(d)} = \langle 1 \rangle$. Thus by 1.1.2 we may pass to $G/u(G)$ and assume that G is completely reducible. By induction we may assume that G is irreducible.

Let A be an abelian normal subgroup of G that is maximal subject to $A \subseteq G'$ and set $R = F[A] \leq D^{n \times n}$. By 1.1.7 and 1.1.12 a) the commutative F-algebra R is semisimple. Let $E = \{ e_1, ..., e_r \}$ be its set of primitive idempotents. Then G normalizes E and $\Sigma_{x \in G} Ve_1{}^x$ is a D-G sub-module of $V = D^n$. But G is irreducible and $V = \oplus_i Ve_i$. Thus G acts transitively on E.

Let $N = C_G(E)$ and $Q = G/N$. Since $G/\zeta_1(G)$ is a q-group, Q is isomorphic to a transitive q-subgroup of Sym(r). Thus $r = q^s$ for some $s \geq 0$ and Q has drived length at most s (Huppert [1], 3.15.3). Set $m = \dim_D Ve_1$. Then $n = mq^s$ and $[\log_q n] = [\log_q m] + s$.

Suppose $s = 0$. Then R is a field and A is a locally cyclic q-group. Hence Aut(A) is abelian and G' centralizes A. But A is equal to its centralizer in G' by the initial choice of A, so $G' = A$ and G has derived length at most 2. Note that if $n < q$ then necessarily $s = 0$.

Now suppose $s > 0$, so $m < n$. By definition N leaves each Ve_i invariant and so by induction the given function, with m in place of n, bounds the derived length of N, and since the derived length of G/N is by the above at most $s = [\log_q n] - [\log_q m]$, the result follows except when $m < q$ and $p + q > 2$.

Suppose we are in this anomalous case. Then $(N \cap G')/(C_{N \cap G'}(Ve_i))$ is isomorphic to a q-subgroup of GL(m,D). Since $p + q > 2$ this group is isomorphic to a linear group of degree $m < q$ and characterisitic p by 2.3.1 and 2.5.3. As such it is abelian, and this is for each i. Thus $N \cap G'$ is abelian. Also $G'/(N \cap G') \cong Q'$ has derived length at most $s - 1$ and consequently G has derived length at most

$$1 + (s - 1) + 1 = 1 + [\log_q n]$$

as required.

We now come to the examples. The example 3.2.8 b) gives a q-primary skew linear group of degree 1 and derived length 2. The only restriction on the characteristic p is $p \neq q$. A Sylow q-subgroup of Sym(nq) has derived length $1 + [\log_q n]$ and is isomorphic to even a linear group of degree n and characteristic $p \neq q$ (use monomial matrices). We are left with the case $p = 0$, $q = 2$. Let D be the real quaternion algebra, Q a quaternion subgroup of D^* of order 8 and S a Sylow 2-subgroup of Sym(n). Then the permutational wreath product $Q \wr S$ has derived length $2 + [\log_2 n]$ and is isomorphic to a monomial subgroup of GL(n,D). □

3.3.8. THEOREM. (Wehrfritz [15]). *Let G be a locally nilpotent subgroup of GL(n,D). Then G is soluble with derived length at most*

$$\begin{cases} 2 + [\log_2 n] & \textit{if} \quad n = 1 \quad \textit{or} \quad p = 0 \\ 1 - [-\log_2 n] & \textit{if} \quad n > 1 \quad \textit{and} \quad p > 0. \end{cases}$$

The bound is attained for all n and p.

For $p > 0$ if F is required to be perfect then G has derived length at most

$$\begin{cases} 2 & \textit{if} \quad n = 1 \quad \textit{or if} \quad n = p = 2 \\ -[-\log_2 n] & \textit{if} \quad n > 2 \quad \textit{and} \quad p = 2 \\ 1 + [\log_2 n] & \textit{if} \quad n > 1 \quad \textit{and} \quad p > 2. \end{cases}$$

This bound is also attained for all n and $p > 0$.

For comparison note that the maximum derived length of a locally nilpotent linear group of degree n and characteristic $p \geq 0$ is

$$\begin{cases} 1 + [\log_2 n] & \textit{if} \quad n = 1 \quad \textit{or} \quad p \neq 2 \\ -[-\log_2 n] & \textit{if} \quad n > 1 \quad \textit{and} \quad p = 2. \end{cases}$$

which superficially is so similar to the above that in some ways it is surprisingly different.

Proof: If $p = 0$ then $G \leq G_u \times G_d \leq GL(n,D)$ by 3.1.7, and so by 3.3.7 and 3.2.4 the best bound for the derived length of G is

$\max\{ -[-\log_2 n], 2 + [\log_2 n], 2, 1 + [\log_q n] : q > 2 \} = 2 + [\log_2 n]$.

If $p > 0$ then by 3.2.4, 3.3.6 and 3.3.7 the best bound for the derived length of G is

$\max\{ 1 - [-\log_2 n], 2, 1 + [\log_q n] : q \neq p \} = \max\{ 1 - [-\log_2 n], 2 \}$.

If also F is perfect then the same argument yields the bound

$$\max\{\ 1,\ -[-\log_2 n],\ 2,\ 1 + [\log_q n] : q \neq p\ \}$$

$$= \begin{cases} 2 & \text{if} \quad n = 1 \quad \text{or if} \quad n = p = 2 \\ -\left[-\log_2 n\right] & \text{if} \quad n \geq 2 \quad \text{and} \quad p = 2 \\ 1 + \left[\log_2 n\right] & \text{if} \quad n \geq 1 \quad \text{and} \quad p \geq 2. \end{cases}$$

□

We need the following simple result at one place only in Chapter 5.

3.3.9. *If* G *is a subgroup of* GL(n,D) *then the following are equaivalent:*

a) G *is locally soluble-by-finite.*

b) G *is (locally soluble)-by-(locally finite).*

c) G^+ *is locally soluble.*

Proof: Clearly b) implies a), and c) implies b) by 3.1.3. Assume a) holds. If Y is any finitely generated subgroup of GL(n,D) and $X = G \cap Y$ then X is locally soluble-by-finite. Hence X^o is soluble by the linear case (Wehrfritz [2], 10.12). But $G^+ = \bigcup_X X^o$. Consequently G^+ is locally soluble as required. □

3.4 THE UPPER CENTRAL SERIES

In this section we consider centrality questions relating to arbitrary subgroups of GL(n,D). Not unexpectedly, unipotent elements frequently present difficulties. Thus we have to start this section with a rather long and technical discussion of unipotent Engel elements. We hope the reader will bear with us. For a first reading, the reader should read the following definitions and the statements of 3.4.2, 3.4.4 and 3.4.5, and pass to 3.4.6.

All our commutators are left normed. In particular $[x,_2y] = [x,y,y]$, $[x,_3y] = [x,y,y,y]$, etc. . If S and T are subsets of a group G we write $S\ e\ T$ if for every $s \in S$ and $t \in T$ there is a positive integer r with $[s,_rt] = 1$. If given t, r can be chosen independently of s, we write $S\ |e\ T$.

3.4.1. *Let* $p = 0$ *and let* G *be a completely reducible subgroup of* $GL(n,D)$. *Suppose that* u *is an element of* $D^{n\times n}$ *such that* $u^x - u$ *and* y *commute for all* x *and* y *in* G. *Then* u *and* G *commute.*

Proof: We may assume that G is irreducible. Let $x \in G$ and set $v = u^x - u$, so $ux - xu = xv$. Since x and v commute, an elementary induction yields that

$$ux^i - x^i u = ix^i v \quad \text{for } i = 1, 2, \dots .$$

Let $f(t)$ be the minimum polynomial of x over F. Then

$$0 = uf(x) - f(x)u = f'(x)xv.$$

Since $\mathrm{char}D = 0$ and x is a unit we have $f'(x)x \neq 0$. Thus $W = \{w \in D^n : wv = 0\}$ contains the non-zero subspace $D^n f'(x)x$. But each element of G commutes with v, so W is a D-G submodule of D^n. The irreducibility of G yields that $W = D^n$; whence $v = 0$, as required. □

3.4.2. LEMMA. (Wehrfritz [10]). *Let* U *be a unipotent normal subgroup of the subgroup* G *of* $GL(n,D)$, *where* $p = 0$ *and* $U \trianglelefteq G$. *If* r *is the composition length of* D^n *as* D-G *bimodule, then* $[U,_{r-1}G] = \langle 1\rangle$. *In particular* $[U,_{n-1}G] = \langle 1\rangle$.

This is definitely false if $p > 0$, even if D is a field, see Wehrfritz [4]. It would be useful to have the conclusion $[U,_{n-1}G] = \langle 1\rangle$ in the case $p > 0$, but we have been unable to prove this. Some partial answers are given below.

Proof: Suppose first that $[U,_m G] = \langle 1\rangle$ for some integer m. Replacing G by one of its conjugates we may assume that each $g \in G$ is of the form $g = (g\sigma_{ij})$, an $r\times r$ block matrix, where the σ_{ii} are irreducible representations of G over D and are either inequivalent or actually equal, and where the σ_{ij} are zero-maps for all $i < j$. Since $U\sigma_{ii}$ is a unipotent normal subgroup of the irreducible group $G\sigma_{ii}$, Clifford's theorem and 3.1.1 yield that $U\sigma_{ii} = \langle 1\rangle$ for each i.

If $i > j$ we prove by induction on $i-j$ that $[U,_{i-j}G]\sigma_{ij} = \{0\}$. Since $[U,_{k+1}G] \subseteq [U,_k G]$ for all k, this will show that $[U,_{r-1}G] = \langle 1\rangle$, as required. Hence assume that $[U,_{k-s}G]\sigma_{ks} = \{0\}$ whenever $0 < k-s < i-j < r$. Let $u \in V = [U,_{i-j-1}G]$, and $g \in G$. Then using $\sigma_{ab} = 0$ for all $a < b$, $u^{-1}\sigma_{ks} = 0 = u\sigma_{ks}$ for $0 < k-s < i-j$, $u^{-1}\sigma_{kk} = 1 = u\sigma_{kk}$ for all k, and

$\Sigma_{s=1}^{r}$ $(g^{-1}\sigma_{is})$ $(g\sigma_{sj}) = 0$ we obtain

$$[u, g]\, \sigma_{ij} = (g^{-1}\sigma_{ii})\, (u\sigma_{ij})\, (g\sigma_{jj}) + u^{-1}\sigma_{ij} \quad (*)$$

Choosing $g = 1$ in this formula shows that $u^{-1}\sigma_{ij} = -u\sigma_{ij}$. If $V\sigma_{ij} = 0$ there is nothing to prove, so assume otherwise. By assumption $[U,_m G] = \langle 1\rangle$ for some m, so there exists $z \in V$ with $z\sigma_{ij} \neq 0$ but $[z,g]\sigma_{ij} = 0$ for all $g \in G$. Then (*) yields that $(g\sigma_{ii})(z\sigma_{ij}) = (z\sigma_{ij})(g\sigma_{jj})$ for all $g \in G$, so $z\sigma_{ij}$ defines a homomorphism between the irreducible constituents σ_{ii} and σ_{jj} of G. Also $z\sigma_{ij} \neq 0$ by choice, so Schur's lemma implies that σ_{ii} and σ_{jj} are equivalent, and so equal by construction.

Suppose now that $v \in V$ is such that $[v,g,h]\sigma_{ij} = 0$ for all $g, h \in G$. Then $[v,g]\sigma_{ij}$ commutes with $G\sigma_{ii}$ for all $g \in G$ by (*). It follows from (*) and 3.4.1 that $v\sigma_{ij}$ commutes with $G\sigma_{ii}$ and so $[v,g]\sigma_{ij} = 0$ for all $g \in G$ by (*) yet again. It follows that $[V,G]\sigma_{ij} = \{0\}$. This completes the proof of the special case.

In general let Y be a finitely generated subgroup of G. Then by the linear case (Wehrfritz [2], 4.13, or Wehrfritz [4] for a better result) there exists an integer m with $[U \cap Y,_m Y] = \langle 1\rangle$. By what we have just proved $[U \cap Y,_{n-1} Y] = \langle 1\rangle$. This is for every such Y, and therefore $[U,_{n-1} G] = \langle 1\rangle$. The result follows from the special case proved above. □

Let F be any field of positive characteristic p. In order to prove some sort of analogue of 3.4.2, we need to be able to conjugate maximal unipotent subgroups. Say F has property (*) if for any $m \geq 1$ and any (Zariski) closed subgroup C of GL(m,F), the maximal p-subgroups of C are all conjugate in C. For example algebraically closed fields (Platonov [1]) and locally finite fields (Wehrfritz [2], 9.10) satisfy (*). No imperfect field satisfies (*) (consider the special linear subgroup of $F^* \wr C_p$ in its obvious monomial representation of degree p over F.) We know of no perfect field that does not satisfy (*).

If one restricts oneself to certain closed subgroups C of GL(n,F) then the maximal unipotent subgroups of C are all conjugate (e.g. C = GL(n,F), by Wehrfritz [2], 1.21). We also have the following result.

3.4.3. *Let* F *be a perfect field of characteristic* $p > 0$ *and let* G *be a closed soluble-by-finite subgroup of* GL(n,F). *Then the maximal unipotent subgroups of* G *are all conjugate in* G.

Proof: We only outline the proof since it strays too far from our main theme. By the Lie-Kolchin theorem, the connected component G^o of G is triangularizable over the algebraic closure of F. Set $U = u(G^o)$. Then G^o/U is abelian and has no element of order p. Since F is perfect G^o/U is also p-divisible, the reason being that for every d-element g of G^o there is a d-element h of G^o with $h^p = g$. To see this, direct calculation shows that GL(n,F) contains such an element h. Also if F is algebraically closed, g lies in a torus T of G, and clearly T contains such an element. Finally diagonalizable p-th roots of g are unique (if $h^p = g = k^p$ then in $\langle h,k\rangle$ modulo its centre, the images of h and k are diagonalizable and unipotent, and are hence 1.)

Let P and Q be maximal unipotent subgroups of G. Certainly $U \leq P \cap Q$. By Sylow's theorem there exists x in G such that $\langle P,Q^x\rangle G^o/G^o$ is a p-group. A well-known generalization of the Schur-Zassenhaus theorem applied to $\langle P,Q^x\rangle G^o/U$ completes the proof (cf. Kegel & Wehrfritz [1], 3.9).□

3.4.4. (Wehrfritz [10]). *Let* F *be a perfect field of characteristic* $p \geq 0$ *and suppose that* U *is a unipotent normal subgroup of the subgroup* G *of* GL(n,D) *such that* $U \in G$. *If either* F *satisfies* (*) *or* G *is locally soluble-by-finite, then* $[U,_{n-1}G] = \langle 1\rangle$.

Proof: It suffices to assume that G is finitely generated. Then $R = F[G] \subseteq D^{n\times n}$ has finite dimension over F. Hence we can regard R as a subalgebra of $F^{m\times m}$ for some integer m. Note that the group of units of R is exactly $R \cap GL(m,F)$, and in particular it is closed in GL(m,F). If G is finite then $[U,_cG] = \langle 1\rangle$, where c is the nilpotency class of a Sylow p-subgroup of G. In general G is residually a finite linear group of degree m and characteristic p (Wehrfritz [2], 4.2). Thus $[U,_{m-1}G] = \langle 1\rangle$, i.e. the result in the linear case holds.

Let $\overline{U}$ and $\overline{G}$ denote the Zariski closures of U and G in GL(m,F), and set $C = C_{\overline{G}}(\overline{U})$. Then $[\overline{U},_{m-1}\overline{G}] = \langle 1\rangle$, and C is closed in $\overline{G}$ (Wehrfritz [2], 5.10 and 5.4). Also $\overline{G}/C$ is a nilpotent p-group since it stabilizes a finite series in the p-group $\overline{U}$ of finite exponent. Since F is perfect, $g_u, g_d \in \overline{G}$ for every $g \in \overline{G}$, by 3.1.6 and Wehrfritz [2], 7.3. But $\overline{G}/C$ is a p-group, so $g_d \in C$ for every $g \in G$ (Wehrfritz [2], 6.5). Thus $gC = g_uC$ for every $g \in \overline{G}$.

Let P be any maximal p-subgroup of $\overline{G}$. If G is soluble-by-finite, the same is true of $\bar{G}$. Consequently either by (*) or by 3.4.3 we have $\bar{G} = \bigcup_{x\in\bar{G}} P^x C$. Suppose $PC \neq \bar{G}$. Since $\bar{G}/C$ is nilpotent, there is a proper normal subgroup N of $\bar{G}$ containing PC. But then $\bar{G} = \bigcup P^x C \subseteq N \neq \bar{G}$, a contradiction. Therefore $PC = \bar{G}$. But $\bar{U} \le P$ by the maximality of P, and $\bar{G}$ lies in the group of units of R since the latter is closed in GL(m,F). Hence $P \le \bar{G} \le GL(n,D)$ and P is nilpotent of class less than n. Consequently $[U,_{n-1}G] \le [\bar{U},_{n-1}\bar{G}] = [\bar{U},_{n-1}P] = \langle 1\rangle$, as required. □

We now consider the unipotent-free case.

3.4.5. (Wehrfritz [10]). *Let* N *be a normal subgroup of the subgroup* G *of* GL(n,D) *such that* $N \in G$ *and* $u(N) = \langle 1\rangle$. *Then* $N^+ \le \zeta_1(G)$ *and* $N/C_N(G)$ *and* [N,G] *are locally finite* p'*-groups.*

Proof: N is a locally-linear Engel group. Therefore N is locally nilpotent (Wehrfritz [2], 8.15). In particular $u(N) = \langle 1\rangle$ is exactly the set of unipotent elements of N. Let Y be any finitely generated subgroup of GL(n,D), and set $X = G \cap Y$. By the linear case (ibid. 8.15, 8.13, and 8.9) $N \cap X \le \zeta(X)$ and the index $(\zeta(X/u(X)):\zeta_1(X/u(X)))$ is finite. Moreover by 7.6 and 7.7 of ibid. applied to $\zeta(X/u(X))$, this index is prime to p whenever $p > 0$. Hence there is a normal subgroup Z of X with $Z \le N \cap X$ and $(N \cap X)/Z$ a finite p'-group such that $[Z,X] \le N \cap u(X) = \langle 1\rangle$. Consequently $Z \le \zeta_1(X)$, $(N \cap X/N \cap \zeta_1(X))$ is a finite p'-group, and $(N \cap X)^\circ \le \zeta_1(X)$. Hence if Y_1 is any finitely generated subgroup of Y and $X_1 = G \cap Y_1$ then

$$(N \cap X_1)^\circ \le \bigcap_{X\supseteq X_1} (N \cap X)^\circ \le \bigcap_{X\supseteq X_1} \zeta_1(X) \le \zeta_1(G).$$

Therefore $N^+ = \bigcup_{X_1} (N \cap X_1)^\circ \le \zeta_1(G)$. Also $(N \cap X_1)/(N \cap \bigcap_{X\supseteq X_1} \zeta_1(X))$ is finite and residually a finite p'-group. Therefore $(N \cap X_1)/(N \cap X_1 \cap \zeta_1(G))$ is a finite p'-group and hence $N/(N \cap \zeta_1(G))$ is a locally finite p'-group.

It remains to prove that [N,G] is a locally finite p'-group. It suffices to assume that G is finitely generated, and hence by the linear case (Wehrfritz [2], 8.15) we may assume that N is G-hypercentral. Let T be the maximum periodic subgroup of N. Then N/T is torsion-free and G-hypercentral, so $(N/T)/C_{N/T}(G)$ is also torsion-free (cf. Robinson [1], 2.25). But $N/C_N(G)$ is locally finite, so it follows that $[N,G] \le T$, a locally finite group. Also if $p > 0$ then $O_p(T) = u(T) \le u(N) = \langle 1\rangle$. Therefore T, and hence [N,G], is a p'-group. □

3.4.6. *Let* G *be any subgroup of* GL(n,D), *and put* $N = \zeta(G)$. *Let* S *be the set of* d*-elements of* N^+.

a) S *is a central subgroup of* G. *If* $p \geq 0$ *then* N/S *is locally finite.*

b) *If either* $u(N) = \langle 1 \rangle$ *or* $p \geq 0$ *then* $N/\zeta_1(G)$ *and* [N,G] *are locally finite.*

Using 3.4.6 one can produce a relativized version of Zalesskiĭ's decomposition of suitable locally nipotent subgroups of GL(n,D) into primary groups, see Point 3 of Wehfrfritz [11]. Below we take a different approach.

Proof: a) By 3.2.3 c) the set S is a normal subgroup of G. The representation ρ of 1.1.2 of G with kernel u(G) is locally continuous, so $N^+\rho \leq (N\rho)^+$. Hence 3.4.5 yields that $[N^+,G] \leq u(G)$. Consequently $[S,G] \leq S \cap u(G) = \langle 1 \rangle$. If $p \geq 0$ then N/S is locally finite by 3.2.3 c) again.

b) If $u(N) = \langle 1 \rangle$ this is a special case of 3.4.5. Let $p \geq 0$. Then $N/\zeta_1(G)$ is locally finite by Part a). It follows that [N,G] is also locally finite. For if T is the maximal periodic subgroup of N, then the upper central G-factors of N/T are all torsion-free (cf. Robinson [1], 2.25) and yet $(N/T)/C_{N/T}(G)$ is locally finite. Thus $[N,G] \leq T$, which is locally finite. □

3.4.7. COROLLARY. *Let* G *be any subgroup of* GL(n,D), *and let* $m = \max\{1,n-1\}$.

a) $\zeta(G)/\zeta_m(G)$ *and* $[\zeta(G),_m G]$ *are both locally finite.*

b) *Suppose* $p \geq 0$ *and that any one of the following conditions holds:*

i) F *is perfect and satisfies* (*),

ii) F *is perfect and* G *is locally soluble-by-finite,*

iii) $u(\zeta(G)) = \langle 1 \rangle$.

Then $\zeta(G)/\zeta_m(G)$ *and* $[\zeta(G),_m G]$ *are* p′*-groups.*

Note that the example of 3.2.12 b) shows that if $p \geq 0$ then $\zeta(G)/\zeta_m(G)$ and $[\zeta(G),_m G]$ need not be p′-groups in 3.4.7 a); indeed they can both be infinite p-groups.

Proof: a) If $p \geq 0$ or $u(\zeta(G)) = \langle 1 \rangle$ then this is a trivial consequence of 3.4.6 b). Suppose $p = 0$ and set $\tilde{G} = K_u G$ for $K = \zeta(G)$ and apply 3.1.8. Certainly $[K_{u,n-1}\tilde{G}] = \langle 1 \rangle$ by 3.4.2. Also $K_d/C_{K_d}(\tilde{G})$ and $[K_d,\tilde{G}]$ locally finite by 3.4.5. Part a) follows.

b) Case iii) follows from 3.4.5, so assume that F is perfect. Let K and $\tilde{G}$ be as in the proof of a). Then again $[K_{u,n-1}\tilde{G}] = \langle 1\rangle$, but now by 3.4.4, and the result follows by an application of 3.4.5 to K_d and $\tilde{G}$. □

Our next goal is to describe $\zeta(G)/\zeta_\omega(G)$ for any subgroup G of GL(n,D). First we need a couple of lemmas.

3.4.8. *Let* A *be a divisible abelian normal* p'*-subgroup of the locally nilpotent subgroup* N *of* GL(n,D), *and set* $C = C_N(A)$. *Then there is an exact sequence*

$$\langle 1\rangle \to E \to N/C \to P \to \langle 1\rangle$$

where E *is an elementary abelian 2-group of order at most* 2^n, *and is* $\langle 1\rangle$ *if* $p > 0$, N/C *is isomorphic to a subgroup of* Sym(2n) *and* P *is a subgroup of* Sym(n).

Proof: Clearly $A \cap u(N) = \langle 1\rangle$ and $C = C_N(u(N)A/u(N))$. Thus we may pass to N/u(N) via 1.1.2 and assume that $u(N) = \langle 1\rangle$ and that N is completely reducible. In particular N is now hypercentral by 3.2.11 and N/C is periodic (3.2.4).

Let $R = F[A] \le D^{n\times n}$. Then R is semisimple Artinian by Clifford's Theorem and 1.1.12. Let $\{e_1,\ldots,e_r\}$ be the set of primitive idempotents of R; note that $r \le n$ by 1.1.9. Then N permutes the e_i; let P be the image and M the kernel of the resulting homomorphism of N into Sym(r).

Let $A_i = e_iA$. Since e_iR is a field, A_i is locally cyclic, and is divisible by hypothesis. Let $A_i(2)$ and $A_i(2')$ be the maximal 2- and $2'$-subgroups of A_i. Now M normalizes A_i and N/C is periodic. But A_i has at most one non-trivial automorphism of finite order that acts hypercentrally, namely the one that inverts $A_i(2)$ and centralizes $A_i(2')$. Thus $(M:C_M(A_i)) \le 2$ and E = M/C is elementary abelian of order dividing 2^r.

Let X consist of the following elements of R: for i = 1, 2,..., r, if $A_i(2) \ne \langle 1\rangle$ put the two elements of order 4 of $A_i(2)$ in X; if $A_i(2) = \langle 1\rangle$ put e_i in X. Then $|X| \le 2r \le 2n$. Also X is normalized by N since N permutes the A_i and e_i. Thus we have an embedding of $N/C_N(X)$ into Sym(X) $\le$ Sym(2n). Clearly $C \le C_N(X) \le M$. Also if $g \in C_N(X)$ then g cannot invert any non-trivial $A_i(2)$. Thus g centralizes every A_i, so $C_N(X) = C$. Therefore N/C is isomorphic to a subgroup of Sym(2n). Suppose $E \ne \langle 1\rangle$. Then for some i we have $A_i(2) \ne \langle 1\rangle$ and we have an element x of M inverting $A_i(2)$. If $p>0$

the subfield of e_iR generated by $A_i(2)$ is isomorphic to $K = \lim_{\rightarrow i} GF(p^{2^i})$, which has no subfield of codimension 2. Thus K has no automorphism of order 2 and consequently no element of N can invert $A_i(2)$. Therefore p = 0 when $E \neq \langle 1 \rangle$. □

If D is the real quaternion algebra and Q is the infinite locally quaternion subgroup of D*, then GL(n,D) contains a copy of the permutational wreath product $N = Q \wr S$ where S is a Sylow 2-subgroup of Sym(n). If A is the maximal divisible subgroup of N and $C = C_N(A)$ then we have an exact sequence

$$1 \to (C_2)^n \to N/C \to S \to 1$$

so the rather complicated structure in 3.4.8 (if p = 0) cannot be materially improved.

Let N be a normal subgroup of the subgroup G of GL(n,D) such that $N \le \zeta(G)$ and $N/C_N(G)$ is a p′-group. Then T = [N,G] is also a p′-group (for the upper G-central factors of $N/O_{p'}(N)$ are all p′-torsion-free (Robinson [1], 2.25) so G centralizes $N/O_{p'}(N)$ and $T \le O_{p'}(G)$). As such T is isomorphic to a linear group of degree 2n and characteristic p by 2.3.1 and 2.5.2. Such locally nilpotent p′-groups are monomial. Let A be the maximal divisible subgroup of T. Then A is normal in G and has rank at most n, the latter by 2.5.1. Set $C = C_N(A)$.

3.4.9. $C \le \zeta_\omega(G)$.

Proof: Now $u(N) = N \cap u(G)$ and $u(N) \cap T = \langle 1 \rangle$. Hence $u(N) \le \zeta_1(G)$ and $C = C_N(u(N)A/u(N))$. Thus by 1.1.2 we may assume that $u(G) = \langle 1 \rangle$. Also $N/C_N(G)$, and $C/C_N(G)$, are p′-groups; for each prime q let C_q be the maximal q-subgroup of C modulo $C_N(G)$. Then $[C_q,G]$ is a q-group (Robinson [1], 2.25). Consequently the index $([C_q,G]:[C_q,G] \cap A)$ is finite and $C_q/(A \cap C_q)$ is nilpotent. Therefore C_q is nilpotent and hence $C_q \le \Delta(G)$ by 3.3.1. It follows that $C \le \Delta(G)$, and then Dietzmann's Lemma (Kegel & Wehrfritz [1], 1.A.3) yields that $C/C_N(G)$ is generated by finite normal subgroups of $G/C_N(G)$. Therefore $C/C_N(G) \le \zeta_\omega(G/C_N(G))$, and consequently $C \le \zeta_\omega(G)$, as required. □

3.4.10. **THEOREM.** (Wehrfritz [11, 12]). *Let G be any subgroup of GL(n,D).*

Then $\zeta(G)/\zeta_\omega(G)$ *is locally finite and its* p′*-part is finite of order dividing*

$$\begin{cases} n! & \textit{if} \quad p > 0 \\ 2^n n! & \textit{if} \quad p = 0. \end{cases}$$

Further, the p′*-part of* $\zeta(G)/\zeta_\omega(G)$ *is isomorphic to a subgroup of* Sym(n) *in the first case and to a subgroup of* Sym(2n) *that is an extension of an elementary abelian group of order dividing* 2^n *by a subgroup of* Sym(n), *in the second.*

Proof: Set $M = \zeta(G)$. If $p = 0$ then $u(M) \le \zeta_{n-1}(G)$ by 3.4.2. Also $u(M) = M \cap u(G)$, so in this case we may apply 1.1.2 and assume that $u(M) = \langle 1 \rangle$. We now have that $M/\zeta_1(G)$ is locally finite in all cases, by 3.4.6.

Let $N/\zeta_1(G)$ be the maximal p′-subgroup of $M/\zeta_1(G)$, and let T, A, and C be as in 3.4.9. Then $C \le \zeta_\omega(G)$ by 3.4.9, and N/C has the structure described by 3.4.8. Finally if $H = \zeta_i(G)/\zeta_1(G)$ for some $1 \le i < \infty$ then the split extension A] H is periodic nilpotent and hence H centralizes A by Kegel & Wehrfritz [1], 1.F.1. Consequently $N \cap \zeta_\omega(G) = C$ and, N/C and $N\zeta_\omega(G)/\zeta_\omega(G)$ are isomorphic. The proof is complete. □

The bounds in 3.4.10 are essentially the best possible, even if F is perfect. For let S be a Sylow q-subgroup of Sym(n) where $q \ne p$, and let A be the Prüfer q^∞-subgroup in an algebraically closed field F of characteristic $p \ge 0$. Then the permutational wreath product $G = A \wr S$ is isomorphic to a group of monomial matrices in GL(n,F). Also G is hypercentral, $\zeta_\omega(G)$ is the base group, and $\zeta(G)/\zeta_\omega(G) \cong S$. Now let $q = 2$, and let Q be the infinite locally quaternion group. Then $G = Q \wr S$ is hypercentral and $\zeta(G)/\zeta_\omega(G)$ is isomorphic to $C_2 \wr S$. Thus the lowest common multiple of the p′-parts of the various $(\zeta(G):\zeta_\omega(G))$ is $2^n n!$ if $p = 0$ and the largest p′-divisor of n! if $p > 0$. Hence our claim that the bounds in 3.4.10 are essentially the best possible follows.

In 3.4.10 if $p > 0$ it is not possible to bound the order of the p-part of $\zeta(G)/\zeta_\omega(G)$.

3.4.11. EXAMPLE. (Wehrfritz [11]). *For each prime* p *and any direct product* Z *of cyclic* p*-groups there exists a locally finite-dimensional division algebra* D *of characteristic* p *and a* p*-primary subgroup* G *of*

GL(2,D) *such that* $G = \zeta_{\omega+1}(G)$ *and* $G/\zeta_{\omega}(G) \cong Z$.

The proof given in Wehrfritz [11] assumes that Z is countable. Using the results of Section 1.4 this restriction is unnecessary. We merely sketch the construction. Let $H = \times_{i=1} C_p \wr C_p^i$. Suppose $Z = \times_{j\in J} Z_j$, where Z_j is a cyclic p-group. Let $K = \times_{j\in J} H \wr Z_j$. Let A be the normal subgroup of K generated by the base groups in H, and Y a complement of A in the base group of K which is normalized by Z. Then there is an exact sequence

$$1 \to B \to G [A \to Z\, C_Y(Z)\, A \to 1$$

where G is locally poly-C_∞, B is central in GA, and $A \le \Delta(GA)$. Let E be the rational function field over F_p in a basis of A. Then G [A embeds into GL(2,D), for a suitable division ring D, by 1.4.3, 1.4.11, and 1.4.6 b). Finally $\zeta_\omega(GA) = C_Y(Z)AB$, and $GA/\zeta_\omega(GA) \cong Z$. □

An immediate consequence of 3.4.10 and 3.4.7 b) is the following.

3.4.12. COROLLARY. (Wehrfritz [11]). *Let* G *be a subgroup of* GL(n,D) *and suppose that*

either $p = 0$,

or F *is perfect and satisfies* (*),

or F *is perfect and* G *is locally soluble-by-finite,*

or $u(\zeta(G)) = \langle 1\rangle$.

Then $\zeta(G)/\zeta_\omega(G)$ *is a finite* p'*-group of order dividing* n! *if* $p > 0$ *and* $2^n n!$ *otherwise.* □

3.4.12 enables one to compute very good bounds for the central height of the group G of 3.4.12. Let q be a prime. Let

$$n = a_0 + a_1 q + \dots + a_r q^r,$$

where each a_i satisfies $0 \le a_i < q$ and $a_r \ne 0$. Set $e(n,q) = (q-1)^{-1}(n - \Sigma a_i)$. Then $q^{e(n,q)}$ is the largest power of q to divide n!. Clearly

$$e(n,q) \le (q-1)^{-1}(n-1).$$

Also $\quad q(a_1 + \dots + a_r) \le a_1 q + \dots + a_r q^r = n - a_0,$

so $e(n,q) \ge (q-1)^{-1}(n - a_0 - q^{-1}(n-a_0)) = q^{-1}(n-a_0) \ge q^{-1}(n - q + 1)$. Thus

$$e(n,q) \le (n-1)/2 \le e(n,2) \le n - 1 \qquad \text{if} \quad q \ge 3$$

and

$$e(n,q) \leq [(n-1)/4] \leq (n-2)/3 \leq e(n,3) \text{ if } q \geq 5 \text{ and } n \geq 2.$$

Trivially $e(1,q) = 0$ for all q. The following is now an immediate consequence of 3.4.12.

3.4.13. COROLLARY. (Wehrfritz [11]). *Let* F *and* G *be as in* 3.4.12. *Then* G *has central height at most*

$$\omega + e(2n, 2) \leq \omega + 2n - 1 \qquad \textit{if} \qquad p = 0$$

$$\omega + e(n, 2) \leq \omega + n - 1 \qquad \textit{if} \qquad p > 2$$

$$\omega + e(n, 3) \leq \omega + [(n-1)/2] \qquad \textit{if} \qquad p = 2. \quad \square$$

If G is locally nilpotent one can do even better. Let $c(n,q)$ be the nilpotency class of a Sylow q-subgroup of Sym(n). Then $c(n,q) = [q^{r-1}]$, where $r = [\log_q n]$ is as above. Since $q^r \leq n < q^{r+1}$ we have

$$[n/q^2] \leq c(n,q) \leq [n/q].$$

Then 3.4.10 and 3.4.12 yield the following.

3.4.14. COROLLARY. (Wehrfritz [11]). *Let* G *be a locally nilpotent subgroup of* GL(n,D) *and suppose that either* F *is perfect or* $u(G) = \langle 1 \rangle$. *Then* G *has central height at most*

$$\omega + 2^{[\log_2 n]} \leq \omega + n \qquad \textit{if } p = 0$$

$$\omega + \max\left\{ 2^{[\log_2 n]-1},\ 3^{[\log_3 n]-1} \right\} \leq \omega + [n/2] \qquad \textit{if } p > 2$$

$$\omega + \max\left\{ 3^{[\log_3 n]-1},\ 5^{[\log_5 n]-1},\ 7^{[\log_7 n]-1} \right\} \leq \omega + [n/3] \qquad \textit{if } p = 2. \square$$

These bounds are attained for all n and p, even when F is perfect *and* $u(G) = \langle 1 \rangle$. For the bounds are the same as for the linear case except for the cases

a) $n = 1$ or $n = p = 2$

and

b) $p = 0$

Case a) is settled by the group G constructed in 3.2.8 d), which is a direct product of nilpotent groups of unbounded class, and so is of central height ω. For Case b) take G to be the permutational wreath product of the infinite locally quaternion group Q by the Sylow 2-subgroup S of Sym(n). Then $\zeta_\omega(G)$ is the maximal divisible subgroup of G and $G/\zeta_\omega(G)$ is isomorphic to the Sylow 2-subgroup of Sym(2n). Thus G has central height $\omega + 2^{[\log_2 n]}$. In 3.4.12 none of the terms in the bounds can be omitted; see Wehrfritz [11], p. 483.

The reader may wonder about the finitely indexed terms of the upper central series of subgroups of G. Something positive can be said, but not very much. For a proof of the following we refer the reader to the original paper.

3.4.15. (Wehrfritz [11], Theorem 3). *Let* G *be any subgroup of* GL(n,D). *If* $p \geq 0$ *or* $n \leq 2$ *or* $u(\zeta(G)) = \langle 1 \rangle$ *then* $\zeta(G)/\zeta_1(G)$ *is locally finite and for each prime* q *there is a positive integer* e(q) *such that for* $2 \leq i < \omega$ *the* q*-primary component of* $\zeta_{i+1}(G)/\zeta_i(G)$ *has exponent dividing* $q^{e(q)}$. *If* $p = 0$ *and* $n \geq 2$ *then* $\zeta(G)/\zeta_{n-1}(G)$ *is locally finite and for each prime* q *there is a positive integer* e(q) *such that for* $n-1 \leq i < \omega$ *the* q*-primary component of* $\zeta_{i+1}(G)/\zeta_i(G)$ *has exponent dividing* $q^{e(q)}$. □

If $p \geq 0$, then for $0 \leq i < \omega$ the p-component of $\zeta_{i+1}(G)/\zeta_i(G)$ has exponent dividing p^m, where $p^{m-1} < n \leq p^m$. This is easy to see. Apart from this, examples in Wehrfritz [11] show, amongst other things, that the e(q) can be unbounded and unrelated, and also that the primary components of the $\zeta_{i+1}(G)/\zeta_i(G)$ can be of infinite order.

3.5 ENGEL ELEMENTS AND GENERALIZED NILPOTENT GROUPS

This is rather a technical section, probably only of interest to specialist group theorists working in the general area of generalized nilpotent groups. No other section of this book depends upon it, and the reader can omit it if he wishes.

The basic group theory and definitions that we require are all in Robinson [1]. In the main we assume they are familiar to the reader. Our aim is to correlate the various Engel sets, radicals, and classes of generalized nilpotent groups, when they are restricted to subgroups of GL(n,D). The objective is completely accomplished whenever the unipotent elements in positive characteristic do not cause problems. However, in spite of the difficulties we have encountered in the earlier sections of this chapter, a reasonably complete picture can be given even in the general case.

We start by combining and consolidating the `Engel´ parts of 3.4.2, 3.4.4, and 3.4.5.

3.5.1. *Let* N *be a normal subgroup of the subgroup* G *of* GL(n,D) *and assume that*

either p = 0,

or F *is perfect and satisfies* (*),

or F *is perfect and* G *is locally soluble-by-finite,*

or u(N) = ⟨1⟩.

Then:

a) *If* N e G *then* $N \leq \zeta(G)$;

b) *If* N |e G *then* $N \leq \zeta_\omega(G)$.

Proof: By 3.4.2, 3.4.4, or by hypothesis $[u(N),_{n-1}G] = \langle 1\rangle$. Also $u(N) = N \cap u(G)$ and the groups N/u(N) and Nu(G)/u(G) are G-isomorphic. Thus by 1.1.2 we may assume that u(N) = ⟨1⟩ in all cases.

a) By 3.4.5 the group T = [N,G] is a locally finite p′-group. It is thus isomorphic to a linear group of degree 2n and characteristic p (2.3.1 and 2.5.2), and consequently has an abelian characteristic subgroup A of finite index and finite rank. Thus T has an ascending characteristic series with finite factors. Each of these must be hypercentral in G. Therefore $T \leq \zeta(G)$ and it follows that $N \leq \zeta(G)$.

b) We continue with the above notation. Let $x \in N$ and set $X = \langle x^G\rangle$. Then $X \leq \langle x\rangle T$ and $\bar{X} = X/C_X(G)$ is periodic by 3.4.5. The image of $A \cap X$ in $\bar{X}$ is an abelian normal subgroup of $\bar{X}$ of finite rank and index. But $\bar{X}$ is locally nilpotent and is generated by the conjugates of $\bar{x} = xC_X(G)$. Consequently $\bar{X}$ involves only the primes involved in the order of $\bar{x}$ and therefore is a Černikov group.

Let Y be the maximal divisible subgroup of $\bar{X}$. Now N |e G implies that $\bar{X}$ |e $\bar{X}$ and so $\bar{X}$ is nilpotent, for example by 8.15 ii) of Wehrfritz [2]. Hence Y lies in the centre of $\bar{X}$ (Kegel & Wehrfritz [1], 1.F.1). Also there exists a characteristic conjugacy class of finite subgroups K of $\bar{X}$ with $KY = \bar{X}$ (ibid 3.9). But Y is central, so K is normal and hence characteristic in $\bar{X}$. (In fact K is just the kernel of the transfer homomorphism of $\bar{X}$ into Y.) There exists an integer r such that $\bar{x} \in KY_r$, where $Y_r = \{ y \in Y : y^r = 1 \}$. But KY_r is characteistic in $\bar{X}$. Thus $\bar{X} = \langle \bar{x}^G\rangle \leq KY_r$ and so $\bar{X}$ is finite. Finally $\bar{X}$ is G-hypercentral by Part a), so $[\bar{X},_h G] = \langle 1 \rangle$ for some positive integer h, and then $x \in X \subseteq \zeta_{h+1}(G)$. This proves that $N \leq \zeta_\omega(G)$. □

We now have to introduce rather a lot of notation. Let G be any group.

$L(G) = \{g \in G : G \text{ e } g\}$ is the set of left Engel elements of G.
$\bar{L}(G) = \{g \in G : G \mid\!\text{e } g\}$ is the set of bounded left Engel elements of G.
$R(G) = \{g \in G : g \text{ e } G\}$ is the set of right Engel elemnts of G.
$\bar{R}(G) = \{g \in G : g \text{ e}\!\mid G\}$ is the set of bounded right Engel elements of G.
$\sigma(G) = \{g \in G : \langle g \rangle$ is ascendant in $G\}$ is the Gruenberg radical of G.
$\bar{\sigma}(G) = \{g \in G : \langle g \rangle$ is subnormal in $G\}$ is the Baer radical of G.
$\rho(G) = \{g \in G : \forall\, x \in G, \langle x \rangle \in \sigma(\langle x, g^G \rangle)\ \}$.
$\bar{\rho}(G) = \{g \in G : (\exists\, k \in \mathbb{N})(\forall\, x \in G)\ \langle x \rangle$ is subnormal in $\langle x, g^G \rangle$ in k steps$\}$

Also recall that $\eta(G)$ denotes the Hirsch-Plotkin radical of G and $\eta_1(G)$ the Fitting subgroup of G.

3.5.2. THEOREM. (Wehrfritz [10]). *Let* G *be any subgroup of* GL(n,D) *and assume that*

either p = 0,
or F *is perfect and satisfies* (*),
or F *is perfect and* G *is locally soluble-by-finite,*
or u(G) = ⟨1⟩.

Then

a) $L(G) = \sigma(G) = \eta(G)$
b) $\bar{L}(G) = \bar{\sigma}(G) = \eta_1(G)$
c) $R(G) = \rho(G) = \zeta(G)$

and d) $\bar{R}(G) = \bar{\rho}(G) = \zeta_\omega(G)$.

Thus the situation is precisely the same as that for linear groups. At the end of this section we show that more than half of 3.5.2 holds for arbitrary F and $G \leq GL(n,D)$.

Proof: If X is a finitely generated subgroup of ⟨L(G), R(G)⟩, then there exists a finitely generated subgroup Y of G with ⟨L(Y), R(Y)⟩ ⊇

X. By the linear case (Wehrfritz [2], 8.15) the group X is nilpotent. Thus $\langle L(G), R(G)\rangle$ is locally nilpotent, and consequently soluble by 3.2.13. Then Theorem 1.5 of Gruenberg [2] ensures that $L(G) = \sigma(G) = \eta(G)$, $\bar{L}(G) = \bar{\sigma}(G)$, $R(G) = \rho(G)$, and $\bar{R}(G) = \bar{\rho}(G)$. Also 3.5.1 a) implies that $\rho(G) = \zeta(G)$. If $x \in \bar{\rho}(G)$ then $\langle x^G\rangle$ |e G, and 3.5.1 b) yields that $\bar{\rho}(G) \leq \zeta_\omega(G)$. The reverse inclusion holds for all groups G. Finally $\eta_1(G) \leq \bar{\sigma}(G)$, and always $\bar{\sigma}(G)$ |e $\bar{\sigma}(G)$, so 3.5.1 b) again implies that $\bar{\sigma}(G) \leq \zeta_\omega(\bar{\sigma}(G)) \leq \eta_1(G)$. The proof is complete. □

We now turn our attention to generalized nilpotent groups.

3.5.3. *Let* G *be a locally nilpotent subgroup of* GL(n,D), *where either* F *is perfect or* u(G) = ⟨1⟩. *Then:*

a) G *is hypercentral;*
b) *if* G *is a Baer group then* $G \in (Q D \underline{N})^S$ *and in particular* G *is a Fitting group;*
c) *if* G *is an* $\underline{N}_1$*-group then* G *is nilpotent.*

Linear Baer groups are nilpotent (Wehrfritz [2], 8.15(ii) and 8.2(ii)), so G in a) need not be a Baer group even if G is linear. The group G constructed in 3.2.8 d) lies in $D\underline{N}$ but not in $\underline{N}$, so it satisfies b) above but not c). Thus there are no further relations between the classes in the diagram on page 3 of Robinson [1], Vol. 2, when restricted to groups G as in 3.5.3, other than those indicated in 3.5.3.

Proof: a) This is 3.2.11.

b) Suppose G is q-primary for some prime q. If q = p then G is nilpotent by 3.2.9, and if $q \neq p$ then G is nilpotent-by-finite by 3.2.10. But G is a Fitting group by 3.5.2 b) and consequently G is nilpotent.

In general, we now have that $G \in Q D \underline{N}$ by 3.2.4. But this applies to every subgroup of G, so in fact $G \in (Q D \underline{N})^S$. (Note that G has no non-identity unipotent elements if u(G) = ⟨1⟩, by 3.1.7.) Clearly any $Q D \underline{N}$-group is a Fitting group.

c) For each prime q let $\bar{G}_q$ be the maximal q-subgroup of $\bar{G} = G/\zeta_1(G)$. Now every $\underline{N}_1$-group is a Baer group, so each $\bar{G}_q$ is nilpotent by the first paragraph of the proof of b) above. If X_q is any subgroup of $\bar{G}_q$ then $\langle X_q : q \text{ prime} \rangle$ is subnormal in $\bar{G}$ by hypothesis. It follows that there is an integer k such that for every q, every subgroup of $\bar{G}_q$ has subnormal depth at most k. By a theorem of Roseblade (Robinson [1], 7.42, Corollary) it follows that $\bar{G}$ is nilpotent. Hence G is nilpotent. □

As a corollary of the proof of 3.5.3 we have the following generalization of 3.3.1 (see Wehrfritz [21]).

3.5.4. *Let* N *be a normal Baer subgroup of the subgroup* G *of* GL(n,D), *with* $u(N) = \langle 1 \rangle$. *Then* $N \le \Delta(G)$.

Proof: For each prime q let $N_q/\zeta_1(N)$ be the maximal q-subgroup of $N/\zeta_1(N)$. Then $N = \langle N_q : q \text{ a prime} \rangle$. Also N_q is nilpotent as in the proof of 3.5.3 b), and so by 3.3.1 each N_q is contained in $\Delta(G)$. The result follows. □

Up till now in this section we have, by hypothesis, avoided those situations in characteristic $p > 0$ where the unipotent elements might cause trouble. We conclude this section with an account of what we know of the general situation.

3.5.5. LEMMA. (Wehrfritz [10]). *Let* G *be a locally nilpotent subgroup of* GL(n,D).

a) *If* X *is any finitely generated subgroup of* G *then* X.u(G) *is nilpotent.*

b) $u(G) \le \eta_1(G) = \bar{\sigma}(G) \le \sigma(G) = G$

and $\bar{\sigma}(G)/u(G) = \bar{\sigma}(G/u(G)) = \eta_1(G/u(G))$.

c) $G^+ \le \eta_1(G)$ *and* $G/\eta_1(G)$ *is a locally finite* p′*-group.*

Proof: a) Let H = X.u(G). If p = 0 then $H \le H_u \times H_d \le GL(n,D)$ by 3.1.7. Now H_u is nilpotent and $H_d \cong H/u(G) \cong X/(X \cap u(G))$, so H_d is also nilpotent. Thus H is nilpotent in this case.

Now assume that $p > 0$. Again by 3.1.7 we have $H \le H_u \times H_d$, where, in the notation of 3.1.7, H_u and H_d lie in $\tilde{R}$ rather than R. Also H_u is a p-group since it is unipotent. Thus $H/C_H(u(G))$ is also a p-group. Let $C = C_X(u(G))$. Since X is nilpotent, C is a normal subgroup of H = X.u(G) and lies in some finitely indexed term of the upper central series of H. Also X/C is a finite p-group since X is finitely generated. Thus H/C is a nilpotent-by-finite p-group of finite exponent. Hence H/C is nilpotent (Robinson [1], 6.34). Therefore H is nilpotent.

b) Since u(G) is nilpotent, $u(G) \leq \eta_1(G)$, and always $\eta_1(G) \leq \bar{\sigma}(G) \leq \sigma(G)$. Clearly $\bar{\sigma}(G)/u(G) \leq \bar{\sigma}(G/u(G)) = \eta_1(G/u(G))$ by 3.5.2 and 1.1.2. If $g \in G$ with $gu(G) \in \bar{\sigma}(G/u(G))$ then $\langle g\rangle u(G)$ is subnormal in G, and nilpotent by Part a). Thus $\langle g\rangle$ is subnormal in G, which implies that $g \in \bar{\sigma}(G)$, i.e. $\bar{\sigma}(G)/u(G) = \bar{\sigma}(G/u(G))$. A similar argument, with ascendant in place of subnormal, shows that $\sigma(G)/u(G) = \sigma(G/u(G))$. But $\sigma(G/u(G)) = G/u(G)$ by 3.5.2. Therefore $\sigma(G) = G$. (This fact can be extracted directly from our proof of 3.5.2.)

Set $N = \bar{\sigma}(G)$. It remains to prove that $N \leq \eta_1(G)$, a fact which is certainly not directly deducible from the proof of 3.5.2. By 3.5.2, however, we can assume that $p > 0$, so by 3.2.4 the group $G/\zeta_1(G)$ is periodic. For each prime q let $G_q/\zeta_1(G)$ be the maximal q-subgroup of $G/\zeta_1(G)$. Set $N_q = N \cap G_q$. Clearly $\zeta_1(G) \leq N$, so $N = \langle N_q : q \text{ a prime} \rangle$. Also each N_q is a normal q-primary Baer subgroup of G.

Let $q \neq p$. Then $u(N_q)$ is central in N_q, so N_q is a Fitting group by 3.5.2 b). Hence N_q is nilpotent by 3.2.10. Also $G_p' \leq u(G_p)$ by 3.2.9, and G_p is generated by its normal subgroups $\langle g\rangle u(G_p)$ for $g \in G_p$, each of which is nilpotent by Part a). Finally $G = G_pC_G(G_p)$, so each $\langle g\rangle G_p$ is normal in G. Thus $G_p \leq \eta_1(G)$, and the result follows. (In fact this proves that $G_p = N_p$.)

c) Certainly 3.2.3 a) implies that $G^+u(G)/u(G) \leq \eta_1(G/u(G))$. Hence $G^+ \leq \eta_1(G)$ by Part b). In particular $G/\eta_1(G)$ is locally finite. Finally $G_p \leq \eta_1(G)$ and so $G/\eta_1(G)$ is a p′-group. □

3.5.6. COROLLARY. (Wehrfritz [10]). *Let G be a locally nilpotent subgroup of GL(n,D). Then G is a Gruenberg group. If G is a Baer group then G is a Fitting group.*

The big open question, as we have said before, is whether or not the group G in 3.5.6 is hypercentral.

Proof: By 3.5.5 b) we have $G = \sigma(G)$, so G is a Gruenberg group. If G is a Baer group then $G = \bar{\sigma}(G)$, so $G = \eta_1(G)$ by the same result. Thus G is a Fitting group. □

To conclude this section, we return to the study of Engel elements.

3.5.7. COROLLARY. *Let* G *be any subgroup of* GL(n,D). *Then*

a) $L(G) = \sigma(G) = \eta(G)$,

b) $\bar{L}(G) = \bar{\sigma}(G) = \eta_1(\eta(G)) \geq \eta_1(G)$,

c) $R(G) = \rho(G)$,

d) $\bar{R}(G) = \bar{\rho}(G)$.

Proof: Just as in the proof of 3.5.2 we have that $\langle L(G), R(G)\rangle$ is locally nilpotent, and hence soluble. The theorem of Gruenberg quoted there yields a), c), d), and the equality $\bar{L}(G) = \bar{\sigma}(G)$ of b). By 3.5.5 b) we have $\eta_1(\eta(G)) = \bar{\sigma}(\eta(G)) = \bar{\sigma}(G)$, the last equality being always true. The proof is complete. □

SUPERSOLUBLE AND LOCALLY SUPERSOLUBLE GROUPS.

In view of the results of this chapter, and of Section 3.2 in particular, one might expect the theory of locally supersoluble subgroups of GL(n,D) to be analogous to the linear theory (Wehrfritz [2], Chapter 11). This has been carefully investigated by I. A. Stewart, but he has yet to publish his results. The outcome is that quite a lot can be said, but that the locally supersoluble theory does not follow the linear case as closely as the locally nilpotent theory does. The following summarizes his conclusions.

3.5.8. THEOREM (Stewart [1]). *Let* G *be a locally supersoluble subgroup of* GL(n,D).

a) G *is soluble, locally-nilpotent by periodic-abelian, and nilpotent by locally-finite.* G *need not be locally-nilpotent by finite.*

b) *If* $u(G) = \langle 1\rangle$ *then* G *is hypercyclic, but the paraheight of* G *is not bounded by any ordinal function of* p *and* n. *The* G*-paraheight of* u(G) *is also not bounded by any such function. Moreover for every* $p \geq 0$ *there exists a suitable* D *with a hypercyclic subgroup of* D^* *of paraheight exactly* ω. *Finally if* $p = 2$ *and* F *is perfect then* G *is hypercyclic.* □

Thus although the structure theorems in the linear case

survive in a recognizable form (3.5.8 a), the hypercyclicity theorems collapse almost completely in the skew linear case. It remains open whether a locally supersoluble subgroup of GL(n,D) is always hypercyclic. Of course this is closely related to Zalesskiĭ's problem as to whether a locally nilpotent subgroup of GL(n,D) is hypercentral; indeed a positive solution to the former implies a positive solution to the latter.

Stewart also considers the wider class of division rings E that are locally finite-dimensional over some, not necessarily central, subfield. For these division rings the above local problems have a negative solution. Specifically, for each $p \geq 0$ he constructs such an E of characteristic p and a locally nilpotent subgroup of GL(2,E) that is not hypercentral, and hence not hypercyclic either. Also for each $p \geq 0$ he constructs such an E with a locally supersoluble subgroup of E^* that is not hypercyclic. However he does have quite a number of positive results for such division rings. For example analogues of the following results hold: solubility of locally nilpotent groups, 3.3.3 a) without $\bar{A} \leq \Delta(\bar{G})$; 3.4.7, 3.4.13, 3.5.2 and (most of) 3.5.3 with $p = 0$ or $u(G) = \langle 1 \rangle$ and with somewhat larger bounds in the cases of 3.4.7 and 3.4.13; 3.5.8 a) with, for the last part, $p = 0$ or $u(G) = \langle 1 \rangle$.

4 DIVISION RINGS ASSOCIATED WITH POLYCYCLIC GROUPS

The skew linear groups considered in the previous two chapters have the property that their finitely generated subgroups are isomorphic to linear groups, usually of unbounded degree. Here we consider a class of skew linear groups that do not, in general, have this property.

Specifically we study skew linear groups over division rings D of the form D = F(G), where F is a central subfield of D, G is a polycyclic-by-finite subgroup of D^*, and D is generated *as a division ring* by F and G. Skew linear groups of this kind were first considered by Lichtman ([5] and [6]). It follows immediately from Goldie's theorem (1.4.2) that D is the (classical) ring of quotients of the Noetherian (Passman [2], 10.2.7) subalgebra F[G] of D = F(G). Thus each element of D can be written in the forms $ac^{-1} = d^{-1}b$ for elements a, b, c, d of F[G], with c and d non-zero. This fact plays a crucial role throughout the chapter.

In Section 4.1 below we show that the above class of skew linear groups is a little wider than might be at first apparent. We also indicate examples of such skew linear groups that are finitely generated and soluble but are not (group-theoretically) isomorphic to any linear group, thus confirming the remark above. Section 4.2 studies residual finiteness, from which we derive proofs of Theorems 1 and 2 of Lichtman [6], and in Section 4.3 we consider residual nilpotence.

In Section 4.4 we show how the results of the earlier sections give information about the division rings of quotients of certain group algebras, in Section 4.5 we present some slight evidence to suggest that Tits' Theorem (Wehrfritz [2], 10.16) may extend from fields to the division rings considered here, and in Section 4.6 we summarize Lichtman's work on skew linear groups over the division rings of quotients of the universal enveloping algebras of certain Lie algebras. Universal enveloping algebras of finite-dimensional Lie algebras have much in common with group algebras of poly-C_∞ groups.

4.1 GENERALITIES AND EXAMPLES

We now start to make considerable assumptions about the reader's knowledge of polycyclic groups and their group algebras. For the former see Segal [3]. For the latter only part of what we require is in Passman [2], and we have to refer the reader to original sources. However not much knowledge is required to understand the statements of the results below and the reader should at least look at these. In particular 4.1.3 plays a crucial role in Sections 4.2 and 4.3, as well as in the present section.

Suppose $D = F(G)$ is a division F-algebra where F is a field and G is a polycyclic-by-finite group. There is an obvious homomorphism of the group algebra FG of G over F onto $F[G] \leq D$. Let $\mathfrak{P}$ be the kernel of this map. Then $\mathfrak{P}$ is a prime ideal (indeed a completely prime ideal) of FG, $F[G] \cong FG/\mathfrak{P}$, and, by Goldie's theorem, D is the ring of quotients of F[G].

Now let us start again but going the other way. Thus let F be a field, G a polycyclic-by-finite group, and $\mathfrak{P}$ any *prime* ideal of FG. There is no need now for $\mathfrak{P}$ to be completely prime, but FG is Noetherian and Goldie's Theorem yields that $FG/\mathfrak{P}$ has a simple Artinian ring Q of quotients (Chatters & Hajarnavis [1], 1.28). Then Q is a matrix ring of finite degree m, say, over some division F-algebra E. Thus these division rings E are also associated with the field F and the group G. However if we consider skew linear groups over these division rings E no new groups arise. This is a consequence of the following.

4.1.1. THEOREM (Brown & Wehrfritz [1]). *Let J be a commutative ring, G a polycyclic-by-finite group and $\mathfrak{P}$ a prime ideal of the group algebra JG. Then there exists a division ring $D = F(H)$ generated, as a division ring, by its central subfield F and polycyclic subgroup H, and a positive integer t such that the ring Q of quotients of $JG/\mathfrak{P}$ is isomorphic to a subring of the matrix ring $D^{t\times t}$.*

In 4.1.1 we have $Q = E^{m\times m}$ for some division ring E and positive integer m. Then $E^{m\times m}$ embeds into $D^{t\times t}$ and hence m divides t. Set

$s = t/m$. Then for every positive integer n the group GL(n,E) is isomorphic to a subgroup of GL(ns,D). Hence follows our remark, that no new groups arise. We need some preliminary discussion and results before we can prove 4.1.1.

4.1.2. *Let* J *be a commutative ring,* G *a polycyclic-by-finite group,* $\mathfrak{P}$ *a prime ideal of* JG *and* G_1 *a normal subgroup of* G. *Then there exists a prime ideal* $\mathfrak{P}_1$ *of* JG_1 *such that* $\mathfrak{P} \cap JG_1 = \bigcap_{x \in G} \mathfrak{P}_1^x$ *and* $(G:N_G(\mathfrak{P}_1))$ *is finite.*

Proof: Suppose first that J is a field. Then JG_1 is Noetherian and so $JG_1/(\mathfrak{P} \cap JG_1)$ has only a finite number of minimal prime ideals, say $\mathfrak{P}_i/(\mathfrak{P} \cap JG_1)$ for $i = 1,2,\ldots,r$ (Chatters & Hajarnavis [1], 1.16). Since G permutes the $\mathfrak{P}_i$ the index $(G:N_G(\mathfrak{P}_i))$ is finite for each i. Also for some integer s we have $(\mathfrak{P}_1\mathfrak{P}_2\ldots\mathfrak{P}_r)^s \subseteq \mathfrak{P} \cap JG_1$.

Set $\mathfrak{p}_i = \bigcap_{x \in G} \mathfrak{P}_i^x$. Then each $\mathfrak{p}_i G$ is an ideal of JG and $(\mathfrak{p}_1 G \ldots \mathfrak{p}_r G)^s \subseteq \mathfrak{P}$. But $\mathfrak{P}$ is prime, so for some i we have $\mathfrak{p}_i G \subseteq \mathfrak{P}$. Consequently, $\mathfrak{P} \cap JG_1 \subseteq \mathfrak{p}_i \subseteq \mathfrak{P} \cap JG_1$, which complete the proof in this case.

Now consider the general case. We may assume that $\mathfrak{P} \cap J = \{0\}$, so J is now a domain; let K be its quotient field. Since $\mathfrak{P}$ is prime $JG/\mathfrak{P}$ is J-torsion-free. Hence $K\mathfrak{P} \cap JG = \mathfrak{P}$. Also if $\underline{a}$ is any ideal of KG then $\underline{a} = K(\underline{a} \cap JG)$. Thus $K\mathfrak{P}$ is a prime ideal of KG. If $\mathfrak{P}_1$ is a prime ideal of KG_1 then $JG_1 \cap \mathfrak{P}_1$ is a prime ideal of JG_1. The general case now follows from the special case proved above. □

Following Roseblade [3] we say that a polycyclic group G is *orbitally sound* if whenever H is a subgroup of G with $(G:N_G(H))$ finite then $H/\bigcap_{x \in G} H^x$ is also finite. Every polycyclic group has an orbitally sound normal subgroup of finite index (Roseblade [3], Theorem C2). For example, every connected soluble subgroup of GL(n,$\mathbb{Z}$) is orbitally sound (and polycyclic). The following result, in many ways the main result of Roseblade [3], is crucial for all of this chapter, but its proof strays too far from our subject of skew linear groups to be included here. For an alternative account of this work of Roseblade see Sections 9, 10 and 11 (and especially results 9.1, 9.3 and 11.1) of Passman [3].

4.1.3. ROSEBLADE'S THEOREM. (Roseblade [3]). *Let* J *be a commutative ring,* G *an orbitally sound polycyclic group and* $\mathfrak{P}$ *any prime ideal of* JG.

Set $P = G \cap (1+\mathfrak{P})$ *and define* D *by* $D/P = \Delta(G/P)$. *Then* $\mathfrak{P} = (\mathfrak{P} \cap JD)G$. □

4.1.4. (Brown & Wehrfritz [1]). *Let* J *be a commutative ring,* G *a polycyclic-by-finite group and* $\mathfrak{P}$ *a prime ideal of* JG. *Then there is a polycyclic fully invariant subgroup* G_o *of* G *of finite index and a completely prime ideal* $\mathfrak{P}_o$ *of* JG_o *such that*

$$\mathfrak{P} \cap JG_o = \bigcap_{x \in G} \mathfrak{P}_o^{\,x} \quad \textit{and} \quad (G:N_G(\mathfrak{P}_o)) < \infty .$$

Proof: It suffices to construct a normal polycyclic subgroup G_1 of G and a completely prime ideal $\mathfrak{P}_1$ of JG_1 such that $\mathfrak{P} \cap JG_1 = \bigcap_{x \in G} \mathfrak{P}_1^{\,x}$ and $(G:N_G(\mathfrak{P}_1)) < \infty$, since then we can set $G_o = G^{(G:G_1)}$ and $\mathfrak{P}_o = \mathfrak{P}_1 \cap JG_o$. Thus it suffices to prove the result for $G/(G \cap (1+\mathfrak{P}))$ in place of G. From now on assume that $G \cap (1+\mathfrak{P}) = \langle 1 \rangle$.

We induct on the Hirsch number h(G) of G. If $h(G) = 0$ then G is finite and we set $G_o = \langle 1 \rangle$ and $\mathfrak{P}_o = J \cap \mathfrak{P}$. Let $h(G) \neq 0$. There exists an orbitally sound normal subgroup G_1 of G of finite index with $G_1 \leq C_G(\Delta(G))$ and $G_1/\Delta(G_1)$ poly-C_∞. By 4.1.2 there is a prime ideal $\mathfrak{P}_1$ of JG_1 with $\mathfrak{P} \cap JG_1 = \bigcap_{x \in G} \mathfrak{P}_1^{\,x}$ and $(G:N_G(\mathfrak{P}_1)) < \infty$. If $G_1 \cap (1+\mathfrak{P}_1)$ is finite then Roseblade's Theorem yields that $\mathfrak{P}_1 = (\mathfrak{P}_1 \cap J\Delta(G_1))G_1$; that is,

$$(JG_1/\mathfrak{P}_1,\ J\Delta(G_1)/(\mathfrak{P}_1 \cap J\Delta(G_1)),\ G_1,\ G_1/\Delta(G_1))$$

is a crossed product. Now $\Delta(G_1) = G_1 \cap \Delta(G)$ is the centre of G_1 and hence $\mathfrak{P}_1 \cap J\Delta(G_1)$ is completely prime. Since $G_1/\Delta(G_1)$ is poly-C_∞ , $JG_1/\mathfrak{P}_1$ is a domain by 1.4.1 and consequently $\mathfrak{P}_1$ is completely prime.

Now suppose that $G_1 \cap (1+\mathfrak{P}_1)$ is infinite. Apply induction to JG_1 and $\mathfrak{P}_1$ modulo $G_1 \cap (1+\mathfrak{P}_1)$. Thus there is a normal subgroup G_2 of G (of G note) of finite index and a completely prime ideal $\mathfrak{P}_2$ of JG_2 with $G_2 \subseteq G_1$,

$\mathfrak{P}_1 \cap JG_2 = \bigcap_{x \in G_1} \mathfrak{P}_2^{\,x}$, and $(G_1:N_{G_1}(\mathfrak{P}_2)) < \infty$. Then

$$\bigcap_{x \in G} \mathfrak{P}_2^{\,x} = \bigcap_{x \in G} \mathfrak{P}_1^{\,x} \cap JG_2 = \mathfrak{P} \cap JG_1 \cap JG_2 = \mathfrak{P} \cap JG_2$$

and $(G:N_G(\mathfrak{P}_2))$ is finite. The proof is complete. □

4.1.5. *Let* J *be a commutative ring,* G *a polycyclic-by-finite group,* G_o *a normal subgroup of* G *and* $\mathfrak{P}$ *a prime ideal of* JG. *Then*

$$\underline{C}_{JG_o}(\mathfrak{P} \cap JG_o) \subseteq \underline{C}_{JG}(\mathfrak{P}).$$

Recall that $\underline{C}_{JG}(\mathfrak{P})$ is the set of elements of JG that are regular modulo $\mathfrak{P}$.

Proof: Suppose first that J is a field, so JG and JG_o are both Noetherian. Set $R = JG/(P \cap JG_o)G$ and let S and $\mathfrak{p}$ be the images of JG_o and P respectively in R. We have to prove that $C = \underline{C}_S(0) \subseteq \underline{C}_R(\mathfrak{p})$.

By 4.1.2 we have that $P \cap JG_o$, and hence S, is semiprime. Therefore by Goldie's Theorem C is an Ore set in S. Now C is G-invariant and R is a crossed product of S by an image of G, so C is Ore in R and $C \subseteq \underline{C}_R(0)$. Hence we can form the ring $RC^{-1} = C^{-1}R$ of quotients. Consider $Q = \mathfrak{p}C^{-1} \cap R$. Then Q is a left ideal of R and as such is finitely generated. A finite set of generators of Q have a common right denominator in RC^{-1} by the Ore condition. Then $Qd \subseteq \mathfrak{p}$ for some $d \in C$. But $\mathfrak{p}$ is prime, Q is a right ideal of R, and $d \notin \mathfrak{p}$. Consequently $Q \subseteq \mathfrak{p}$.

Let $x \in R$ and $c \in C$. If $xc \in \mathfrak{p}$ then $x = (xc)c^{-1} \in \mathfrak{p}C^{-1} \cap R = Q \subseteq \mathfrak{p}$. Similarly if $cx \in \mathfrak{p}$ then $x \in \mathfrak{p}$. Hence $c \in \underline{C}_R(\mathfrak{p})$, which completes the proof of the case where J is a field.

In general $P \cap J$ is a prime ideal of J and we may factor out by it and assume that J is a domain. Let K be its quotient field. Since J is central and P is prime JG/P is J-torsion-free. Hence $KP \cap JG = P$. If $\underline{a}$ is an ideal of KG then $\underline{a} = K(\underline{a} \cap JG)$. It follows easily that $\underline{C}_{JG}(P) = JG \cap \underline{C}_{KG}(KP)$. Further, $KP \cap JG_o = P \cap JG_o$ and hence $\underline{C}_{JG_o}(P \cap JG_o) = JG_o \cap \underline{C}_{KG_o}(KP \cap KG_o)$. The general case follows from the special case proved above. □

Let R be a semiprime Goldie ring. By Goldie's Theorem R has a semisimple Artinian ring Q(R) of right quotients. If P is a prime ideal of R then $Q(R/P)$ is simple Artinian by the same result. We use below the well-known fact that

$$Q(R) \cong \oplus\, Q(R/P)$$

where P ranges over all the minimal prime ideals of R. To see this let $Q(R) \cong \oplus\, S_i$ where the S_i are simple. Let $\pi_i : Q(R) \twoheadrightarrow S_i$ be the natural projection. Then $S_i = Q(R\pi_i)$ and so $P_i = R \cap \ker\pi_i$ is prime (Goldie's Theorem again). Also $\bigcap_i P_i = \{0\}$, so every minimal prime is one of the P_i. Suppose $P_1 \supseteq P_2$. Then $\bigcap_{i>1} P_i = \{0\}$, $\oplus_{i>1}\, \pi_i$ embeds R into $\oplus_{i>1}\, S_i$ and $Q(R) \cong \oplus_{i>1}\, S_i$. This contradiction proves that the P_i are exactly the minimal primes of R.

Proof of 4.1.1: Pick G_o and P_o as in 4.1.4. Set $R = JG/P$ and $S = JG_o/(P \cap JG_o)$. Then S is semiprime and both R and S have rings Q_R and Q_S of quotients, by Goldie's Theorem. Also Q_S is naturally embedded in

Q_R since $\underline{C}_{JG_o}(\mathfrak{P} \cap JG_o) \subseteq \underline{C}_{JG}(\mathfrak{P})$ by 4.1.5. Further Q_R is finitely generated as left Q_S-module, the number of generators required being at most $(G:G_o)$.

Now $\mathfrak{P} \cap JG_o = \bigcap_{i=1}^{m} \mathfrak{P}_i$, say, where each $\mathfrak{P}_i$ is a conjugate of $\mathfrak{P}_o$ (see 4.1.4). Choose m minimal. Then each $\mathfrak{P}_i$ is completely prime, $S_i = JG_o/\mathfrak{P}_i$ is a domain, and its ring D_i of quotients is a division ring. Also by the minimal choice of m the $\mathfrak{P}_i$ are exactly the minimal primes of JG_o over $\mathfrak{P} \cap JG_o$ and $Q_S \cong \oplus_{i=1}^{m} D_i$. But all the D_i are isomorphic to $F(H) = D$, say, where F is the quotient field of $J/(\mathfrak{P} \cap J)$ and $H = G_o/(G_o \cap (1+\mathfrak{P}_o))$. Therefore if Q_R has r generators as left Q_S-module then $t = \dim_D Q_R \leq mr$ and Q_R is isomorphic to a subring of $\mathrm{End}_D(Q_R) \cong D^{t \times t}$. □

4.1.6. REMARKS. In the above proof of 4.1.1 the field F is just the quotient field of $J/(\mathfrak{P} \cap J)$ and H is some section of G. Specifically $H = K/L$ where K has finite index in G, L is normal in K, and $L_G = \bigcap_{x \in G} L^x = K \cap (1+\mathfrak{P})$. Note first that by increasing t we can replace K by any of its subgroups of finite index, so for example K can always be chosen to be fully invariant in G. Frequently L/L_G is finite, for example if G is orbitally sound. In this case K has a subgroup M of finite index with $M \cap L = L_G$, since G/L_G is residually finite. If we then replace K by M it follows that we can choose H to be a subgroup of $G/(G \cap (1+\mathfrak{P}))$ of finite index. We do not know whether this is true for all G.

We now describe some finitely generated skew linear groups over division rings of the type we are considering in this section, that are not isomorphic to any linear group. The details are excessively long and complicated and we omit them, proving only a very special case.

Let G be a finitely generated group with an abelian normal subgroup A such that G/A is polycyclic-by-finite. Suppose that p is zero or a prime and that A is a p-group (recall that a 0-group is a torsion-free group). It is an easy consequence of 1.4.4 and 1.4.5 that G is isomorphic to a skew linear group of characteristic p over a division ring $D = E(H)$ where H is some polycyclic subgroup of D^* and E is a subfield of D normalized by H. It would be very interesting, especially in connection with the theory of abstract finitely generated soluble groups, if E could be chosen to be centralized by H, in which case D is of the type considered above. Unfortunately, only very special cases of this are known. The following,

Theorem 2 of Wehrfritz [17], we state without proof, since the proof is very technical and gives no clue to a general solution of the problem.

4.1.7. **(Wehrfritz [17]).** ***Let*** **H** ***be a polycyclic-by-finite group,*** **F** ***a field of characteristic*** $p \geq 0$ ***and*** **M** ***an*** **FH*-module. Under each of the following three conditions there is an integer*** n ***and a division ring*** **D*, generated as such over its centre by a polycyclic group, such that the split extension*** $G = H [M$ ***is isomorphic to a subgroup of*** **GL(n,D).**

- **a)** **C** ***is a normal subgroup of*** **H,** $\mathfrak{P}$ ***is a prime ideal of*** **FC** ***normalized by*** **H,** ***and*** $M = FH/\mathfrak{P}H$.
- **b)** **C** ***is an abelian normal subgroup of*** **H,** $\mathfrak{P}$ ***is a prime ideal of*** **FC** ***and*** $M = FH/\mathfrak{P}H$.
- **c)** **C** ***is any subgroup of*** **H** ***and*** $M = FH/(C-1)FH$. □

Finitely generated abelian-by-polycyclic-by-finite groups are not usually isomorphic to linear groups. The precise conditions for their linearity are known (Wehrfritz [8]). Not suprisingly these are rather complicated, so we do not state them here. A special case of c) in 4.1.7 is where G is the standard wreath product $F^+ \wr H$**, this being the case** $C = \langle 1 \rangle$**. Now 10.21 of Wehrfritz [2] gives the conditions for a wreath product to be isomorphic to a linear group. It implies the following.**

4.1.8. ***Let*** **H** ***and*** **F** ***be as in*** **4.1.7,** ***and let*** **X** ***be any non-trivial subgroup of*** F^+***. Then*** $X \wr H$ ***is isomorphic to a linear group if and only if*** **H** ***is abelian-by-finite.*** □

Thus for example let X be a cyclic group of infinite or prime order and H a polycyclic group that is not abelian-by-finite, $H = Tr_1(3,\mathbb{Z})$ **for example. Then** $G = X \wr H$ **is isomorphic to a skew linear group over a division ring of the type we are considering here (by 4.1.7c)), but not isomorphic to any linear group (by 4.1.8). Trivially G is finitely generated and soluble.**

Let us prove that $G = X \wr H$ **, for X and H as above, is a skew linear group as in 4.1.7. H has a torsion-free normal subgroup K of finite index n, say. Let P be a prime field. Then PK is an Ore domain by 1.4.8; let D be its division ring of quotients. Clearly D = P(K). If charP = 0**

let X be infinite cyclic. Otherwise let X be cyclic of order charP. Then

$$\left\langle \begin{bmatrix} a & 0 \\ b & 1 \end{bmatrix} \quad : \quad a, b \in K \right\rangle \leq GL(2,D)$$

is isomorphic to $X \wr K$. Hence $X \wr H$ is isomorphic to subgroup of $GL(2n,D)$.□

4.2 RESIDUAL FINITENESS

Every finitely generated linear group is residually finite by a theorem of Mal'cev (Wehrfritz [2], 4.2). We extend this result to the class of skew linear groups considered in the previous section. As a by-product we prove Lichtman's theorem which establishes the equivalence of unipotence and stability for such groups. The whole of this section depends upon the following reduction theorem, variants of which appear in Lichtman [5] and [6].

4.2.1. REDUCTION THEOREM. *Let* D *be a division ring generated as such by the infinite polycyclic-by-finite subgroup* G *of* D^*, *and let* n *be a positive integer. If* $d_1,\ldots,d_m$ *are non-zero elments of* $D^{n\times n}$ *then there exists a division ring* D_0, *generated as such by a polycyclic group* G_0 *of Hirsch number* $h(G_0) < h(G)$, *an integer* t, *and a ring homomorphism* Φ *of the subring generated by* $d_1,\ldots,d_m$ *into* $D_0^{t\times t}$ *such that* $\bigcap_i (\ker\Phi)^i = \{0\}$ *and each* $d_j\Phi \neq 0$. *If* $\mathrm{char}D = 0$ *then for all but a finite number of primes* p *(depending on* $d_1,\ldots,d_m$*) we can choose* D_0 *of characteristic* p. *Further if* G *is nilpotent-by-cyclic-by-finite we can choose* t *independently of* $d_1,\ldots,d_m$, *and if* G *is torsion-free nilpotent we can choose* t = n.

Note that if $\mathrm{char}D > 0$ then necessarily $\mathrm{char}D_0 = \mathrm{char}D$. Before we prove 4.2.1 we show how it can be used. We require the following simple result.

4.2.2. *Let* $D = F(G)$ *be a division algebra over the field* F, *where* G *is a polycyclic-by-finite subgroup of* D^*. *Then any finitely generated subring of* D *lies in a division subring of* D *generated by a polycyclic-by-finite group.*

Proof: Let R be a finitely generated subring of D. Then R lies in a division subring of D generated by a subring of the form J[G], where J is a finitely generated subring of F. We may enlarge J so that J is generated by a finitely generated subgroup U of its group of units. (In fact its whole group of units is finitely generated - see Samuel [1].) Clearly UG is polycyclic-by-finite and it generates J[G] as a ring. This proves the result. □

The following is our main application of 4.2.2.

4.2.3. THEOREM. *Let* F *be a field of characteristic* $p \geq 0$, n *a positive integer, and* D = F(G) *a division* F*-algebra generated by the locally polycyclic-by-finite group* G.

a) (Lichtman [6]). *Any unipotent subgroup* U *of* GL(n,D) *is a stability group.*

b) *If* H *is a subgroup of the group of units of a finitely generated subring* R *of* $D^{n\times n}$ *(e.g. if* H *is a finitely generated subgroup of* GL(n,D)), *then* H *is residually finite.*

c) *Let* H *be as in Part b). If* p = 0 *or* n = 1 *then* H *is torsion-free by finite. If* $p > 0$ *then* H *has a subgroup of finite index whose only torsion is* p*-torsion.*

d) (Lichtman [6]). *A periodic subgroup of* GL(n,D) *is locally finite.*

The next section contains alternative proofs of Parts b), c), and consequently d) above. The approach adopted below is shorter but gives less information.

Proof: Suppose H is as in Part b). By 4.2.2 we may assume that D is generated by the polycyclic-by-finite group G. Thus we can apply the Reduction Theorem to the finitely generated subrings of $D^{n\times n}$. We then use induction on h(G). If h(G) = 0 then G is finite. It follows that $d = \dim_F F[G]$ is finite, that F[G] = D, and that $D^{n\times n}$ is isomorphic to subring of $F^{nd\times nd}$. Thus in what follows, the start of each induction on h(G) is covered by the case of linear groups.

a) Let T be the natural image of the augmentation ideal of **Z**U in $D^{n\times n}$. By Levitzki's Theorem (1.3.9) it suffices to prove that T is a nil ring. To do this we may assume that U is finitely generated, so our initial remarks are applicable.

Let $x \in T$. The set $\{ x^i : i = 1,2,\ldots\}$ is left linearly dependent over D, say $\lambda_1 x^{n_1} + \ldots + \lambda_r x^{n_r} = 0$, where $\lambda_1,\ldots,\lambda_r \in D^*$ and $1 \leq n_1 < \ldots < n_r$. We may assume that $h(G) > 0$. Then the Reduction Theorem produces D_0, G_0, t, and a ring homomorphism Φ from the subring generated by U and the $\lambda_i^{\pm 1} 1_n$ into $D_0^{t \times t}$ as in that theorem. By induction $x\Phi$ is nilpotent. Pick s minimal with $x^s\Phi = 0$. If $s > n_1$ then

$(\lambda_1 1_n)\Phi.(x^{s-1})\Phi = (-\Sigma_2^r \lambda_i x^{n_i - n_1 - 1})\Phi.(x^s)\Phi = 0$. But $(\lambda_1 1_n)\Phi$ is a unit, so $(x^{s-1})\Phi = 0$, a contradiction of the choice of s. Consequently $s \leq n_1$, whence $x^{n_1}\Phi = 0$. But this is for all possible choices of Φ. It follows from 4.2.1 that $x^{n_1} = 0$, and therefore T is a nilring.

b) By the linear case, induction on h(G), and the Reduction Theorem we have $H \in R\underline{RF} = R\underline{F}$.

c) Let q be a prime, and let $\underline{F}_q$ denote the class of finite q-groups. Clearly every periodic $(PRL\underline{F}_q)$-group is a q-group. Part c) is therefore a consequence of the following statement (a much stronger version of which will be proved in Section 4.3).

If $p = 0$ *then* $H \in (PRL\underline{F}_q)\underline{F}$ *for almost all primes* q. *If* $p > 0$ *then* $H \in (PRL\underline{F}_p)\underline{F}$.

For by the Reduction Theorem there exists a ring homomorphism $\Phi : R \longrightarrow D_0^{t \times t}$ such that $H\Phi$ has the required properties and $\bigcap_i (\ker\Phi)^i = \{0\}$. Now $K = H \cap (1+\ker\Phi)$ is equal to $C_H(R/\ker\Phi)$, the action being right multiplication. Thus if $q = \mathrm{char}D_0$ then $K \in R(\underline{N} \cap L\underline{F}_q)$, for example by Wehrfritz [2], 4.6. The result follows.

d) Let H be a finitely generated periodic subgroup of GL(n,D). We have to prove that H is finite. If $p = 0$ then H is finite by Part c). Assume that $p > 0$. Then H has a normal p-subgroup K of finite index, by Part c). Therefore K is finitely generated. But K is also unipotent, and hence nilpotent by Part a). Consequently K, and so H, is finite. □

Not surprisingly, not every finitely generated skew linear group is residually finite.

4.2.4. (Wehrfritz [16]). *Let* p *be zero or a prime. Then there exists a division ring* D *of characteristic* p *and a 3-generator, nilpotent-of-class-2-by-cyclic subgroup* G *of* D^* *such that* G *is not residually finite. Moreover every subgroup of* G *of finite index contains* p'*-torsion.*

Proof: Let $r \neq p$ be any prime. For each integer i let

$$H_i = \langle x_i, y_i : [x_i,y_i,x_i] = [x_i,y_i,y_i] = [x_i,y_i]^r = 1 \rangle.$$

Let H be the central product of the H_i amalgamating the $[x_i,y_i]$ to z, say. Define the automorphism g of H by

$$x_i^g = x_{i+1} \quad , \qquad y_i^g = y_{i+1}$$

Let $G = H] \langle g \rangle$. Now $Z = \langle x_i^r , y_i^r , z : i \in \mathbb{Z} \rangle$ is the centre of H. It follows that H is nilpotent of class 2. Also G/H is cyclic, and $G = \langle x_0, y_0, g \rangle$ is a 3-generator group. Suppose N is normal subgroup of G of finite index which does not contain z. Now $H' = \langle z \rangle$, so $[H, N \cap H] \subseteq H' \cap N = \langle 1 \rangle$, i.e. $N \cap H \subseteq Z$. But H/Z is infinite and $H/(N \cap H)$ is finite. Therefore no such N exists. It follows that G is not residually finite, and that G satisfies the final statement of 4.2.4.

Let F be a field of characteristic p containing a primitive r-th root ς of unity. Identify ς and z, and let G act trivially on F. Since $G/\langle z \rangle \in PLPC_\infty$ we can apply 1.4.3 to the crossed product

$$(F[G], F, G, G/\langle z \rangle).$$

Thus G embeds into the multiplicative group of a division ring of characteristic p. □

The above example depends upon the existence of torsion elements. The following example of a non residually-finite group (due originally to P. Hall) shows that this is not necessary.

4.2.5. (Wehrfritz [16]). *Let* p *be zero or a prime. Then there exists a division ring* D *of characteristic* p *and a 3-generator torsion-free soluble subgroup* G *of* D^* *of derived length* 3 *which is not residually finite.*

Proof: Let A be the direct product of copies A_i ($i \in \mathbb{Z}$) of the rationals, written multiplicatively. Let $i \longmapsto p_i$ be a bijection from $\mathbb{Z}$ to the set of all positive primes. Let x be the automorphism of A that permutes the A_i cyclicly (so $A_i^x = A_{i+1}$), and let y be the automorphism of A that for each i raises A_i to the p_i-th power. Then $H = \langle x,y \rangle \cong C_\infty \wr C_\infty$, the base group being the direct product of the conjugates of y by the powers of x, and the top group being $\langle x \rangle$. Therefore $G = A] H$ is torsion-free and soluble of derived length 3. Also A is infinite and irreducible as an H-module, so G is 3-generator and not residually finite. Finally G is poly torsion-free abelian, so G is isomorphic to a skew linear group of characteristic p and degree 1, by 1.4.4. □

We now turn to the proof of the Reduction Theorem. For this some preliminary lemmas are required. 4.2.6 and 4.2.8 also play a key role in Section 4.3. The reader will have no difficulty in generalizing 4.2.7; we prove what we need.

An ideal $\underline{a}$ of a ring R is *right (weak) Artin-Rees* if for any submodule N of any finitely generated right R-module M there is a positive integer m such that $N \cap M\underline{a}^m \subseteq N\underline{a}$. There is a corresponding notion of left Artin-Rees. An *Artin-Rees ideal* is one that is both left and right Artin-Rees.

4.2.6. *Let* R *be a right Noetherian ring and* $\underline{a}$ *a semiprime ideal of* R *such that* $\underline{C}_R(\underline{a}) \subseteq \underline{C}_R(\underline{a}^i)$ *for each* $i \geq 1$. *Then:*

a) $\underline{C}_R(\underline{a})$ *is right Ore in* R *modulo* $\underline{a}^i$ *for each* $i \geq 1$;

b) *if* $\underline{a}$ *is right Artin-Rees then* $\underline{C}_R(\underline{a})$ *is right Ore in* R;

c) *if* $\underline{a}$ *is right Artin-Rees and* R *is a left Noetherian domain then in the ring* RQ^{-1} *of quotients of* R *by* $Q = \underline{C}_R(\underline{a})$ *we have* $\bigcap_i(\underline{a}Q^{-1})^i = \langle 0\rangle$.

Proof: a) We may assume that $\underline{a}^i = \{0\}$. Then $\underline{a}$ is the nilpotent radical of R. By hypothesis $\underline{C}_R(\underline{a}) \subseteq \underline{C}_R(0)$. Hence by Small's theorem $\underline{C}_R(\underline{a}) = \underline{C}_R(0)$ and $\underline{C}_R(0)$ is right Ore in R, see Chatters & Hajarnavis [1], 2.3.

b) This is an immediate consequence of Part a) and a lemma of P. F. Smith (ibid. 11.9).

c) Here $\underline{a}Q^{-1}$ is an ideal of RQ^{-1} since if $c^{-1}a = bd^{-1}$ where $a \in \underline{a}$, $b \in R$ and $c, d \in Q$, then $cb = ad \in \underline{a}$ and $c \in \underline{C}_R(\underline{a})$, so $b \in \underline{a}$ and $Q^{-1}\underline{a} \subseteq \underline{a}Q^{-1}$. Therefore $(\underline{a}Q^{-1})^i = \underline{a}^iQ^{-1}$ for all $i \geq 1$. Also if $ac^{-1} = b \in R$ where $a \in \underline{a}^i$ and $c \in Q$, then $bc \in \underline{a}^i$, $c \in \underline{C}_R(\underline{a}^i)$ by hypothesis, and we find that $b \in \underline{a}^i$. Consequently $R \cap \underline{a}^iQ^{-1} = \underline{a}^i$ for all i. Thus

$$R \cap \bigcap_i (\underline{a}Q^{-1})^i = \bigcap_i \underline{a}^i = \langle 0\rangle$$

by another result of P. F. Smith (Passman [2], 11.2.13). Therefore

$$\bigcap_i (\underline{a}Q^{-1})^i = (R \cap \bigcap_i (\underline{a}Q^{-1})^i)Q^{-1} = \langle 0\rangle,$$

as required. □

We need to produce rings and ideals satisfying the hypotheses of 4.2.6.

4.2.7. *Let* (R,S,G,H) *be a crossed product, where* S *is commutative, semiprime, and finitely generated as a ring, and* H *is poly-*C_∞. *Let* $\underline{m}$ *be a maximal ideal of* S *and set* $\underline{m}_G = \bigcap_{x\in G} \underline{m}^x$ *and* $\underline{n} = \underline{m}_G R$. *Then:*

a) $\underline{n}$ *is a prime ideal of* R;

b) $\underline{C}_R(\underline{n}) \subseteq \underline{C}_R(\underline{n}^i)$ *for all* $i \geq 1$;

c) $\underline{n}$ *is Artin–Rees.*

d) *if* $a = t_1\alpha_1 + \ldots + t_r\alpha_r \in R$ *where* $\{t_1,\ldots,t_r\}$ *is part of a transversal of* $S \cap G$ *to* G *and* $\alpha_1,\ldots, \alpha_r \in S \setminus (\bigcup_{x\in G} \underline{m}^x)$, *then* $a \in \underline{C}_R(\underline{n})$.

The group of units of S is finitely generated (Samuel [1]), so G is actually a polycyclic group. For our applications we do not need to quote this since there G will be polycyclic by assumption.

Proof: a) Clearly $\underline{n} = \underline{m}_G R = R\underline{m}_G$ is an ideal of R. Let T be a transversal of $S \cap G$ to G. The group H is orderable since it is poly-C_∞. Transfer this order to T.

Let $\underline{a}$ and $\underline{b}$ be ideals of R with $\underline{ab} \subseteq \underline{n} \subseteq \underline{a} \cap \underline{b}$. Let $\underline{a}'$ be the set of minimal coefficients of $\underline{a}$; that is, if $a \in \underline{a}\setminus\{0\}$ then $a = t_1\alpha_1+\ldots+t_r\alpha_r$ for some $t_1 < t_2 < \ldots < t_r$ in T and α_i in $S\setminus\{0\}$, and then $\underline{a}'$ consists of 0 and all the elements α_1. Similarly define $\underline{b}'$ for $\underline{b}$. Then $\underline{a}'$ and $\underline{b}'$ are ideals of S containing $\underline{m}_G$ and normalized by G. Also since $\underline{ab} \subseteq \underline{n} = \oplus_{t\in T}\, t\underline{m}_G$ we have $\underline{a}'\underline{b}' \subseteq \underline{m}_G \subseteq \underline{m}$. But $\underline{m}$ is prime, so one of $\underline{a}'$ or $\underline{b}'$, say $\underline{a}'$, lies in $\underline{m}$. The G-invariance of $\underline{a}'$ then yields $\underline{a}' \subseteq \underline{m}_G$. But then $\underline{a}'R \subseteq \underline{n} \subseteq \underline{a}$ and so $\underline{a} = \underline{a}'R \subseteq \underline{n}$. Therefore $\underline{n}$ is prime.

b) Since S is finitely generated and commutative, $S/\underline{m}$ is finite by Hilbert's Nullstellensatz, and thus $\bar{S} = S/\underline{m}_G$ is finite and semisimple. We induct on i. Suppose that $\underline{C}_R(\underline{n}) \subseteq \underline{C}_R(\underline{n}^i)$; this certainly holds if i = 1. Let $x \in R$ and $c \in \underline{C}_R(\underline{n})$, and suppose that $xc \in \underline{n}^{i+1}$. Then $x \in \underline{n}^i$. Now $\underline{n}^i = (\underline{m}_G)^i R$ and hence $\underline{n}^i/\underline{n}^{i+1} \cong (\underline{m}_G{}^i/\underline{m}_G{}^{i+1}) \otimes_S R$. But $\bar{S}$ is semisimple, so $\underline{m}_G{}^i/\underline{m}_G{}^{i+1}$ is isomorphic, as $\bar{S}$-module, to a submodule of some free $\bar{S}$-module, $\bar{S}^m$ say. Then $\underline{n}^i/\underline{n}^{i+1}$ is isomorphic, as right R-module, to a submodule of $\bar{S}^m \otimes_S R \cong \bar{R}^m$, where $\bar{R} = R/\underline{n}$. Clearly $c+\underline{n}$ is not a zero divisor in $\bar{R}$ and hence is not a zero divisor on $\underline{n}^i/\underline{n}^{i+1}$. Therefore $x \in \underline{n}^{i+1}$. In the same way $cx \in \underline{n}^{i+1}$ implies that $x \in \underline{n}^{i+1}$. Thus $c \in \underline{C}_R(\underline{n}^{i+1})$, as required.

c) Since S is commutative $\underline{m}_G$ is trivially centrally generated in S. Also as remarked above G is polycyclic. Part c) is thus a very special case of a theorem of Roseblade (Passman [2], 11.2.3).

d) As above $S/\underline{m}_G$ is finite and semisimple. The hypothesis on the α_i ensures that each $\alpha_i \in \underline{C}_S(\underline{m}_G)$. Also H is poly-$C_\infty$, so Higman's Theorem (1.4.1) implies that $a \in \underline{C}_R(\underline{n})$. □

Finally we need a good supply of elements a satisfying the hypotheses of 4.2.7 d). The next result, due to Roseblade [1], provides us with precisely this. We need the following definition.

Let G be any group. A *plinth* for G is a G-module A that is free of finite rank, n say, as $\mathbb{Z}$-module, such that A is rationally irreducible (that is, $\mathbb{Q} \otimes_{\mathbb{Z}} A$ is irreducible) for every subgroup of G of finite index. Let $\rho : G \longrightarrow GL(n,\mathbb{Z})$ be afforded by the action of G on A. Then A is a plinth for G if and only if the connected component $(G\rho)^o$ is irreducible as a subgroup of $GL(n,\mathbb{Q})$. For the proof of the following result we refer the reader either to Roseblade's original paper or Passman [2], Section 12.3, especially 12.3.6.

4.2.8. (Roseblade [1], Theorem E). *Let* A *be a plinth for the polycyclic-by-finite group* G, *let* k *be a locally finite field, and let* $a \in kA \setminus \{0\}$. *Then there is a maximal ideal* $\underline{m}$ *of* kA *such that* $a \notin \bigcup_{x \in G} \underline{m}^x$. □

Proof of 4.2.1: We need to reduce G so that we can apply Roseblade's Theorem (4.1.3). Now G has a normal subgroup G_1 of finite index that is orbitally sound (see the preamble to 4.1.3). By the standard properties of polycyclic groups we can choose G_1 to be torsion-free with $\zeta_1(G_1) = \Delta(G_1)$ and $G_1/\zeta_1(G_1)$ a poly-C_∞ group (see Segal [3], Chapter 1). Then D has finite left dimension $n_1 \leq (G:G_1)$ over the division subring D_1 of D generated by G_1 and thus right multiplication embeds $D^{n \times n}$ into ${D_1}^{k \times k}$, where $k = nn_1$. If G is torsion-free nilpotent we can choose $G = G_1$ and $n = k$. To simplify the notation we assume that $G = G_1$.

Let $\mathfrak{P}$ be the kernel of the natural map of $\mathbb{Z}G$ into D and let R be the subring of D generated by G. By 4.1.3 (Roseblade's Theorem) $\mathfrak{P} = (\mathfrak{P} \cap \mathbb{Z}\zeta_1(G))G$. Let Z be the subring of R generated by $\zeta_1(G)$. Then R is a crossed product of Z by $G/\zeta_1(G)$. Let $\underline{m}$ be any maximal ideal of Z, put $R_o = R/\underline{m}R$, and denote the natural image of G in R_o by G_o. By P. Hall's theorem

(Passman [2], 10.2.7) the ring R_0 is Noetherian. Also R_0 is the crossed product of the field $Z/\underline{m}$ by the poly-C_∞ group $G/\zeta_1(G)$. Thus R_0 is a domain by 1.4.1. Hence Goldie's theorem yields that R_0 has a ring D_0 of quotients that is a division ring. By the same result D is the ring of quotients of R.

Let $c_1,\dots,c_k$ be the non-zero entries of the matrices $d_1,\dots,d_m$. There exists $r \in R$ such that $R[r^{-1}] \subseteq D$ contains $c_1,\dots,c_k$ and their inverses. Now $r = \Sigma\, t_i z_i$ for some distinct members t_i of a transversal of $\zeta_1(G)$ to G and some non-zero elements z_i of Z. Pick $\underline{m}$ not containing $z = \Pi_i\, z_i$. This is possible since the Jacobson radical of a finitely generated integral domain is always zero (e.g. Wehrfritz [2], 4.1). Moreover if charD $= 0$ we can choose $\underline{m}$ such that, with a finite number of exceptions, the characteristic of $Z/\underline{m}$ is any prime, since the set of primes q such that q is a unit of $Z[z^{-1}]$, is finite. Note that $r \notin \underline{m}R$.

Let $Q = \underline{C}_R(\underline{m}R)$. By 4.2.7 and 4.2.6 the set Q is right Ore in R and we can form the ring RQ^{-1} of quotients and regard it as a subring of D. There is a natural map, θ say, of RQ^{-1} onto D_0, and 4.2.6 yields that $\bigcap_i (\ker\theta)^i = \{0\}$. Now $r \in Q$, so $R[r^{-1}] \subseteq RQ^{-1}$. Thus each c_i is a unit of RQ^{-1} and hence each $c_i\theta \neq 0$. Let Φ be the map of $(RQ^{-1})^{n\times n}$ onto $D_0{}^{n\times n}$ induced by θ. Then each $d_j\Phi \neq 0$. Further $((\ker\theta)^{n\times n})^i \subseteq ((\ker\theta)^i)^{n\times n}$, so $\bigcap_i (\ker\Phi)^i = \{0\}$.

If $h(G_0) < h(G)$ the proof is complete. This happens, for example, if initially (i.e., before assuming that $G = G_1$) we have $\zeta_1(G_1) \neq \langle 1\rangle$. In this case we also have $t = nn_1$, which is independent of the d_j. In particular if G is nilpotent-by-finite then we can choose G_1 nilpotent. The conclusion, therefore, follows in this special case.

We are left with the case $h(G_0) = h(G)$. Then as G is torsion-free we have $\zeta_1(G) = \langle 1\rangle$, and R is just the group ring of G over a prime image J of $\mathbb{Z}$. With a slightly smaller choice of G_1 we may assume that G has a non-trivial abelian normal subgroup A such that G/A is poly-C_∞ and A is a plinth for G. Let S be the subring of R generated by A. Then S is a finitely generated commutative domain. We now essentially repeat the above argument with Z replaced by S.

Choose $c_1,\dots, c_k$ and r as before. Then $r = \Sigma_i\, t_i s_i$ for some distinct members t_i of a transversal of A to G and some non-zero elements s_i of S. If $p = \mathrm{char}D > 0$ then $J = \mathbb{Z}/p\mathbb{Z}$ and by 4.2.8 there is a maximal ideal $\underline{m}$ of S such that $s = \Pi_i\, s_i \notin \bigcup_{x\in G} \underline{m}^x$. If not then $J = \mathbb{Z}$ and $\bigcap qS = \{0\}$, where q ranges over any infinite set of primes. Thus for all but a finite set of primes q we have $s \notin qS$, and by 4.2.8 applied to S/qS there is a maximal ideal $\underline{m}$ of S such that $q \in \underline{m}$ and $s \notin \bigcup_{x\in G} \underline{m}^x$.

Set $\underline{m}_G = \bigcap_{x\in G} \underline{m}^x$, $\underline{n} = \underline{m}_G R$, $R_0 = R/\underline{n}$, $S_0 = S/\underline{m}_G \subseteq R_0$, and let G_2 be the image of G in R_0. Set $Q = \underline{C}_R(\underline{n})$ and $Q_0 = \underline{C}_{R_0}(0)$. By Goldie's Theorem, 4.2.7 and 4.2.6 we can form the rings $R_0Q_0^{-1}$ and RQ^{-1} of quotients. There is an obvious ring homomorphism θ of RQ^{-1} onto $R_0Q_0^{-1}$, and $\bigcap (\ker\theta)^i = \{0\}$. Also by 4.2.7 we have $r \in Q$, so $c_1,\dots,c_k$ are units of RQ^{-1}. Let $\phi : (RQ^{-1})^{n\times n} \twoheadrightarrow (R_0Q_0^{-1})^{n\times n}$ be the induced map. Then each $d_j\phi$ is non-zero and $\bigcap (\ker\phi)^i = \{0\}$. Also if $\mathrm{char}D = 0$ we can choose $\mathrm{char}R_0$ to be any prime with a finite number of exceptions.

If initially G is nilpotent-by-cyclic-by-finite we can choose G_1 so that $G_1/C_{G_1}(A)$ is cyclic. Using Theorem 1 of Wehrfritz [6] instead of 4.2.8 we can choose G_1 to satisfy $\underline{m} = \underline{m}_G$. Then R_0 is a domain by 1.4.1 and $D_0 = R_0Q_0^{-1}$ is a division ring generated by G_2. Clearly $h(G_2) \leq h(G)$, so we can set $G_0 = G_2$. This deals with this special case, and again we have $t = nn_1$.

In general S_0 is a direct sum of a finite number of copies of the finite field $S/\underline{m}$. Let k be the diagonal copy of $S/\underline{m}$ in S_0. Now G_2, being polycyclic, is residually finite, and S_0 is finite. Hence there is a normal subgroup G_0 of G_2 of finite index such that $S_0 \cap G_0 = \langle 1\rangle$ and G_0 centralizes S_0. Thus $k[G_0] \subseteq R_0$ is just the group algebra of G_0 over k. Also $k[G_0]\setminus\{0\} \subseteq Q_0$ by 4.2.7 d) since G_0 is poly-C_∞, and hence G_0 generates a division subring D_0 of $R_0Q_0^{-1}$, by Goldie's Theorem. Also $R_0[D_0]$ has finite left dimension $d \leq (G_2:G_0)$ over D_0, and right multiplication embeds $R_0[D_0]$ into $D_0^{d\times d}$. Further $R_0[D_0]$ is Artinian and therefore $R_0[D_0] = R_0Q_0^{-1}$. This completes the proof. Note that in this final case we can take $t = dnn_1$, where d depends on the choice of $\underline{m}$ and hence on $d_1,\dots,d_m$. □

The Reduction Theorem can only be indirectly used to prove results that apply to characteristic zero only (e.g. 4.2.3 c)) since D_0 in 4.2.1 can (and in our proof of 4.2.1 will) have positive characteristic. What would be required is an algebra version of the Reduction Theorem. If G is nilpotent-by-finite then the proof of 4.2.1 does not use 4.2.8 or the related results of Wehrfritz [6]. It does require a version of 4.2.7 in which the term F-algebra replaces ring, but only in the special case where S is central. Then $\underline{m}_G = \underline{m}$ and $\bar{S}$ is a field. The proof of 4.2.7 then goes through as before. As a result we obtain the following variant of the Reduction

Theorem.

4.2.9. *Let* $D = F(G)$ *be a division ring generated over its centre* F *by the infinite, finitely generated nilpotent-by-finite group* G, *and let* n *be a positive integer. If* $d_1,\ldots,d_m$ *are non-zero elements of* $D^{n\times n}$ *there is a division* F-*algebra* $D_o = F(G_o)$ *generated as such by the finitely generated nilpotent group* G_o *of Hirsch number* $h(G_o) < h(G)$, *an integer* t *depending only on* n *and* G, *and an* F-*algebra homomorphism* Φ *of the* F-*subalgebra generated by* $d_1,\ldots,d_m$ *into* $D_o^{t\times t}$ *such that* $\bigcap_i (\ker\Phi)^i = \{0\}$ *and each* $d_j\Phi \neq 0$. □

As a corollary that follows easily from 4.2.9 but not (apparently) from 4.2.1 consider the following.

4.2.10. *Let* F, G *and* D *be as in 4.2.9 with* charF = 0. *Suppose that* P *is a periodic subgroup of the group of units of a finitely generated* F-*subalgebra of* $D^{n\times n}$. *Then* P *is isomorphic to a linear group over* $\mathbb{C}$. *In particular* P *is abelian-by-finite.*

Proof: If G is finite then $n_1 = \dim_F D$ is finite and P embeds into $GL(nn_1,F)$. Since P is countable (Wehrfritz [2], 9.5) it follows that P embeds into $GL(nn_1,\mathbb{C})$. Suppose G is infinite. By hypothesis we have $P \subseteq F[d_1,\ldots,d_m]$ for some $d_j \in D^{n\times n}$. Apply 4.2.9. Since charF = 0 stability theory shows that $P \cap (1+\ker\Phi)$ is torsion-free and hence trivial. Thus P embeds into $D_o^{t\times t}$. The proof is completed by induction on h(G). □

4.3 RESIDUAL NILPOTENCE

The main result of this section is the following generalization of a well-known theorem for linear groups (Wehrfritz [2], 4.7).

4.3.1. THEOREM. (Wehrfritz [16]). *Let* $D = F(G)$ *be a division algebra over the field* F *generated as such by the locally polycyclic-by-finite subgroup* G *of* D^*, *and let* n *be a positive integer. Suppose that* X *is any subgroup of the group of units of a finitely generated subring* R *of* $D^{n\times n}$. *If* charD = 0 *then for all but a finite number of primes* p *there is a normal subgroup of* X *of finite index that is residually a finite* p-*group. If* charD =

$p \geq 0$ *there is a normal subgroup of* X *of finite index that is residually a finite* p*-group.*

REMARKS. a) If X is any finitely generated subgroup of GL(n,D) then X is a subgroup of the group of units of the subring of $D^{n\times n}$ generated by X. Thus the conclusion of 4.3.1 applies to all finitely generated subgroups of GL(n,D).

b) Trivially a periodic residually finite-p group is a p-group. Thus the group X is torsion-free by finite if charD = 0, and has a normal subgroup Y of finite index whose only torsion is p-torsion if charD = $p \geq 0$. In the latter case as D^* has no p-torsion, Y is torsion-free if n = 1. That is, 4.2.3 c) follows at once from 4.3.1. Then 4.2.3 d) follows from 4.3.1 and 4.2.3 a). Finally 4.2.3 b) is trivially a consequence of 4.3.1.

Let p be a prime. A residually finite-p group is clearly residually nilpotent. Thus the following result, which is *not* a consequence of Section 4.2, is immediate from 4.3.1.

4.3.2. COROLLARY. *Let* X *be as in* 4.3.1. *Then* X *is a finite extension of a residually nilpotent group.* □

Even a finitely generated linear group need not be residually nilpotent; for example it could be a finite simple group. Let π be any set of primes, and let e be a positive integer. A group G is *centrally π-erimitic of eccentricity* e if for every subset S and element g of G, if g^m centralizes S for some π-number m then g^e centralizes S, and if e is minimal with this property. (We use this concept only in the next result.) If π is the set of all primes we simply say that G is centrally erimitic. The following is a simple consequence of 4.3.1 (cf. Wehrfritz [1], 2.2).

4.3.3. COROLLARY. *Let* X *be as in* 4.3.1. *If* charD = 0 *then* X *is centrally erimitic, and contains a normal subgroup of finite index with eccentricity* 1. *If* charD = $p \geq 0$ *then* X *is centrally* p′*-erimitic and contains a normal subgroup of finite index of eccentricity* 1. □

An obvious question is whether X in 4.3.3 is always centrally erimitic even if charD ≠ 0, since this is the case if X is linear (see

Wehrfritz [1], or [2], 4.24). In view of the known properties of linear groups there are other obvious questions. For example if X is as in 4.3.1, is the Frattini subgroup of X nilpotent, is the Hirsch-Plotkin radical of X nilpotent, does X have finite central height, is X centrally stunted, and what are the chief factors of X like? None of these questions are likely to have easy answers.

We now embark upon the proof of 4.3.1. This requires considerable preliminary development.

4.3.4. *Let R be a ring, J a subring of R such that R is finitely generated as a right J-module, and J_1 a ring direct summand of J that is a right Noetherian ring. If $a \in R$ is right regular in R then $aR \cap J_1 \neq 0$.*

Proof: R is a (not necessarily unital) J_1-module via right multiplication and $R = A \oplus B$ as right J_1-module, where J_1 annihilates A, and B is unital and finitely generated. Then B contains a free J_1-submodule M of finite maximal rank, m say (Chatters & Hajarnavis [1], 1.9). Now $M \cong aM$ under the obvious map. If $aR \cap J_1 = \{0\}$ then $aM + J_1$ is a free J_1-submodule of B of rank m+1. This contradicts the choice of m. The result follows. □

4.3.5. *Let A be a plinth for the polycyclic-by-finite group G. For i = 1,…,r let k_i be a locally finite field, and let $J = \oplus k_iA = (\oplus k_i)A$. Let G act on the ring J via the given action on A and the trivial action on $\oplus k_i$. If ν is a non-zero element of J then there is a maximal ideal $\underline{m}$ of J with $\nu^G \cap \underline{m} = \emptyset$.*

This is a very mild generalization of Roseblade's theorem 4.2.8. Here we use the notation $\nu^G = \langle\ \nu^x : x \in G\ \rangle$ for brevity.

Proof: Let $\nu = \Sigma\, \nu_i$, where each $\nu_i \in k_iA$. For at least one i we have $\nu_i \neq 0$, say $\nu_1 \neq 0$. By 4.2.8 there is a maximal ideal $\underline{m}_1$ of k_1A with $\nu_1{}^G \cap \underline{m}_1 = \emptyset$. Let $\underline{m} = \underline{m}_1 \oplus \Sigma_{i>1}\, k_iA$. Clearly $\underline{m}$ is a maximal ideal of J. If $x \in G$ then $\nu^x = \Sigma\, \nu_i{}^x$ and each $\nu_i{}^x \in k_iA$. Since $\nu_1{}^x \notin \underline{m}_1$ we have that $\nu^x \notin \underline{m}$ and the lemma follows. □

If X is a subset of a ring R then rg{X} denotes the subring

(with 1) generated by X. Let $\mathrm{rad}_R X$ denote the intersection of the prime ideals of R containing X (meaning R if no such ideals exist).

4.3.6. *Let (R, S, H, K) be a crossed product where S is finite and semisimple. Suppose that G is a polycyclic-by-finite group of automorphisms of R normalizing S and H such that K is a plinth for G. Let m be a positive integer and a a regular element of R. Then there is a G-invariant ideal* $\underline{a} \neq R$ *of R of finite index such that a is a unit modulo* $\underline{a}$ *and*

$$\underline{a} = \mathrm{rad}_R(\underline{a} \cap \mathrm{rg}\{H^m\}).$$

Proof: Clearly $B = H^{|S|!}$ centralizes S and $H' \leq S \cap H$, which has order dividing $|S|!$. Thus B stabilizes the series $H \supseteq S \cap H \supseteq \langle 1 \rangle$ and $B^{|S|!}$ centralizes H. Set $A = H^t$ where $t = m(|S|!)^3$. Then A is a free abelian subgroup of H of finite index lying in H^m that is normalized by G and central in R. Note that A is also a plinth for G.

The subring k of R generated by its identity element has the form $k = \oplus_i k_i$, where each k_i is a finite field. Set $J_i = k_i A = k_i[A] \subseteq R$, so $J = \mathrm{rg}\{A\} = \oplus J_i$; we have here used the fact that $S \cap A = \langle 1 \rangle$. Trivially J is central in R and normalized by G, and R is a finitely generated J-module. By 4.3.4 there exists $\lambda \in aR \cap J_1 \setminus \{0\}$. Interchanging right and left there also exists $\mu \in Ra \cap J_1 \setminus \{0\}$. Let $\nu = \lambda\mu \in J$. By 4.3.5 there exists a maximal ideal $\underline{m}$ of J with $\nu^G \cap \underline{m} = \emptyset$. Since J is a finitely generated commutative ring, $J/\underline{m}$ is finite. Consequently $R/\underline{m}R$ is also finite. Trivially $\underline{m}R = R\underline{m}$.

Let $x \in G$. Then $a^{x^{-1}}R + \underline{m}.R \supseteq (\nu^{x^{-1}}J + \underline{m})R = JR = R$. Thus $aR + \underline{m}^x R = R$, and similarly $Ra + \underline{m}^x R = R$. Consequently a is a unit modulo $\underline{m}^x R$ and therefore also modulo $\mathrm{rad}_R(\underline{m}^x)$. The set $\langle \underline{m}^x R : x \in G \rangle$ is finite since $R/\underline{m}R$ is finite and R is finitely generated. Let $\underline{a} = \bigcap_{x \in G} \mathrm{rad}_R(\underline{m}^x) = \mathrm{rad}_R(\bigcap_{x \in G} \underline{m}^x R)$. Then $R/\underline{a}$ is a finite semisimple ring and each $\mathrm{rad}_R(\underline{m}^x)/\underline{a}$ is a direct sum of the simple components of $R/\underline{a}$. Therefore a is a unit modulo $\underline{a}$.

Now $R = \oplus_{t \in T} SAt$, where T is any transversal of $(S \cap H)A$ to H. Also S is a direct sum of irreducible k-modules, so SA is a direct sum of cyclic J-modules, each isomorphic to a direct summand of J and one being J itself. If $J = Je \oplus Jf$ where $1 = e + f$ then $\bigcap_{x \in G} \underline{m}^x e = (\bigcap_{x \in G} \underline{m}^x)e$, and therefore $\bigcap \underline{m}^x R = (\bigcap \underline{m}^x)R \subseteq (\underline{a} \cap J)R \subseteq (\underline{a} \cap \mathrm{rg}\{H^m\})R \subseteq \underline{a}$. Consequently $\underline{a} = \mathrm{rad}_R(\underline{a} \cap \mathrm{rg}\{H^m\})$. Finally $R = J \oplus K$ as J-module for some K, and so

$\underline{m}R \subseteq \underline{m} \oplus K \neq R$. The proof is complete. □

For brevity call a group G *polyplinthic* if G has a series of finite length of normal subgroups each of whose factors is a plinth for G. Note that every subgroup of finite index of a polyplinthic group is polyplinthic, as follows directly from the definition of plinth. Also every polycyclic-by-finite group has a polyplinthic subgroup of finite index. Clearly a polyplinthic group is poly-C_∞.

4.3.7. *Let* (R, S, G, K) *be a crossed product, where* S *is finite and semisimple, and* K *is polyplinthic. Let* P *be a normal subgroup of* G *of finite index and let* $a = \Sigma_{t\in T}\, ta_t \in R\setminus\{0\}$, *where* T *is a transversal of* S ∩ G *to* G *and the* a_t *are zero or units of* S. *Then there is an ideal* $\underline{a} \neq R$ *of* R *of finite index such that* a *is a unit modulo* $\underline{a}$ *and* $\underline{a} = \mathrm{rad}_R(\underline{a} \cap \mathrm{rg}\{P\})$.

Proof: We induct on the length of a plinth series for K. Let H/(S ∩ G) be a normal subgroup of K such that H/(S ∩ G) is a plinth for G and G/H is polyplinthic. If U is a transversal of H to G then $R = \oplus_{u\in U}\, uS[H]$. Let $a = \Sigma\, ub_u$ where each $b_u \in S[H]$. By 1.4.1 each non-zero b_u is a regular element of S[H]. Consequently so is $b = \Pi_{b_u \neq 0} b_u$ (multiplied in any fixed order). Let m = (H:H ∩ P). Then by 4.3.6 there is a G-invariant ideal $\underline{b}$ of S[H] of finite index such that b is a unit modulo $\underline{b}$ and $\underline{b} = \mathrm{rad}_{S[H]}(\underline{b} \cap \mathrm{rg}\{H \cap P\})$. By an elementary property of finite semisimple rings each non-zero b_u is also a unit modulo $\underline{b}$.

Now $\underline{b}R = R\underline{b} = \oplus_{u\in U}\, u\underline{b}$ since $\underline{b}$ is G-invariant. By induction applied to $R/\underline{b}R$ there is an ideal $\underline{a} \supseteq \underline{b}$ of R of finite index with $\underline{a} \neq R$ such that a is a unit modulo $\underline{a}$ and $\underline{a}/\underline{b}R = \mathrm{rad}_{R/\underline{b}R}(\underline{a}/\underline{b}R \cap \mathrm{rg}\{P \text{ modulo } \underline{b}R\})$. Then

$$\begin{aligned}\underline{a} &= \mathrm{rad}_R(\underline{a} \cap (\mathrm{rg}\{P\} + \underline{b}R))\\ &= \mathrm{rad}_R((\underline{a} \cap \mathrm{rg}\{P\}) + \underline{b}R)\\ &= \mathrm{rad}_R((\underline{a} \cap \mathrm{rg}\{P\}) + \mathrm{rad}_{S[H]}(\underline{b} \cap \mathrm{rg}\{H \cap P\}))\\ &= \mathrm{rad}_R(\underline{a} \cap \mathrm{rg}\{P\})\end{aligned}$$

since if $\underline{c}$ is a G-invariant ideal of S[H] of finite index then for some integer s we have $(\mathrm{rad}_{S[H]}\underline{c})^s \subseteq \underline{c}$, and so $((\mathrm{rad}_{S[H]}\underline{c})G)^s \subseteq \underline{c}G$ and $\mathrm{rad}_{S[H]}\underline{c} \subseteq \mathrm{rad}_R\underline{c}$. □

4.3.8. *Let (R, Z, G, K) be a crossed product where R is a domain, Z is central and finitely generated as a ring by* $Z \cap G$, *and K is polyplinthic. For each prime p let* G_p *be a normal subgroup of G of finite index containing* $Z \cap G$. *Let a be a non-zero element of R.*

a) *If* $\text{char}R = 0$ *then for all but a finite number of primes p there is an ideal* $\underline{a} \neq R$ *of R of finite index such that a is a unit modulo* $\underline{a}$ *and* $p \in \underline{a} = \text{rad}_R(\underline{a} \cap \text{rg}\{G_p\})$.

b) *If* $\text{char}R = p > 0$ *there is an ideal* $\underline{a} \neq R$ *of R of finite index such that a is a unit modulo* $\underline{a}$ *and* $\underline{a} = \text{rad}_R(\underline{a} \cap \text{rg}\{G_p\})$.

Proof: Let T be a transversal of $Z \cap G$ to G, and let $a = \Sigma_T\, ta_t$ where each $a_t \in Z$. Let $b = \Pi_{a_t \neq 0}\, a_t$. Now $Z[b^{-1}]$ is a finitely generated integral domain since Z is. If $\underline{n}$ is a maximal ideal of $Z[b^{-1}]$ then $\underline{n}$ has finite index and $Z \cap \underline{n}$ is a maximal ideal of Z. If $\text{char}R = 0$, set $\pi = \langle \text{char}(Z[b^{-1}]/\underline{n}) : \underline{n} \text{ as above}\rangle$. Then π is cofinite.

If $\text{char}R = 0$ let $p \in \pi$. Otherwise set $p = \text{char}R$. The above shows that there is a maximal ideal $\underline{m}$ of Z, necessarily of finite index, with $p \in \underline{m}$ and $b \notin \underline{m}$. It follows that each non-zero a_t is a unit modulo $\underline{m}$. Also $R/\underline{m}R \cong \oplus_{t \in T}\, t(Z/\underline{m})$ and $\underline{m}R$ is an ideal of R since $\underline{m}$ is central. By 4.3.7 there is an ideal $\underline{a} \neq R$ of R containing $\underline{m}R$ such that a is a unit modulo $\underline{a}$ and

$$\begin{aligned}\underline{a} &= \text{rad}_R(\ \underline{a} \cap (\text{rg}\{G_p\} + \underline{m}R)) \\ &= \text{rad}_R((\ \underline{a} \cap \text{rg}\{G_p\}) + \underline{m}) \\ &= \text{rad}_R(\ \underline{a} \cap \text{rg}\{G_p\})\end{aligned}$$

since $Z = \text{rg}\{Z \cap G\} \subseteq \text{rg}\{G_p\}$. □

4.3.9. *Let G be a polycyclic group, J a commutative Noetherian ring, and* $\underline{a}$ *an ideal of the group algebra R = JG of finite index. Suppose that G is p-nilpotent for every prime p dividing the characteristic of* $R/\underline{a}$. *Then* $\underline{a}$ *is Artin-Rees.*

For the definition of Artin-Rees see the preamble to 4.2.6. If p is a prime then a group G is p-nilpotent if every finite image of G is an extension of a p′-group by a p-group. This lemma is a slight generalization of a theorem of Roseblade and Smith [1], their result covering the case where J is a field.

Proof: Since R is Noetherian it suffices to consider a finitely generated (say right) R-module M and an essential submodule N of M killed by $\underline{a}$, and to prove that some power of $\underline{a}$ kills M, see Chatters & Hajarnavis [1], 11.2.

N is a finitely generated module over the finite ring $R/\underline{a}$, so N is finite. Set $\underline{b} = J \cap \underline{a}$. Then $\underline{b}G$ is an Artin-Rees ideal of the Noetherian ring R by Passman [2], 11.2.2. Hence $N \cap M\underline{b}^t \subseteq N\underline{b} = \{0\}$ for some positive integer t. But $\underline{N}$ is essential and $M\underline{b}^t$ is an R-submodule of M, so $M\underline{b}^t = \{0\}$. Clearly $J/\underline{b}$ is finite. Thus M is killed by a product $\underline{m}_1 \dots \underline{m}_n$ of maximal ideals of J containing $\underline{b}$, not necessarily all distinct. We induct on n.

Let $\underline{m}$ be a maximal ideal of J containing $\underline{b}$ with $M\underline{m} < M$. Clearly $N \cap M\underline{m}$ is essential in $M\underline{m}$ (even if $M\underline{m} = \{0\}$) so by induction $M.\underline{m}.\underline{a}^r = \{0\}$ for some positive integer r. Then $M\underline{a}^r\underline{m} = \{0\}$ and so $M\underline{a}^r$ is a finitely generated $(J/\underline{m})G$-module. Also $J \cap \underline{a} \subseteq \underline{m}$, so G is p-nilpotent for p = char$(J/\underline{m})$. By the theorem of Roseblade & Smith [1] every ideal of $(J/\underline{m})G$ is Artin-Rees, so there is a positive integer s with $N \cap M\underline{a}^r \cap M\underline{a}^{r+s} \subseteq N\underline{a} = \{0\}$. Therefore $M\underline{a}^{r+s} = \{0\}$. □

4.3.10. *Let* $R = S[G]$ *be a ring where* S *is a subring of* R *and* G *is a subgroup of the group of units of* R *normalizing* S. *Suppose that* P *is a normal subgroup of* G *of finite index with* $P \subseteq S$, *and let* $\underline{a}$ *be a* G*-invariant right (resp. left) Artin-Rees ideal of* S *such that* $S/\underline{a}$ *is right (resp. left) Noetherian. Then* $\underline{b} = \mathrm{rad}_R\underline{a}$ *is a right (resp. left) Artin-Rees ideal of* R.

Proof: We prove the right version. Now $\underline{a}R = \Sigma_{x \in G}\, \underline{a}x$ is an ideal of R and $\underline{b}/\underline{a}R$ is the radical of $R/\underline{a}R$. Also R is finitely generated as right S-module and therefore $R/\underline{a}R$ is right S-Noetherian and consequently right Noetherian. Thus some power of $\underline{b}$, say $\underline{b}^r$, lies in $\underline{a}R$, Jacobson [1], p. 196, Theorem 1, and Chatters & Hajarnavis [1], 1.8.

Let M be a finitely generated right R-module and N a submodule of M. Then M is also finitely S-generated so for some positive integer s we have $N \cap M\underline{a}^s \subseteq N\underline{a}$. Then $N \cap M\underline{b}^{rs} \subseteq N \cap M(\underline{a}R)^s = N \cap M\underline{a}^s \subseteq N\underline{a} \subseteq N\underline{b}$ since $(\underline{a}R)^s = (\underline{a}G)^s = G\underline{a}^s$ as $\underline{a}$ is G-invariant. The result follows. □

4.3.11. *Let* R *be a ring and* $\underline{a}$ *an ideal of* R *with* $R/\underline{a}$ *semisimple Artinian. Then* $\underline{C}_R(\underline{a}) \subseteq \underline{C}_R(\underline{a}^i)$ *for each* $i \geq 1$.

The point of 4.3.11 is so that we can apply 4.2.6.

Proof: We induct on i. We may assume that $\underline{a}^{i+1} = \{0\}$. Suppose $xc = 0$ where $x \in R\setminus\{0\}$ and $c \in \underline{C}_R(\underline{a})$. By induction $x \in \underline{a}^i$. Now $\underline{a}^i$ is an $R/\underline{a}$-module and $R/\underline{a}$ is semisimple. Thus $\underline{a}^i$ is a direct sum of irreducible $R/\underline{a}$-modules and there exists an irreducible $R/\underline{a}$-module V and an element $v \in V\setminus\{0\}$ with $vc = 0$. But V is isomorphic to a submodule of $R/\underline{a}$, so c is a right zero-divisor on $R/\underline{a}$. This contradiction shows that c is left regular in R. In the same way c is also right regular in R. □

4.3.12. THEOREM. (Wehrfritz [16]). *Let* $D = F(G)$ *be a division algebra over the field* F, *generated as such by the polycyclic-by-finite group* G. *Let* n *be a positive integer and* R *a finitely generated subring of* $D^{n\times n}$. *Then there exists an ideal* $\underline{a}$ *of* R *of finite index with* $\bigcap_{i\geq 1} \underline{a}^i = \{0\}$. *Moreover if* charD = 0 *then for all but a finite number of primes* p *we can choose* $\underline{a}$ *as above with* $p \in \underline{a}$.

4.3.12 is the crux of the proof of 4.3.1. Note that if charD = $p > 0$ then necessarily $p1_R \in \underline{a}$.

Proof: If $R \subseteq S$ are rings and $\underline{b}$ is an ideal of S of finite index containing the rational prime p and satisfying $\bigcap \underline{b}^i = \{0\}$ then $\underline{a} = R \cap \underline{b}$ is an ideal of R of finite index containing p and satisfying $\bigcap \underline{a}^i = \{0\}$. Also $\underline{a}^{n\times n}$ is an ideal of the matrix ring $R^{n\times n}$ of finite index containing p and satisfying $\bigcap_i (\underline{a}^{n\times n})^i = \{0\}$. There exists a finitely generated subring R_1 of D with $R \subseteq R_1{}^{n\times n}$. Thus we may assume that n = 1, and by 4.2.2 we may assume that F is a prime field. Since F[G] is Noetherian, D is the classical ring of quotients of F[G] and so $R \subseteq J[G,a^{-1}]$ for some prime image J of **Z** and some element a in $J[G]\setminus\{0\}$.

G has a normal subgroup H of finite index such that H is orbitally sound (for the definition see before 4.1.3), $\zeta_1(H) = \Delta(H)$, and $H/\zeta_1(H)$ is polyplinthic. By 4.1.3 we have that J[H] is the crossed product of $Z = J[\zeta_1(H)]$ by $H/\zeta_1(H)$. Clearly F is the quotient field of J in D. Also D has finite dimension, d say, as left F(H)-space. Thus $J[G,a^{-1}] \subseteq D$ is isomorphic to a subring of $F(H)^{d\times d}$ and therefore $J[G,a^{-1}]$ is isomorphic to a subring of $J[H,b^{-1}]^{d\times d}$ for some $b \in J[H]\setminus\{0\}$. We may therefore assume that $H = G$.

For each prime p choose a p-nilpotent normal subgroup G_p of G of finite index containing $\zeta_1(G)$. By 4.3.8 there is a prime p and an ideal $\underline{a}$ of $J[G] = Z[G]$ of finite index with $p \in \underline{a}$ such that a is a unit modulo $\underline{a}$ and $\underline{a} = \mathrm{rad}_R(\underline{a} \cap J[G_p])$. By 4.3.9 the ideal $\underline{a} \cap J[G_p]$ is Artin-Rees. Hence $\underline{a}$ is Artin-Rees by 4.3.10. Let $Q = \underline{C}_{J[G]}(\underline{a})$. Then by 4.2.6 and 4.3.11 we have that Q is right Ore in $J[G]$, that $R \subseteq J[G,a^{-1}] \subseteq J[G]Q^{-1} = T$ say, and that $\underline{a}Q^{-1}$ is an ideal of T of finite index with $p \in \underline{a}Q^{-1}$ and $\bigcap_i (\underline{a}Q^{-1})^i = \{0\}$. If $\mathrm{char}D = 0$ then by 4.3.8 we can choose p to be any prime with at most a finite number of exceptions. The result follows. □

4.3.13. *Let* $\underline{a}$ *be an ideal of finite index in the finitely generated ring* R. *Then* $R/\underline{a}^i$ *is finite for each positive integer* i.

Proof: Let $u_1,\ldots,u_h$ generate R modulo $\underline{a}$ as $\mathbb{Z}$-module and let $1 = r_1,\ldots,r_k$ generate R as a ring. Set $M = \Sigma_1^h u_i\mathbb{Z} + \Sigma_1^k r_j\mathbb{Z}$, so in particular $\underline{a} + M = R$. If m_1 and m_2 are elements of M then $m_1m_2 = x + m$ for some $x \in \underline{a}$ and $m \in M$. Clearly $x \in \underline{a} \cap (M + M_2)$, where $M_2 = \Sigma_{m_i \in M} m_1m_2\mathbb{Z}$. Let $\underline{b}$ be the ideal of R generated by $\underline{a} \cap (M+M_2)$. Trivially $\underline{b} \subseteq \underline{a}$. Also $\underline{b}+M$ is multiplicatively closed by the construction of $\underline{b}$ and contains the r_j, so $\underline{b}+M = R$. Thus $R/\underline{b}$ is finitely generated as $\mathbb{Z}$-module, so $\underline{a}/\underline{b}$ is too. But $\underline{b}$ is a finitely generated ideal since $M+M_2$ is finitely $\mathbb{Z}$-generated. Therefore $\underline{a}$ is a finitely generated ideal.

$R/\underline{a}$ is finite. Suppose $R/\underline{a}^i$ is finite for some i. By the previous paragraph $\underline{a}/\underline{a}^i$ is finitely generated as $R/\underline{a}^i$-$R/\underline{a}^i$ bimodule. Since $R/\underline{a}^i$ is finite it follows that $\underline{a}/\underline{a}^{i+1}$ is finite. Therefore $R/\underline{a}^{i+1}$ is finite. The proof is complete. □

Proof of 4.3.1: We may assume that G is polycyclic-by-finite. By 4.3.12 for a suitable prime p we can find an ideal $\underline{a}$ of R of finite index with $p \in \underline{a}$ and $\bigcap_i \underline{a}^i = \{0\}$, and by 4.3.13 each $R/\underline{a}^i$ is finite. Regard R as an R-X bimodule via left and right multiplication, and set $C_i = C_X(R/\underline{a}^i)$. Then C_i is a normal subgroup of X of finite index, each C_i/C_{i+1} is a finite p-group, and $\bigcap_i C_i = \langle 1 \rangle$, see Wehrfritz [2], 4.6. □

4.4 DIVISION RINGS OF QUOTIENTS OF GROUP ALGEBRAS

4.4.1. Let $\underline{X}$ denote the class of groups G with a normal series

$$\langle 1\rangle = G_0 \subseteq G_1 \subseteq \dots \subseteq G_m \subseteq G \qquad (*)$$

of finite length, such that G/G_m is torsion-free polycyclic-by-finite, each G_{i+1}/G_i is abelian, and each G/G_i is residually torsion-free polycyclic-by-finite. It is elementary that a torsion-free nilpotent group of derived length d has a normal series of length d with torsion-free abelian factors. Thus for example $\underline{X}$ contains all soluble-by-finite groups that are residually finitely generated torsion-free nilpotent; for if G is such a group then G is in fact soluble, of derived length m say, since the nilpotent images of G have bounded derived length, and then we set $G_i = \bigcap N$, where N ranges over all normal subgroups of G such that G/N is torsion-free nilpotent of derived length at most m - i. This is the case covered in Lichtman [9].

Let F be any field and let G be an $\underline{X}$-group. Then R = FG is an Ore domain by 1.4.23, as is $F(G/G_i)$ for each i, the notation being as in the previous paragraph. Let D be the division ring of quotients of R and suppose that X is a finite subset of GL(n,D). There exists $r \in R\backslash\{0\}$ such that $Xr \subseteq R^{n\times n}$ and $X \subseteq GL(n,R[r^{-1}])$. By hypothesis there is a normal subgroup K of G such that G/K is torsion-free and polycyclic-by-finite, and such that the natural projection of FG onto F(G/K) does not map r to zero and induces a one-to-one map on Xr.

Let $A = G_1 \cap K$ where G_1 is as in (*), and let $\underline{a}$ be the augmentation ideal of A in FA. Then $r \notin \underline{a}G$. We claim that the natural map of G onto G/A extends to a homomorphism Φ of $R[r^{-1}]$ into the division ring D_1 of quotients of F(G/A), which exists by 1.4.23 again. Assuming that this is so, the induced map Φ_n of $GL(n,R[r^{-1}])$ into $GL(n,D_1)$ is one-to-one on X. Consequently GL(n,D) is locally super-residually of the same type, but with m replaced by m - 1. (Recall that a group Y is super-residually a $\underline{Y}$-group if for every finite subset X of Y there is a homomorphism of Y onto a $\underline{Y}$-group that is one-to-one on X.) We therefore have the opportunity for an inductive proof. The base of the induction, namely the case m = 0, is where G is torsion-free and polycyclic-by-finite. This case is covered by the results of Sections 4.2 and 4.3.

It remains to construct the homomorphism Φ. We require several preliminary lemmas, some of which will also be useful later in a different context.

4.4.2. *Let* A *be a torsion-free abelian group, let* F *be a field, and let* Q *be the quotient field of* FA. *Denote the augmentation ideal of* A *in* FA *by* $\underline{a}$ *and let* H *be a group of automorphisms of* A *and hence of* Q. *Then there exists a discrete valuation domain (that is, a local principal ideal domain)* J *with maximal ideal* $\underline{m}$ *such that* $FA \leq J \leq Q$, $FA \cap \underline{m} = \underline{a}$, *and* H *normalizes* J *and* $\underline{m}$.

Proof: It is elementary that $\bigcap_{i=1}^{\infty} \underline{a}^i = \{0\}$, although this does in fact follow immediately from the Krull Intersection Theorem. Define the map $\lambda : FA \longrightarrow \mathbb{Z} \cup \{\infty\}$ by $x\lambda = i$ whenever $x \in \underline{a}^i \backslash \underline{a}^{i+1}$; of course $\underline{a}^0 = FA$ and $0\lambda = \infty$. If $x, y \in FA$ then clearly $(x + y)\lambda \geq \min\{x\lambda, y\lambda\}$, with equality whenever $x\lambda \neq y\lambda$. We show below that $(xy)\lambda = x\lambda + y\lambda$. Assuming that this is so, extend λ to Q by defining $(xy^{-1})\lambda = x\lambda - y\lambda$ for $x, y \in FA\backslash\{0\}$. This is then a well-defined discrete valuation on Q. Moreover H normalizes $\underline{a}^i$ for each i, and therefore $x^h\lambda = x\lambda$ for all $x \in Q$ and $h \in H$.

Let J be the valuation ring of λ; that is, set

$$J = \{ x \in Q : x\lambda \geq 0\} \supseteq FA.$$

The maximal ideal of J is $\underline{m} = \{ x \in Q : x\lambda > 0 \}$. By definition of λ we have $FA \cap \underline{m} = \underline{a}$ and clearly H normalizes J and $\underline{m}$.

Let $x \in \underline{a}^i\backslash\underline{a}^{i+1}$, $y \in \underline{a}^j\backslash\underline{a}^{j+1}$. It remains to show that $(xy)\lambda = x\lambda + y\lambda$, so all we need is that $xy \notin \underline{a}^{i+j+1}$. Here we can assume that A is finitely generated, so A is now free abelian, say on $a_1,\dots, a_t$. Let $x_i = a_i - 1$. Then $S = F[a_1,\dots,a_t] = F[x_1,\dots,x_t]$ is a polynomial ring in the x_i and $\underline{a} = \Sigma x_i FA$. Suppose $xy \in \underline{a}^{i+j+1}$. There exists $a, b \in A$ such that $ax, by \in S$ and $axby \in (\Sigma x_k S)^{i+j+1}$. Then either $ax \in (\Sigma x_k S)^{i+1}$ or $by \in (\Sigma x_k S)^{j+1}$. Consequently $x \in \underline{a}^{i+1}$ or $y \in \underline{a}^{j+1}$. This contradiction proves the point. □

4.4.3. *Let* R *be a right Ore domain with division ring* D *of right quotients, and let* S *be a subring of* D *containing* R. *Then* S *is right Ore.*

Proof: Let ax^{-1}, cy^{-1} be non-zero elements of S, where $a, c, x, y \in R$. By hypothesis there are $b, d \in R$ with $ad = cb \neq 0$. Clearly $xd, yb \in R \subseteq S$ and $ax^{-1}.xd = cy^{-1}.yb \neq 0$. The result follows. □

4.4.4. *Let* (R,J,G,H) *be a crossed product, where* J *is a commutative ring with nilpotent maximal ideal* $\underline{m}$, *necessarily normalized by* G. *Set* $P = \underline{m}G$ *and assume that* R/P *is a domain. Then* $\underline{C}_R(0) = \underline{C}_R(P) = R\backslash P$.

Proof: Suppose $\underline{m}^{d+1} = \{0\}$. Then $\mathfrak{P}^{d+1} = \underline{m}^{d+1}G = \{0\}$ and hence

$$\underline{C}_R(0) \subseteq R\backslash\mathfrak{P} = \underline{C}_R(\mathfrak{P})$$

by hypothesis. The reverse inclusion we prove by induction on d, the case $d = 0$ being trivial. Let $d \geq 0$ and suppose $ac = 0$, where $a \in R$ and $c \in R\backslash\mathfrak{P}$. By induction $a \in \mathfrak{P}^d$. Now $\mathfrak{P}^d \cong \underline{m}^d \otimes_J R$ as right R-module, and $J/\underline{m}$ is a field. Thus $\underline{m}^d$ is $(J/\underline{m})$-free and consequently $\mathfrak{P}^d$ is free as right $R/\mathfrak{P}$-module. Therefore $a = 0$ and so c is left regular. In a similar way c is right regular, and the proof is complete. □

4.4.5. *Let J be a discrete valuation domain with maximal ideal* $\underline{m} = \pi J$ *and suppose* (R,J,G,H) *is a crossed product with* R *a right Ore domain. Set* $\mathfrak{P} = \underline{m}G$ *and assume that* $R/\mathfrak{P}$ *is a domain. Then* $\underline{C}_R(\mathfrak{P})$ *is a right Ore subset of* R.

Note that necessarily G normalizes $\underline{m}$, so $\mathfrak{P}$ is an ideal of R.

Proof: Let $a \in R\backslash\{0\}$ and $c \in \underline{C}_R(\mathfrak{P})$, $= R\backslash\mathfrak{P}$ by hypothesis. Then there exists $b_1, d_1 \in R$ with $ad_1 = cb_1 \neq 0$. Now $d_1 = \pi^t d$ for some $t \geq 0$ and $d \in R\backslash\mathfrak{P}$. Hence π^t divides cb_1 and $cb_1 \equiv 0 \bmod \mathfrak{P}^t$. By 4.4.4 applied to $R/\mathfrak{P}^t$ we have $c \in \underline{C}_R(\mathfrak{P}^t)$, so $b_1 \in \mathfrak{P}^t$. Consequently $b_1 = \pi^t b$ for some $b \in R$ and the assumption that R is a domain yields that $ad = cb$. Therefore $\underline{C}_R(\mathfrak{P})$ is right Ore in R. □

4.4.6. *Let* F *be a field and* G *an* $\underline{O}_1$*-group (see Section 1.4) with an abelian normal subgroup* A *such that* G/A *is also an* $\underline{O}_1$*-group. Let* $\underline{a}$ *denote the augmentation ideal of* A *in* FA *and let* $r \in FG\backslash\underline{a}G$. *Denote by* D *and* D_1 *the division rings of quotients of* FG *and* F(G/A) *respectively. Then the natural map of* FG *onto* F(G/A) *extends to a homomorphism of* $FG[r^{-1}] \subseteq D$ *into* D_1.

Note that the claim left unsubstantiated in 4.4.1 follows from 4.4.6.

Proof: Necessarily A is torsion-free. By 4.4.2 there is a discrete valuation domain J with maximal ideal $\underline{m} = \pi J$ lying between FA and its quotient field $Q \leq D$ and normalized by G, such that $FA \cap \underline{m} = \underline{a}$. Set $R = J[G] \leq D$ and $\mathfrak{P} = \underline{m}G$. Then (R,J,G,G/A) is a crossed product and $\mathfrak{P}$ is an ideal of R. Now R is an Ore domain by 4.4.3. Also R/P is a skew group ring of $G/A \in \underline{O}_1$ over $J/\underline{m}$, so R/P is certainly a domain; even an Ore one. Then $\underline{C}_R(P)$ is Ore in R by 4.4.5 and we can form the ring $S = R\underline{C}_R(P)^{-1}$ of quotients and regard it as a subring of D. Moreover $FG[r^{-1}] \subseteq S$. Now $S/PS = D_2$ is the quotient division ring of R/P. Also $F(G/A)$ is embedded in R/P in the obvious way, so D_2 contains D_1 and the natural projection of S onto D_2 maps $FG[r^{-1}]$ into D_1. The result follows. □

4.4.7. *With the notation of the proof of 4.4.6 let*

$$L = GL(n,S) \cap (1_n + (PS)^{n \times n}).$$

If $\mathrm{char}F = 0$ *then* L *is residually torsion-free nilpotent.*

If $\mathrm{char}F = p > 0$ *then* L *is residually a nilpotent* p*-group.*

Note that L is just the kernel of the obvious map of $GL(n,S)$ into $GL(n,D_2)$.

Proof: Now $\mathrm{char}F = \mathrm{char}(J/\underline{m}) = \mathrm{char}D_2$. Also $\bigcap_{i=1}^{\infty} \underline{m}^i = \{0\}$, so $\bigcap P^i = \{0\}$. It follows from this and 4.4.4 that $(PS)^i = P^iS$ and $\bigcap_{i=1}^{\infty} P^iS = \{0\}$, cf. the proof of 4.2.6 c). Now L stabilizes the series $\{P^iS^n\}_{i \geq 1}$. If $\mathrm{char}F = 0$ the factors of this series, being D_2-modules, are torsion-free abelian and stability theory then yields that L is residually torsion-free nilpotent. If $\mathrm{char}F = p > 0$ the factors of the series are elementary abelian p-groups and L is residually a nilpotent p-group (see Kegel & Wehrfritz [1], § 1.C). □

We now come to the main result of this section. Parts e) and f) are slight generalizations of Lichtman [9], Theorem 4.1, Parts 2 and 4.

4.4.8. THEOREM. *Let* F *be a field,* G *a locally* $\underline{X}$*-group,* $\underline{X}$ *as in* 4.4.1 *and* D *the division ring of quotients of* FG. *Let* T *be any subgroup of the group of units of a finitely generated subring of* $D^{n \times n}$.

a) T *is super-residually a skew linear group of degree* n *over division rings of quotients of group algebras over* F *of torsion-free polycyclic-by-finite groups.*

b) T *is residually finite.*

c) *If* $\mathrm{char}F = 0$ *then* T *is torsion-free by finite.*

d) *If* charF = p > 0 *then* T *has a subgroup of finite index whose elements of finite order are all* p-*elements. In particular* T *is torsion-free by finite if* n = 1.

e) *Any unipotent subgroup* U *of* GL(n,D) *is a stability subgroup.*

f) *Any periodic subgroup of* GL(n,D) *is locally finite.*

Note that T could be any finitely generated subgroup of GL(n,D).

Proof: a) Since T lies in a finitely generated subring of $D^{n\times n}$ we may assume that G ∈ $\underline{X}$. Then a) follows from 4.4.1.

b) This is an immediate consequence of a) and 4.2.3 b).

c) This follows from 4.4.7, induction as in 4.4.1, and 4.2.3 c) (or 4.3.2 if you prefer).

d) This follows from 4.4.7, 4.4.1, and 4.2.3 c) again.

e) U is locally residually a stability subgroup of degree n by 4.4.1 and 4.2.3 a). Thus U is locally residually nilpotent-of-class-less-than-n. Therefore U is nilpotent, and consequently U is a stability subgroup of GL(n,D), by 1.3.4 for example.

f) If charF = 0 then this follows at once from c). Let charF = p > 0 and suppose that T is a finitely generated periodic subgroup of GL(n,D). Then $T/O_p(T)$ is finite by d). Also $O_p(T)$ is nilpotent by e). But then $O_p(T)$ is periodic, nilpotent, and finitely generated. Therefore $O_p(T)$ is finite and the proof is complete. □

By a famous theorem of Jordan there is a function $\beta(n)$ of n only such that a finite linear group of degree n over the complex numbers has an abelian normal subgroup of index at most $\beta(n)$. See comments after Wehrfritz [2], 9.2 for examples of such $\beta(n)$. We call such a function a Jordan function. For a proof of Jordan's Theorem see Curtis & Reiner [1], 36.13 or Dixon [1], 5.6. Recall also Dickson's Theorem, that a finite linear group of degree n of order not divisible by the characteristic is isomorphic to a subgroup of GL(n,ℂ); see Dixon [1], Corollary 3.8.

4.4.9. THEOREM. (*Lichtman* [9], *Proposition* 4.2). *Let* F *be a field of characteristic zero, let* G *be any* $\underline{O}_1$-*group, and let* D *be the division ring of quotients of* FG. *If* P *is any locally finite subgroup of* GL(n,D) *then* P *is* *isomorphic to a subgroup of* GL(n,$\mathbb{C}$). *In particular for any Jordan function* β *there is an abelian normal subgroup of* P *of index at most* β(n).

Proof: By Mal'cev's Local Theorem (Kegel & Wehrfritz [1], § 1.L, Appendix) we may assume that P is finite. There is a finitely generated subring J_1 of F and a non-zero element r of $R_1 = J_1G$ such that $Pr \subseteq R_1{}^{n\times n}$ and $P \subseteq GL(n,R_1[r^{-1}])$. Moreover we may assume, since charF = 0, that $|P|^{-1} \in J_1$, and that F is the quotient field of J_1.

Let J_2 be the integral closure of J_1 in F. Since J_1 is Noetherian, so J_2 is a Krull domain (Nagata [1], 33.10, but at the end of this proof we include an elementary proof of the special case of this deep result that we are effectively using). Hence there is a rank-1 prime $\underline{q}$ of J_2 such that if $\phi : R_2 = J_2G \twoheadrightarrow (J_2/\underline{q})G$ is the natural map, with ϕ_n from $R_2{}^{n\times n}$ to $(R_2\phi)^{n\times n}$ the induced map, then $r\phi \neq 0$ and ϕ_n is one-to-one on Pr.

Let J be the localization of J_2 at $\underline{q}$ and let $\underline{m}$ be its maximal ideal. Then J is a discrete valuation domain by the definition of Krull domain. Clearly ϕ extends to a map, still called ϕ, of R = JG onto $R_0 = (J/\underline{m})G$. Since $G \in \underline{O}_1$ there is a division ring D_0 of quotients of R_0. Set $P = \underline{m}G = \ker\phi$. Now FG is an Ore domain, so R is too. Then by 4.4.5 the set $\underline{C}_R(P)$ is Ore in R and ϕ extends to a map of $R\underline{C}_R(P)^{-1} \subseteq D$ to D_0. Then ϕ maps $R_1[r^{-1}]$ into D_0 and ϕ_n embeds P into $GL(n,D_0)$.

Suppose $\mathrm{char}R_0 = p > 0$. Then p and |P| are coprime since the latter is a unit of J. Thus 2.3.1 and Dickson's Theorem yield an embedding of P into GL(n,$\mathbb{C}$). Suppose $\mathrm{char}R_0 = 0$. Now $P\phi_n \leq GL(n,R_1\phi[r^{-1}\phi])$, $R_1\phi \cong J_1\phi G$, and the Krull dimension $k(J_1)$ of J_1 satisfies

$$k(J_1) = k(J_2) > k(J_2/\underline{q}) = k(J_2\phi) = k(J_1\phi)$$

(Kaplansky [1], Theorem 44 and p. 110). Therefore by induction on the Krull dimension of J_1 we deduce that $P\phi_n \cong P$ is isomorphic to a subgroup of GL(n,$\mathbb{C}$). Note that $k(J_1) \geq 1$ always since $\mathrm{char}J_1 = 0$, and if $k(J_1\phi) = 0$ then $\mathrm{char}R_0 > 0$. Thus the induction does start.

We give an elementary proof that we can assume that J_2 is a Krull domain. By Noether's Normalization Lemma (Zariski & Samuel [1], Vol 1, p. 266) there is a subring J_3 of F of the form $\mathbb{Z}[h^{-1},x_1,\dots,x_t]$ where h is a positive integer and $x_1,\dots, x_t \in J_1$ are indeterminates, with J_1 integral over J_3. By replacing J_1 by $J_1[h^{-1}]$ we may assume that $J_3 \subseteq J_1$. Then J_2 is also the integral closure of J_3 in F. Now J_3 is a unique factorization domain (ibid. p. 38, Theorem 13) and hence is integrally closed (ibid. p. 261). Then

J_2 is finitely generated as J_3-module (ibid. p. 265, Corollary 1) and in particular is a Noetherian domain. Therefore J_2 is a Krull domain (Kaplansky [2], p. 82). □

4.4.10. COROLLARY. *Let* F *be a field of characteristic zero,* G *a locally* $\underline{X}$*-group,* $\underline{X}$ *as in* 4.4.1, *and* D *the division ring of quotients of* FG. *If* P *is any periodic subgroup of* GL(n,D) *then* P *is isomorphic to a subgroup of* GL(n,$\mathbb{C}$). *In particular* P *has an abelian normal subgroup of finite index at most* $\beta(n)$ *for any Jordan function* β. □

This is a slight generalization of Lichtman [9], Theorem 4.1, Part 5. It is an immediate consequence of 4.4.8 f) and 4.4.9. Note that we have already seen (2.5.10) that for arbitrary division rings D of characteristic zero, the conclusion of 4.4.10 is far from true. Note also that a locally finite skew linear p'-group of degree n and characteristic $p \geq 0$ is always isomorphic to a subgroup of GL(n,$\mathbb{C}$), whatever the coefficient ring. This is an easy consequence of 2.3.1 and Dickson's Theorem.

We could not derive 4.4.10 from the Reduction Theorem (and 4.4.1). For let D = F(G) be a division F-algebra of characteristic zero generated as such by the poly-C_∞ group G. There does *not* exist a function $\alpha(n)$ of n only such that a finite subgroup of GL(n,D), for any such D, has an abelian normal subgroup of index at most $\alpha(n)$.

For let p and q be primes with $q|(p-1)$; for any prime q there is such a p by Dirichlet's theorem on primes in an arithmetic progression (e.g. Serre [2], p. 104). By 2.1.5 there is a division ring $D = \mathbb{Q}(a,b)$ where a has order p, b has order q^e for e a sufficiently large integer, and b normalizes $\langle a\rangle$ with $(\langle b\rangle : C_{\langle b\rangle}(a)) = q$. Set $F = C_{\mathbb{Q}(a)}(b)$. Clearly F is central in D. There exists $x \in \mathbb{Q}(a)$ with $y = x^{2p} \notin F$. The group $\langle x^{\langle b\rangle}\rangle$ is finitely generated and abelian, and the group of roots of unity of $\mathbb{Q}(a)$ has order 2p. Therefore $\langle y^{\langle b\rangle}\rangle$ is free abelian of finite rank. Let $c \in \mathbb{Q}\backslash\{0,\pm1\}$ with $\langle c\rangle \cap \langle y^{\langle b\rangle}\rangle = \langle 1\rangle$. Then $G = \langle bc, y\rangle \leq D^*$ is poly-C_∞ of Hirsch length at most q + 1. Since $(\mathbb{Q}(a):F) = q$ is a prime and $y \in \mathbb{Q}(a)\backslash F$ we have D = F(G). Finally D^* contains the finite group $\langle a,b\rangle$, whose only maximal abelian normal subgroup $\langle a,b^q\rangle$ has index q. Thus if α exists then $\alpha(n) \geq q$ for all primes q and integers $n \geq 1$. Consequently no such α exists.

This section raises as many questions as it answers. Are periodic subgroups of GL(n,D) locally finite for D as in 4.4.9, with or without characteristic zero? In the example above G is not nilpotent. Are there examples in which G is nilpotent? Now 4.2.10 suggests otherwise, but alternatively can the hypotheses on G in 4.2.10 be weakened to G polycyclic-by-finite? The class L$\underline{X}$ is really very artificial. Is there a wider and more natural class to replace it, for which we can derive all or at least some of the conclusions of 4.4.8? In the main in this section we have been extending the results of Section 4.2. What about the results of Section 4.3; do they extend?

4.5 FREE SUBGROUPS OF NORMAL SUBGROUPS OF GENERAL LINEAR GROUPS

A famous theorem of Tits states that a finitely generated linear group either is soluble-by-finite or contains non-cyclic free subgroups. We have already seen (1.4.9) that this does not extend to skew linear groups in general. However Tits' Theorem may extend to skew linear groups over division rings of the type we are considering in this chapter. There is some evidence for this, much of which is due to Lichtman.

We look not at arbitrary subgroups of GL(n,D) but just at normal ones. The case $n \geq 1$ is easily dealt with. Let D be any division ring and $n \geq 1$ an integer. Denote the standard matrix units in $D^{n\times n}$ by e_{ij} for $1 \leq i,j \leq n$. The subgroup of GL(n,D) generated by all matrices of the form $1_n + \delta e_{ij}$, $\delta \in D$, $1 \leq i,j \leq n$, $i \neq j$, is denoted by SL(n,D). Suppose that N is a non-central normal subgroup of GL(n,D). By a well-known theorem of Dieudonné (see Artin [1], Theorem 4.9, p. 165) either $SL(n,D) \leq N$ or $n = 2$ and $|D| \leq 3$. The following is now easy.

4.5.1. *Let D be a division ring that is not a locally finite field and let $n \geq 1$ be an integer. If N is any non-central normal subgroup of GL(n,D) then N contains a non-cyclic free subgroup.*

Proof: If charD = 0 then N contains $SL(n,\mathbb{Q}) \leq SL(n,D)$ by the above remarks. Clearly then N contains non-cyclic free subgroups. Let charD = $p \geq 0$, and suppose first that D contains an element t that is trancendental over the field $\mathbb{F}_p$ of p elements. Then $E = \mathbb{F}_p(t)$ is a subfield of D, N contains SL(n,E) and again N contains non-cyclic free subgroups.

Now assume that D is algebraic over $\mathbb{F}_p$. Then $\mathbb{F}_p(x)$ is finite for every $x \in D$ and D^* is periodic. In this case D is a locally finite field (see Jacobson [3], p. 185, Corollary, or Scott [1], 14.1.6) and we have excluded this possibility by hypothesis, for the excellent reason that then GL(n,D) is locally finite and so contains no non-identity free subgroups. □

Note that we have not used Tits' Theorem in the proof of 4.5.1. However we do need it in the difficult case $n = 1$, to which we now turn. Notice that if D is actually a field in 4.5.1 then for trivial reasons we can include the case $n = 1$.

4.5.2. THEOREM.(Lichtman [5]). *Let* $D = F(G)$ *be a division algebra over the field* F, *generated as such by the locally polycyclic-by-finite group* G. *Assume that* D *is not a locally finite field. If* N *is a normal subgroup of* GL(n,D), *then either* N *is central or* N *contains non-cyclic free subgroups.*

Unfortunately for this result we have to work with a more general class of rings than hitherto. The reason is that in the proof of the Reduction Theorem (4.2.1), $R_0Q_0^{-1}$, in the second case, need not be simple. It turns out that we can choose θ so that θ maps the group $U(RQ^{-1})$ of units of RQ^{-1} onto that of $R_0Q_0^{-1}$, so assuming $n = 1$, $(N \cap U(RQ^{-1}))\theta$ is a normal subgroup of $U(R_0Q_0^{-1})$. But then there seems to be no useful way of obtaining a normal subgroup of $GL(d,D_0)$ on which to apply induction.

4.5.3. Rings with a normalizing basis.

We need to consider rings of the form $L = \oplus_{t \in T} tR$, where R is a subring of L and T is a *subset* of the units U(L) of L normalizing R (meaning that $R = R^t = t^{-1}Rt$ for all $t \in T$). Call such a T a ***normalizing basis*** *of* L *over* R. The obvious example of such is a crossed product (L,R,G,H), where T is chosen to be any transversal of $R \cap G$ to G. We can replace T by $t_0^{-1}T$ for any $t_0 \in T$. That is, we may choose to have $1 \in T$ if we wish. There are a number of weaker versions of the notion of a normalizing basis in the literature; for example some authors would replace "t is a unit of L normalizing R" by "$Rt = tR$ and $rt \neq 0 \neq tr$ for all $r \in R\backslash\{0\}$". The above suffices for our purposes.

The pertinent example from our point of view is the following. Let $L = R[G]$ be a ring where R is a division subring of L and G is a subgroup of the group of units of L normalizing R. Then G spans L as say right R-space, and so G contains a basis T of L as right R-space. T is then a normalizing basis of L over R. Note that necessarily $R \cap G$ is a normal subgroup of G.

We need to work with certain finitely generated subrings of such a ring $L = R[G]$. Suppose that G is finitely generated and that $(G:R \cap G)$ is finite. Let X be any finite subset of L containing a transversal of $R \cap G$ to G. If $x \in X$ then $x = \Sigma\, tx_t$ for some $x_t \in R$. Let R_1 be the subring of R generated by $R \cap G$ and all the x_t^g for $x \in X$, $t \in T$, and $g \in G$. Since T and $(G:R \cap G)$ are finite and $R \cap G$ is a finitely generated group, both R_1 and $L_1 = R_1[G]$ are finitely generated rings. We claim that $T \subseteq G$ is a normalizing basis of L_1 over R_1. Clearly G normalizes R_1 and $X \subseteq \oplus_{t \in T}\, tR_1$. Since X contains a transversal of $R \cap G$ to G, it suffices to prove that $\oplus\, tR_1$ is a subring of L. Clearly it is additively closed. If $s, t \in T$ then $st = xr$ for some $x \in X$ and $r \in R \cap G$. Hence

$$sR_1.tR_1 = stR_1 = xR_1 = (\Sigma_{u \in T}\, ux_u)R_1 \subseteq \oplus\, uR_1.$$

This proves the claim. More generally if R_2 is any G-invariant subring of L containing R_1 the same argument shows that T is a normalizing basis of $R_2[G]$ over R_2.

We collect some properties of rings with normalizing bases. Thus let $L = \oplus\, tR$ be a ring with a normalizing basis T over its subring R.

Ideals: If $\underline{a}$ is an ideal of L then $\underline{a}$ is normalized by T and $\underline{b} = \underline{a} \cap R$ is a T-invariant ideal of R. Conversely suppose $\underline{b}$ is a T-invariant ideal of R and set $\underline{a} = \oplus\, t\underline{b}$. We claim that $\underline{a} = \underline{b}L = L\underline{b}$ is an ideal of L. For certainly $\underline{a} + \underline{a} \subseteq \underline{a}$. Also $\underline{a} \subseteq L\underline{b} \subseteq \Sigma\, tR\underline{b} \subseteq \Sigma\, t\underline{b} \subseteq \underline{a}$, so $\underline{a} = L\underline{b}$. Also $\underline{a} = \oplus\, \underline{b}t$ as $\underline{b}$ is T-invariant and so $\underline{a} = \underline{b}L$. Thus $\underline{a}$ is an ideal of L. We can therefore form the ring $L/\underline{a}$. Set $\bar{t} = t + \underline{a} \in L/\underline{a}$ for $t \in T$. Identify $R/\underline{b}$ with a subring of $L/\underline{a}$ via $R/\underline{b} = R/R \cap \underline{a} \cong (R + \underline{a})/\underline{a} \subseteq L/\underline{a}$. Then $L/\underline{a} = \oplus_{t \in T}\, \bar{t}(R/\underline{b})$ and $\{\bar{t}: t \in T\}$ is a normalizing basis of $L/\underline{a}$ over $R/\underline{b}$.

Regular Elements: Let c be a left regular element of R and suppose $\Sigma\, (ta_t)c = 0$, where the $a_t \in R$. Then each $a_tc = 0$ and so each $a_t = 0$. Hence c is also left regular as an element of L. In a similar way right regular elements of R are right regular in L, using that $L = \oplus Rt$. Therefore $\underline{C}_R(0) \subseteq \underline{C}_L(0)$.

Ore Subsets: Let Q be a T-invariant right divisor set (i.e. Q is a multiplicative submonoid not containing zero and satisfying the right Ore condition) in R. We claim that Q satisfies the right Ore condition in L. For let $c \in Q$ and $a = \Sigma\, ta_t \in L\backslash\{0\}$, where the $a_t \in R$. By hypothesis $c_t = c^t \in Q$. Thus $a_t d_t = c_t b_t$ for some $d_t \in Q$ and $b_t = R$. Again by the Ore condition there exists $d \in Q$ and, for each $a_t \neq 0$, an element $e_t \in R$ with $d = d_t e_t$. Then

$$ad = \Sigma\, ta_t d_t e_t = \Sigma\, tc_t b_t e_t = c(\Sigma\, tb_t e_t),$$

where the summations are over all $t \in T$ with $a_t \neq 0$. The claim is proved. There is an analogous result for the left Ore subsets.

Rings of Quotients: Suppose Q is a T-invariant right divisor set in R with $Q \subseteq \underline{C}_R(0)$. Certainly Q is a right divisor set of L in $\underline{C}_L(0)$ by the above. Thus we can form the rings LQ^{-1} and RQ^{-1} of right quotients. Moreover we can identify RQ^{-1} with the subring $R[Q^{-1}]$ of LQ^{-1}. Then $LQ^{-1} = \oplus_T\, tRQ^{-1}$ and T is also a normalizing basis of LQ^{-1} over RQ^{-1}.

The Relevant Special Case: Let $\underline{b}$ be a T-invariant ideal of R, let $\underline{a} = L\underline{b}$, $L_o = L/\underline{a}$ and let $\Phi : L \twoheadrightarrow L_o$ be the natural map. After the obvious identifications Φ maps R onto $R_o = R/\underline{b}$. Set $Q_o = \underline{C}_{R_o}(0)$ and $Q = R \cap Q_o\Phi^{-1} = \underline{C}_R(\underline{b})$. Assume that R is a domain and that Q is a (right and left) Ore subset of R. Trivially $Q \subseteq \underline{C}_R(0)$ and Q_o is an Ore subset of R_o. By the above we can form the rings

$$LQ^{-1} = Q^{-1}L \supseteq RQ^{-1} = Q^{-1}R$$

$$\text{and} \qquad L_oQ_o^{-1} = Q_o^{-1}L_o \supseteq R_oQ_o^{-1} = Q_o^{-1}R_o$$

of quotients and extend Φ to a map $ac^{-1} \longmapsto a\Phi(c\Phi)^{-1}$, still to be called Φ, of LQ^{-1} onto $L_oQ_o^{-1}$.

Suppose also that T is finite. We claim that Φ maps the group $U(LQ^{-1})$ of units of LQ^{-1} onto that of $L_oQ_o^{-1}$. (This will imply that $N\Phi$ is a normal subgroup of $U(L_oQ_o^{-1})$ whenever N is a normal subgroup of $U(LQ^{-1})$.) Assume for the moment that $\underline{b}Q^{-1}$ lies in the Jacobson radical $J(RQ^{-1})$ of RQ^{-1}. Then $\underline{b}Q^{-1}.LQ^{-1} = \underline{a}Q^{-1}$ lies in $J(LQ^{-1})$ (Passman [2], 7.2.5). Choose a_o in $L_o \cap U(L_oQ_o^{-1})$. Then $a_ob_od_o^{-1} = 1$ for some $b_o \in L_o$ and $d_o \in Q_o$. Pick $a, b \in L$ with $a\Phi = a_o$ and $b\Phi = b_o$, and set $d = ab$. Then

$d\Phi = d_o$ and $d \in Q_o\Phi^{-1} = Q + \underline{a} \subseteq (1 + \underline{a}Q^{-1})Q$. Since $\underline{a}Q^{-1} \subseteq J(LQ^{-1})$ so d, and hence a, has a right inverse in LQ^{-1}. Interchanging right and left we obtain a left inverse for a. This proves that $a \in U(LQ^{-1})$ and hence that $L_o \cap U(L_oQ_o^{-1}) \subseteq U(LQ^{-1})\Phi \subseteq U(L_oQ_o^{-1})$. But

$$U(L_oQ_o^{-1}) = (L_o \cap U(L_oQ_o^{-1}))Q_o^{-1}$$

and so the claim will follow. It remains to prove that $\underline{b}Q^{-1} \subseteq J(RQ^{-1})$. This follows from the following lemma.

4.5.4. *Let* R *be a ring,* $\underline{b}$ *an ideal of* R, *and set* $Q = \underline{C}_R(\underline{b})$.

a) *Suppose* Q *is right Ore in* R *and lies in* $\underline{C}_R(0)$. *Then* $\underline{b}Q^{-1}$ *is an ideal of* RQ^{-1} *and* $\underline{C}_{RQ^{-1}}(\underline{b}Q^{-1}) \subseteq U(RQ^{-1})$.

b) *If* $Q \subseteq U(R)$ *then* $\underline{b} \subseteq J(R)$.

Proof: a) That $\underline{b}Q^{-1}$ is an ideal of RQ^{-1} we have effectively seen before. For if $c^{-1}a = bd^{-1}$ where $a \in \underline{b}$, $b \in R$ and $c, d \in Q$ then $cb = ad \in \underline{b}$ and $c \in \underline{C}_R(\underline{b})$, so $b \in \underline{b}$ and $Q^{-1}\underline{b} \subseteq \underline{b}Q^{-1}$. Therefore $\underline{b}Q^{-1}$ is an ideal. Also $\underline{b} = R \cap \underline{b}Q^{-1}$, for if $r = ac^{-1} \in R$, where $a \in \underline{b}$ and $c \in Q$, then $rc = a$ belongs to $\underline{b}$ and $c \in \underline{C}_R(\underline{b})$, so $r \in \underline{b}$.

Suppose $ac^{-1} \in \underline{C}_{RQ^{-1}}(\underline{b}Q^{-1})$ where $a \in R$, $c \in Q$, and let $r \in R$. If $ra = 0$ then $r(ac^{-1}) = 0$ and $r \in R \cap \underline{b}Q^{-1} = \underline{b}$. If $ar = 0$ then $ac^{-1}.cr = 0$, $cr \in \underline{b}Q^{-1}$, and $r \in R \cap \underline{b}Q^{-1} = \underline{b}$. This proves that $a \in \underline{C}_R(\underline{b}) = Q$ and therefore $ac^{-1} \in QQ^{-1} \subseteq U(RQ^{-1})$.

b) Clearly $1 + \underline{b} \subseteq \underline{C}_R(\underline{b}) = Q \subseteq U(R)$, so $\underline{b} \subseteq J(R)$. □

Before we can indicate the proof of the fundamental step in the proof of 4.5.2 we need one further result.

4.5.5. *Let* D *be a division ring with centre* F *and* N *a normal subgroup of* D^* *that is soluble by locally-finite and contains* F^*. *Then* $N = F^*$.

Proof: If N is soluble then $N = F^*$ by Scott [1], 14.4.1. In particular N/F^* is locally finite. Therefore so is N' by Schur's Theorem and it suffices to prove that N' is soluble. If $\mathrm{char}D \neq 0$ then N' is abelian (2.1.1). Suppose $\mathrm{char}D = 0$. If N' is not soluble then it is finite, by 2.5.9. In this case $(D^*:C_{D^*}(N'))$ is finite and Scott [1], 14.2.1 yields that $D = C_D(N')$, which is a contradiction. Thus N' is soluble and the proof is complete. □

4.5.6. (Lichtman [5]).*Let* $E = D[G]$ *be a ring, where* D *is an infinite division subring of* E *generated as division ring by* $D \cap G$, *and* G *is a polycyclic-by-finite subgroup of* U(E) *normalizing* D *such that the index* $(G:D \cap G)$ *is finite. Let* N *be a normal subgroup of* U(E). *Then either* $[N,U(E)]$ *is unipotent or* N *contains a non-cyclic free subgroup.*

Proof: Assume that N contains no non-cyclic free subgroups and suppose first that G has an abelian normal subgroup B of finite index. Let K be the subfield of D generated by $D \cap B$. By Galois theory K has finite dimension over $F = C_K(G)$ and E is a finite-dimensional F-algebra. Clearly F is not locally finite. Suppose for the moment that $E = D^{n \times n}$. Then N is soluble-by-locally finite by the linear case (Tits' Theorem in effect, see Wehrfritz [2], 10.17). If $n > 1$ then N is central by 4.5.1 and if $n = 1$ then N is central by 4.5.5. This proves the special case. In general $U(E/\underline{m}) = U(E) + \underline{m}/\underline{m}$ for every maximal ideal $\underline{m}$ of E, and so by the special case $[N,U(E)] \subseteq 1 + \underline{n}$ for $\underline{n}$ the nilradical of E. But $\underline{n}$ is nilpotent and therefore $[N,U(E)]$ is unipotent as required.

Now assume that G is not abelian-by-finite. In particular G has positive Hirsch number. We apply the proof of the Reduction Theorem (4.2.1), but to E rather than D. Pick G_1 as before, but also with $G_1 \subseteq D$. We cannot now assume that $G = G_1$, but we may assume that G_1 generates D as a division ring. Let R be the subring of D generated by G_1. Now G contains a normalizing basis T of E over D. By 4.5.3 for all large enough finitely generated G-invariant subrings R_1 of D, the set T is also a normalizing basis of $R_1[G]$ over R_1. Now $R_1 \subseteq R[r^{-1}]$ for some $r \in R\backslash\{0\}$. If X is a transversal of G_1 to G then

$$R_2 = R[\ r^{-x} : x \in X\ ] \supseteq R_1$$

is G-invariant. Suppose Q is a right Ore subset of R containing r and normalized by G. Then $R \leq R_2 \leq RQ^{-1}$, Q is right Ore in R_2, and $RQ^{-1} = R_2Q^{-1}$. By the above T is also a normalizing basis of $L = R_2[G]$ over R_2 and, by 4.5.3, we can form LQ^{-1}.

As in the proof of 4.2.1 let Z be the subring of R generated by the centre of G_1, and pick $\underline{m}$ as before. Now G may not normalize $\underline{m}$ so set $\underline{m}_G = \bigcap_{x \in G} \underline{m}^x$ and $\underline{n} = \underline{m}_G R$. Now $\underline{n}$ need not be prime, but $\underline{m}R$ is by 4.2.7 and $(G:G_1)$ is finite. Thus at least $\underline{n}$ is semiprime. Also $\underline{n}$ still satisfies properties b) to d) of 4.2.7, so 4.2.6 is again applicable. Thus with $Q = \underline{C}_R(\underline{n})$ we can form RQ^{-1} and LQ^{-1}as above. Set $R_0 = R/\underline{n}$, $Q_0 = \underline{C}_{R_0}(0)$, and $E_0 =$

$LQ^{-1}/\underline{n}LQ^{-1}$ and let G_o be the image of G in E_o. Then we can form $R_oQ_o^{-1}$, which in fact is isomorphic to $RQ^{-1}/\underline{n}Q^{-1}$ and $E_o = R_oQ_o^{-1}[G_o]$. Now $R_oQ_o^{-1}$ is semisimple Artinian and contains a division subring D_o generated as such by a normal subgroup of G_o of finite index. Also $R_oQ_o^{-1} \subseteq D_o[G_o]$, so $E_o = D_o[G_o]$. The natural map ϕ of LQ^{-1} onto E_o maps $U(LQ^{-1})$ onto $U(E_o)$ by 4.5.3 since $LQ^{-1} = \underline{LC}_{R_2}(R_2 \cap \underline{n}Q^{-1})^{-1}$. Since G is not abelian-by-finite, G_o is not finite and D_o is infinite. Thus if $h(G_o) < h(G)$ then by induction $[N \cap U(LQ^{-1}), U(LQ^{-1})]\phi$ is unipotent. The result in this case follows as in the proof of 4.2.3 a).

Suppose now that $h(G_o) = h(G)$. Here R is just the group ring of G_1 over a prime image of $\mathbb{Z}$. We continue applying the proof of 4.2.1, but now we have a problem. We cannot in general pick G_1 and a plinth $A_1 \leq G_1$ of G_1 such that A is normal in G. The best we can do is to pick G_1 and $A \leq G_1$ such that A is normal in G, $A = A_1 \times \ldots \times A_h \leq G_1$ is a direct product of a finite number of G_1-plinths A_i and G_1/A is poly-C_∞. Let S be the subring of R generated by A. Suppose the following is true.

4.5.7. *Let* k *be a locally finite field,* G *a polycyclic-by-finite group and* $A_1,\ldots,A_h$ *a finite number of plinths for* G. *If* $a \in S = kA_1 \times\ldots\times A_h$ *is non-zero, then there is a maximal ideal* $\underline{m}$ *of* S *with* $a^G \cap \underline{m} = \emptyset$.

Then we can choose a maximal ideal $\underline{m}$ of S exactly as in the proof of 4.2.1. Set $\underline{n} = \underline{m}_G R$. As in the previous case $\underline{n}$ is now only semiprime, but Parts b) to d) of 4.2.7 hold and again 4.2.6 is applicable. Thus we can complete the proof as in the previous case.

It remains to prove 4.5.7. (Note that we only use 4.5.7 where k is finite, a case that avoids part of the argument below.) We induct on h, the case h = 1 being 4.2.8. Now $a = \Sigma t\alpha_t$, where the $\alpha_t \in S_1 = kA_2 \times\ldots\times A_h$ and the summation is over all $t \in A_1$. By induction applied to $\alpha = \Pi_{\alpha_t \neq o}\alpha_t$, there is a maxiaml ideal $\underline{m}_1$ of S_1 with $\alpha^G \cap \underline{m}_1 = \emptyset$. Thus $a^x \notin \underline{m}_1 S$ for all $x \in G$.

Set $\underline{m}_{1G} = \bigcap_{x \in G} \underline{m}_1^x$ and $C = C_G(S_1/\underline{m}_{1G})$. We claim that (G:C) is finite, a fact that is obvious if k is finite. In general $S_1/\underline{m}_1$ is a locally finite field, so $B^s \subseteq 1 + \underline{m}_1$ for $B = A_2 \times\ldots\times A_h$ and some integer $s > 0$. Then $B^s \subseteq 1 + \underline{m}_{1G}$ and $S_1/\underline{m}_{1G}$ is semisimple, say $S_1/\underline{m}_{1G} = k_1 \oplus\ldots\oplus k_t$ where each k_i is a field. Then each k_i is a finite extension of k and Galois theory yields that $(G:C) \leq t!.\Pi_i(k_i:k) < \infty$.

Let $g_1,\ldots, g_m$ be a transversal of C to G and set $b = \Pi_i\ a^{g_i}$. Clearly $b \notin \underline{m}_1S$, since this ideal is prime (e.g. by 1.4.1). Also $S/\underline{m}_1S \cong k_1A_1$ for $k_1 = S_1/\underline{m}_1$. Consequently by 4.2.8 there exists a maximal ideal $\underline{m} \supseteq \underline{m}_1S$ of S such that $b^C \cap \underline{m} = \emptyset$. Suppose $a^x \in \underline{m}$ for some $x \in G$. Then $x = g_ic$ for some i and some $c \in C$. Hence $b^c \in Sa^xS \subseteq \underline{m}$, a contradiction. Therefore $a^G \cap \underline{m} = \emptyset$. □

Proof of 4.5.2: By 4.5.1 we may assume that n = 1. Assume also that N contains no non-cyclic free subgroups. Suppose there exists $x \in N$ and $g \in G$ with $[x,g] \neq 1$. There is a polycyclic-by-finite subgroup G_1 of F^*G such that the division subring D_1 of D generated by G_1 contains x and g and is not locally finite. By 4.5.5, with D_1 playing the roles of both E and D, the element [x,g] is unipotent. But D is a division ring, so we obtain the contradiction [x,g] = 1. This proves that N lies in the centre of D. □

The following is a slight generalization of Lichtman [9], Theorem 4.1, Part 3.

4.5.8. COROLLARY. *Let* F *be a field,* G *locally an* $\underline{X}$*-group where* $\underline{X}$ *is as in* 4.4.1, *and* D *the division ring of quotients of* FG. *Suppose that either* F *is not locally finite or* $G \neq \langle 1\rangle$. *If* N *is any normal subgroup of* GL(n,D), *then either* N *is central or* N *contains a non-cyclic free subgroup.*

Proof: By 4.5.1 we may assume that n = 1. Suppose that N does not contain a non-cyclic free subgroup. By means of a trivial localization we assume that $G \in \underline{X}$. We now induct on m, where m is as in 4.4.1. If m = 0 the result follows immediately from 4.5.2, so let $m \geq 1$. In the proof of 4.4.6 the natural map Φ of S onto $S/\mathbb{P}S = D_2$ maps $\underline{C}_R(\mathbb{P})$ onto $(R/\mathbb{P})\backslash\{0\}$ and hence maps $U(S) \supseteq \underline{C}_R(\mathbb{P})\underline{C}_R(\mathbb{P})^{-1}$ onto $D_2{}^*$. Consequently $(N \cap U(S))\Phi \cap D_1$ is a normal subgroup of $D_1{}^*$. Induction easily yields that N is centralized by G.□

There is some further positive evidence that Tits' Theorem extends to the division rings we have been considering here. Bachmuth & Mochizuki [1] prove the following, see Theorem A and Remark 1. Let F be any field, let G be a poly-C_∞ group and let D be the division ring of quotients of the group algebra FG, which exists by 1.4.4. Suppose that H is a finitely generated subgroup of SL(2,D) such that for some b ∈ FG\{0} we have $Hb^{-1} \subseteq (FG)^{2\times 2}$. Then H either contains a non-cyclic free subgroup or is soluble-by-finite. The proof uses Ihara's Theorem, see for example Serre [3], p. 82, to express SL(2,D) as a non-trivial amalgamated free product, and then the subgroup theorem for such free products is applied.

EXERCISE: In 4.5.1, 4.5.2, 4.5.5, 4.5.6, and 4.5.8, replace "N is a normal subgroup" by "N is a subnormal subgroup".

4.6 UNIVERSAL ENVELOPING ALGEBRAS OF LIE ALGEBRAS

Let F be a field, L a Lie algebra over F and A = A(L) the universal enveloping algebra of L. It is a well-known consequence of the Poincaré-Birkhoff-Witt Theorem (Jacobson [2], p. 159) that A is a domain (ibid. p. 166, Theorem 6). In fact A can always be embedded into a division ring D by a theorem of P. M. Cohn (Cohn [1], Theorem 4.3, Corollary and Theorem 5.1), but it is difficult to make more than superficial use of this fact. Specifically one would like D to be the ring of quotients of A; that is, one would like A to be an Ore domain.

Let $\underline{O}_F$ denote the class of Lie F-algebras L such that A(L) is an Ore domain, and let $\underline{Z}_F$ denote the class of Lie F-algebras of dimension at most 1. The local, poly, and subalgebra operators L, P and S on classes of Lie F-algebras can be defined in the obvious way; see Amayo & Stewart [1], §1.5.

4.6.1. a) $\underline{O}_F$ *is* ⟨L,S⟩*-closed.*

b) $\underline{O}_F.\langle L,P\rangle\underline{Z}_F = \underline{O}_F$.

c) $\underline{O}_F$ *contains all locally finite-dimensional Lie* F*-algebras.*

d) $\underline{O}_F$ *contains all locally soluble Lie* F*-algebras.*

Proof: a) The L-closure of $\underline{O}_F$ is trivial. If H is a subalgebra of the Lie F-algebra L then by the Poincaré-Birkhoff-Witt Theorem A(H) is a subalgebra of A(L), and A(L) is free as both right and left A(H)-module, (see Jacobson [2], p. 161, Corollary 2). If A(L) is Ore it follows easily that A(H) is too; cf. the proof of 1.4.16.

b) If $\underline{X}$ is a class of Lie algebras with $\underline{O}_F.\underline{X} = \underline{O}_F$ then $\underline{O}_F.L\underline{X} = \underline{O}_F$ by a) and $\underline{O}_F.P\underline{X} = \underline{O}_F$ by an elementary induction. Thus $\underline{O}_F.\langle L,P\rangle\underline{X} = \underline{O}_F$.

It is therefore sufficient to prove that $\underline{O}_F.\underline{Z}_F = \underline{O}_F$. This we postpone for the moment.

c) If L is a finite-dimensional Lie F-algebra then A(L) is Noetherian (Jacobson [2], p. 166, Theorem 6 again) and hence Ore by Goldie's Theorem. Thus $L \in \underline{O}_F$, and c) follows from a).

d) Clearly b) implies that $\langle L,P\rangle\underline{Z}_F \subseteq \underline{O}_F$. Also $\langle L,P\rangle\underline{Z}_F$ contains all locally soluble Lie F-algebras since $LP\underline{Z}_F$ contains all abelian Lie F-algebras.

It remains to prove that $\underline{O}_F.\underline{Z}_F \subseteq \underline{O}_F$. For this some preliminary development is needed. Suppose that R is a right Ore domain with ring Q of right quotients, and let $\delta : R \longrightarrow R$ be a derivation; that is, δ is an additive map satisfying $(xy)^\delta = x^\delta y + xy^\delta$ for all $x, y \in R$. Extend δ to a map of Q into itself by defining

$$(ac^{-1})^\delta = a^\delta c^{-1} - ac^{-1}c^\delta c^{-1}.$$

We have first to check that δ is well-defined. Suppose $ac^{-1} = bd^{-1}$. By the Ore condition $cx = dy \neq 0$ for some $x, y \in R$, and then also $ax = by$. Consequently

$$(cx)^\delta = c^\delta x + cx^\delta = d^\delta y + dy^\delta,$$

and $$(ax)^\delta = a^\delta x + ax^\delta = b^\delta y + by^\delta.$$

Then $a^\delta c^{-1} - ac^{-1}c^\delta c^{-1}$

$$= (b^\delta y + by^\delta - ax^\delta)x^{-1}xy^{-1}d^{-1} - ac^{-1}(d^\delta y + dy^\delta - cx^\delta)x^{-1}xy^{-1}d^{-1}$$

$$= b^\delta d^{-1} - bd^{-1}d^\delta d^{-1}$$

since $ac^{-1} = bd^{-1}$. Thus δ is well-defined.

We now check that δ is a derivation. Any two elements of Q can be taken to have the same denominator, and

$$(ac^{-1} + bc^{-1})^\delta = ((a+b)c^{-1})^\delta = (a+b)^\delta c^{-1} - (a+b)c^{-1}c^\delta c^{-1}$$

$$= (ac^{-1})^\delta + (bc^{-1})^\delta.$$

Now consider $x = (ac^{-1})(bd^{-1})$. Then $c^{-1}b = ef^{-1}$ for some e and f. Hence $bf = ce$ and so

$$b^\delta f + bf^\delta = c^\delta e + ce^\delta. \qquad (*)$$

Then $(ac^{-1})^\delta(bd^{-1}) + (ac^{-1})(bd^{-1})^\delta$

$$= (a^\delta c^{-1} - ac^{-1}c^\delta c^{-1})bd^{-1} + ac^{-1}(b^\delta d^{-1} - bd^{-1}d^\delta d^{-1})$$

and

$$x^\delta = (ae)^\delta f^{-1}d^{-1} - x(df)^\delta f^{-1}d^{-1}$$

$$= a^\delta ef^{-1}d^{-1} + ae^\delta f^{-1}d^{-1} - xd^\delta d^{-1} - xdf^\delta f^{-1}d^{-1}$$

$$= a^\delta c^{-1}bd^{-1} + ac^{-1}(ce^\delta - bf^\delta)f^{-1}d^{-1} - ac^{-1}bd^{-1}d^\delta d^{-1}$$

$$= a^\delta c^{-1}bd^{-1} + ac^{-1}(b^\delta f - c^\delta e)f^{-1}d^{-1} - ac^{-1}bd^{-1}d^\delta d^{-1} \quad \text{(by (*))}$$

$= (ac^{-1})^{\delta}(bd^{-1}) + (ac^{-1})(bd^{-1})^{\delta}.$

Therefore δ is a derivation of Q.

We can now form the skew polynomial ring Q[x] where $ax-xa= a^{\delta}$ for all $a \in Q$ (see Cohn [3], §1.3).

Now let us return to our original problem. Let L be a Lie F-algebra and let H be an ideal of L such that $H \in \underline{O}_F$ and $\dim_F L/H = 1$. We have to prove that A(L) is Ore. Now $L = Fx \oplus H$ for some $x \in L$ and then $A(L) = B[x]$ for $B = A(H)$ by the Poincaré-Birkhoff-Witt Theorem. Also $\delta : a \longmapsto ax - xa$ is a derivation of B into itself. By hypothesis B has a ring of quotients Q. Extend δ to Q and form the corresponding skew polynomial ring $S = Q[x]$. The subring B[x] of S is simply a copy of A(L). Now S, being a skew polynomial ring over a division ring, is a Noetherian domain, and is therefore Ore by 1.4.2.

Let $a, b \in A(L) \setminus \{0\}$. Then there exist $r, s \in S$ with $ar = bs \neq 0$. Now $r = \Sigma\, x^i r_i$ and $s = \Sigma\, x^i s_i$ for some $r_i, s_i \in Q$. As elements of Q we can choose a common denominator h for them. Then $rh, sh \in B[x] = A(L)$ and $a(rh) = b(sh) \neq 0$. Therefore A(L) is right Ore. In the same way it is left Ore, and $L \in \underline{O}_F$ as required. The proof of 4.6.1 is complete. □

Lichtman has proved, for the algebras A(L), analogues of certain of the results of the earlier sections of Chapter 4. We state his main results without proof.

Let L *be a Lie* F*-algebra, and assume that one of the following holds:*

1) $L \in \underline{O}_F$ *and* $\bigcap_{m \geq 0} L^m = \{0\}$.

2) L *is metabelian.*

3) L *is locally finite-dimensional over* F.

Then $L \in \underline{O}_F$ in each case by 4.6.1. Let D be the division ring of quotients of A(L).

4.6.2. THEOREM. (Lichtman [8]). *With the notation above, the following hold:*

a) *Unipotent subgroups of* GL(n,D) *are stability subgroups.*

b) *Periodic subgroups of* GL(n,D) *are locally finite.*

c) *If* char F = 0 *and* $\beta(n)$ *is any Jordan function, then periodic subgroups of* GL(n,D) *have abelian normal subgroups of index at most* $\beta(n)$.

d) *A normal subgroup of* GL(n,D) *is either central or contains a non-cyclic free subgroup (apart from the trivial case where* L = {0} *and* F *is locally finite.)*

e) *Suppose* L *satisfies* 1) *and is soluble. Then nilpotent subgroups of* D^* *are abelian. If also* charF = 0 *then nilpotent-by-finite subgroups of* D^* *are abelian.* □

Recall that a Jordan function is any integer-valued function $\beta(n)$ such that every finite subgroup of $GL(n,\mathbb{C})$ has an abelian normal subgroup of index at most $\beta(n)$. If L satisfies 3) then easy localizing arguments reduce one to the finite-dimensional case, and then Lichtman [8] applies.

5 NORMAL SUBGROUPS OF ABSOLUTELY IRREDUCIBLE GROUPS

We have asserted, more than once, that absolute irreducibility is a much stronger condition than irreducibility. In this chapter, at long last, we really substantiate this claim. In general terms our contention is that a reasonable absolutely irreducible group should be abelian by locally-finite at least. Further, the same conclusion should hold for a reasonable normal subgroup H of an absolutely irreducible group G. Moreover H should be more or less rigid, in that $G/C_G(H)$ should be much smaller than one might at first expect. What groups are reasonable in this context ? Clearly a group with a non-cyclic free subgroup is unreasonable. We are a long way from proving the converse of this, if indeed it is actually true.

The main sections of this chapter are Section 5.4, where we consider locally finite normal subgroups, and Section 5.6, where soluble normal subgroups are considered. We need to start, in Section 5.1, with a close look at the linear case and, since it is required for Section 5.6, in Section 5.5 we study what happens when the ground division ring is locally finite-dimensional over its centre. Sections 5.2 and 5.3 are service sections, of a mainly ring-theoretic nature, for what comes later. In the final Section 5.7 we consider generalized nilpotent and generalized soluble normal subgroups. In fact much of it is devoted to the locally nilpotent case.

Where possible we relax the absolute irreducibility condition on G to at least irreducibility, and sometimes even to just $u(G) = \langle 1 \rangle$. This does not happen very often, and since our counterexamples are irreducible, this will justify the claim made in the first sentence of this chapter.

5.1 NORMAL SUBGROUPS OF LINEAR GROUPS

As always the linear case is the basis of the skew linear case.

We need a more detailed analysis of the former in order to prove the general result. Much of the section is therefore a study of periodic normal subgroups of linear groups, and it involves the classification of finite simple groups in the weak form that there are only finitely many sporadic simple groups. We conclude the section with the linear case of the main theorem of this chapter. Although this has stronger conclusions than is possible in the skew linear case, it should give the reader a good idea of what to expect.

We begin with a few general results.

5.1.1. *Let* G *be a group, and* K *a subgroup of* G *that is the central product of finitely many subgroups* $H_1,\ldots,H_r$. *Then* $KC_G(K) = \bigcap_i H_iC_G(H_i)$.

Proof: Clearly if $i \neq j$ then $H_i \subseteq C_G(H_j)$. Thus repeated application of modular law shows that

$$\begin{aligned}\bigcap_i H_iC_G(H_i) &= (H_1\ldots H_{r-1} \textstyle\bigcap_{i<r} C_G(H_i)) \cap H_rC_G(H_r)\\ &= H_1\ldots H_{r-1}(\textstyle\bigcap_{i<r} C_G(H_i) \cap H_rC_G(H_r))\\ &= H_1\ldots H_{r-1}H_r \textstyle\bigcap_i C_G(H_i)\\ &= KC_G(K). \ \square\end{aligned}$$

5.1.2. *Let* $L \leq H$ *be normal subgroups of the group* G *with* Z *the centre of* L, *and set* $C = C_G(L) \cap C_G(H/L)$. *Then* $C/ZC_G(H)$ *is isomorphic to a subgroup of the first cohomology group* $H^1(H/L,Z)$. *In particular if* $m = (H:L)$ *is finite then* $C/ZC_G(H)$ *has finite exponent dividing* m, *and if also* Z *has finite rank* n *then* $C/ZC_G(H)$ *is finite with order dividing* m^{mn}.

Proof: By elementary stability theory (e.g. Kegel & Werfritz [1], § 1.C) there is an exact sequence

$$\langle 1\rangle \longrightarrow C_G(H) \longrightarrow C \longrightarrow \mathrm{Der}(H/L,Z)$$

where the right-hand map sends $c \in C$ to the derivation $Lh \longmapsto [h,c]$. If this derivation is inner there exists $z \in Z$ with $[h,c] = [h,z]$ for every $h \in H$. It follows that $cz^{-1} \in C_G(H)$ and that the inverse image of Ider(H/L,Z) in C is $ZC_G(H)$. Therefore $C/ZC_G(H)$ is isomorphic to a subgroup of $H^1(H/L,Z)$. If $m = (H:L)$ is finite then $H^1(H/L,Z)$ has exponent dividing m and its rank is at most that of Map(H/L,Z). The latter is clearly (H:L).rankZ. Therefore if rankZ = n then the order of $H^1(H/L,Z)$ divides m^{mn}. $\square$

5.1.3. *Let* G *be a periodic group and let* A *be a* G*-module that is finitely* $\mathbb{Z}$*-generated. Then* $H^1(G,A)$ *is periodic.*

Proof: Periodic subgroups of $GL(n,\mathbb{Z})$ are finite (Wehrfritz [2], 4.8). This and elementary stability theory yield that $C = C_G(A)$ has finite index, m say, in G. Now $H^1(C,A) \cong Hom(C,A)$ is certainly periodic. Also the image under co-restriction of $H^1(C,A)$ in $H^1(G,A)$ contains $m.H^1(G,A)$ (restriction followed by co-restriction is multiplication by m). The result follows.□

5.1.4. *Let* H *be a non-trivial periodic linear group with no non-trivial abelian normal subgroups. Then* H *has a non-trivial subnormal simple subgroup.*

Proof: Consider a composition series of H. By Wehrfritz [2], 9.30, such a series has only a finite number of non-abelian factors. Suppose that $\langle 1\rangle \subseteq \ldots \subseteq K_1 \subset H_1 \subseteq \ldots \subseteq K_r \subset H_r \subseteq \ldots \subseteq H$ is the series, where the H_i/K_i are the non-abelian factors. We induct on r. If $r = 0$ then H is locally soluble, and so soluble. This is impossible, so $r \geq 0$.

For some m every soluble section of H has derived length at most m (ibid. 9.21). Thus if X is any finite subgroup of H the m-th term of the derived series of X lies in $X \cap H_r$. Consequently the m-th term L of the derived series of H lies in H_r. Now L has a unique maximal soluble normal subgroup which, being normal in H, is trivial. Thus the maximal soluble normal subgroup of $K_r \cap L$ is also trivial. If $r = 1$ it follows that $K_1 \cap L = \langle 1\rangle$, $K_1L = H_1$, and $L \cong H_1/K_1$ is a normal simple subgroup of H. If $r \geq 1$ then by induction $K_r \cap L$ has a subnormal simple subgroup, which is clearly subnormal in H. □

We also need the following generalization of part of 2.4.7.

5.1.5. *Let* H *be a periodic linear group with Hirsch-Plotkin radical* N*. Then there exists a perfect characteristic subgroup* E *of* H *such that* $[E,N] = \langle 1\rangle$, $C_H(EN) \subseteq N$, *and* E *modulo its centre is a direct product of finitely many simple groups, the number of such being bounded by a function of* n *only.*

Proof: Let $C = C_H(N)$ and $Z = C \cap N$. If $C \subseteq N$ set $E = \langle 1 \rangle$. Suppose $C \not\subseteq N$. If X/Z is the Hirsch-Plotkin radical of C/Z then XN is locally nilpotent; whence $X = Z$. Now N is closed in H (Wehrfritz [2], 8.2 ii)) and so Z is closed in C. Thus C/Z is isomorphic to a linear group (ibid. 6.4) and therefore contains a simple subnormal subgroup by 5.1.4. Let K/Z be the group generated by all the simple subnormal subgroups of C/Z. Then K/Z is a direct product of perfect simple groups (Robinson [1], Vol. 1, 5.42). If $C_{C/Z}(K/Z) \neq \langle 1 \rangle$ then it contains a simple subnormal subgroup. Thus $C_C(K/Z) = Z$.

Let E be the last term of the derived series of K; this exists by Wehrfritz [2], 9.21. Then E is perfect, $E \cap Z$ is central in E, and $EZ = K$, so $E/(E \cap Z) \cong K/Z$ is a direct product of perfect simple groups, and the number of such factors is bounded by a function of n only (ibid. 9.30). Finally $C_H(EN) = C_C(E) \subseteq C_C(K/Z) = Z \subseteq N$.□

NOTE. In fact the number of perfect simple factors of E is at most $3n/4$ if the characteristic is not 2, and $2n$ otherwise, see Wehrfritz [9]. Presumably the bound $2n$ is far too large.

EXERCISE. In 5.1.5 prove that E is the central product of quasisimple groups (possibly $E = \langle 1 \rangle$). Recall that a group is quasisimple if it is perfect, and simple modulo its centre. (Hint: suppose

$$E/B = K_1/B \times \ldots \times K_r/B,$$

where B is the centre of E and each K_i/B is simple. Let E_i be the last term of the derived series of K_i. Prove that E_i is quasisimple and that $E = E_1 \ldots E_r$.)

5.1.6. THEOREM. (Wehrfritz [19]). *Let G be a linear group of degree n and characteristic $p \geq 0$ and let H be a periodic normal subgroup of G. If $p > 0$ assume further that $O_p(H) = \langle 1 \rangle$. Then the order of $G/HC_G(H)$ is finite and n-bounded.*

The term "n-bounded" is an abbreviation for "bounded by an integer-valued function of n only". In a similar way we use the term (m,n)-bounded if the bounding function involves only the two variables m and n.

Easy examples show that no qualitative improvement is possible to 5.1.6. For example if A is a cyclic group of finite order (coprime to p if $p \geq 0$), then the permutational wreath product $G = A \wr Sym(n)$ is isomorphic to a linear group of degree n and characteristic p. If H denotes the base group of G then $(G:HC_G(H)) = n!$. A second type of example arises from adjoining diagonal automorphisms to a group of Lie type. For example let $n \geq 2$ be an integer and let k be any locally finite field. Set $H = SL(n,k)$ and $G = GL(n,k)$. Then $(G:HC_G(H)) = (k^*:(k^*)^n)$, which for suitable k takes the value n. The bound on $(G:HC_G(H))$ arising from the proof below (which differs somewhat from that given in Wehrfritz [19]) is vastly greater than n!, and can almost certainly be reduced.

Proof: Note that $u(H) = \langle 1 \rangle$ in all characteristics. Thus we may pass to $G/u(G)$ and assume that G is absolutely completely reducible. We break the proof up into more managable pieces.

a) *If* $p = 0$ *or* H *is soluble then the conclusion of* 5.1.6 *holds.*

For in either case H has an abelian characteristic subgroup A of rank at most n and n-bounded index, m say, (Wehrfritz [2], 3.5 and 9.4). Now $(G:C_G(A))$ divides n! by ibid. 1.12 and 1.8, and trivially $(G:C_G(H/A))$ divides m!. Finally the order of $C_G(A) \cap C_G(H/A)$ modulo $AC_G(H)$ divides m^{mn} by 5.1.2, and therefore $(G:HC_G(H))$ divides $m!n!m^{mn}$.

b) *If* H *is simple then the conclusion of* 5.1.6 *holds.*

Suppose first that H is finite. Since there are only finitely many sporadic simple groups we may assume that H is not sporadic. Suppose H is alternating of degree m. Then H contains an elementary abelian 2-subgroup of rank $[m/2] - 1$ and an elementary abelian 3-subgroup of rank $[m/3]$. Hence $m \leq 2n + 3$ if $p \neq 2$ and $m \leq 3n + 2$ if $p = 2$. Thus m is n-bounded.

Since infinite periodic simple linear groups are all of Lie type (Thomas [1]) we are left with the case that H is of Lie type (finite or infinite). Using the structure of the automorphism group of a group of Lie type (Steinberg [1], Theorems 30 & 36) and the detailed structure of H (ibid., or Carter [1]) it can be shown that the order of the group of *outer* automorphisms of H induced by G is n-bounded; see Wehrfritz [19], 1.6. The result follows in this case.

c) *Let* N *and* E *be as in* 5.1.5. *Then* $(G:ENC_G(EN))$ *is* n-*bounded.*

Let B be the centre of E. The number of non-abelian composition factors of E is n-bounded. Thus some subgroup of G of n-bounded index normalizes the simple factors of E/B. Also $G/C_G(E)$ is isomorphic to a linear group of degree n^2 (Wehrfritz [2], 6.2). Thus by 5.1.1 and b) above the index $(G:EC_G(E/B))$ is n-bounded. By ibid. 1.12, the index $(G:C_G(B))$ divides n!, and since $Hom(E/B,B) = \langle 1 \rangle$, stability theory shows that $C_G(E/B) \cap C_G(B) = C_G(E)$. Consequently $(G:EC_G(E))$ is n-bounded. Also $(G:NC_G(N))$ is n-bounded by a) above, and since $[E,N] = \langle 1 \rangle$ we have $EC_G(E) \cap NC_G(N) = ENC_G(EN)$. The claim follows.

d) *The completion of the proof of* 5.1.6.

Since $C_H(EN) \leq N$ by 5.1.5 we have that $H \cap ENC_G(EN) = EN$. Thus $(H:EN)$ is n-bounded by c) above. Hence $ENC_G(EN) \cap C_G(H/EN) = EN(C_G(EN) \cap C_G(H/EN))$ has n-bounded index in the group G. Finally $C_G(EN) \cap C_G(H/EN)$ modulo $ENC_G(H)$ has n-bounded order by 5.1.2. Therefore $(G:ENC_G(H))$ is n-bounded. The result follows.□

If p in 5.1.6 is positive then the condition $O_p(H) = \langle 1 \rangle$ cannot be removed. For let k be any locally finite field of characteristic p, and let $n \geq 2$ be an integer. Let $G = Tr(n,k)$ and $H = O_p(G)$. Then $HC_G(H) = k^*H$, and $G/HC_G(H)$ is isomorphic to the direct product of n-1 copies of k^*. Similarly by choosing G to be the normalizer and H the stabilizer of a suitable flag one can obtain examples where

$$G/HC_G(H) \cong GL(n_1,k) \times \ldots \times GL(n_r,k)/k^*$$

and $\Sigma n_i = n$. However something can be salvaged if $O_p(H)$ is finite. We need the following simple lemma.

5.1.7. *Let* K *be a periodic group and* p *a prime such that* $O_p(K)$ *is finite of order* p^e *and* $H = K/O_p(K)$ *is isomorphic to a linear group of degree* n *and characteristic* p. *Let* Z *denote the centre of* $O_p(K)$.

a) *Any* p-*image of* H *is finite of* n-*bounded order, and any* p-*image of* K *is finite of* (n,p^e)-*bounded order.*

b) Der(H,Z) *is finite of* (n,p^e)-bounded order.

If K is a linear group of degree n and characteristic p then so is H, for example by 1.1.2.

Proof: a) Let E and N be subgroups of H as in 5.1.5. Clearly the p-images of N and E are trivial. But (H:EN) is n-bounded by Part c) of 5.1.6. Consequently any p-image of H has n-bounded order. Part a) follows.

b) Let $C = C_H(Z)$. Clearly $(H:C) \leq p^e!$. Suppose $k_1 \ldots, k_r \in H$ generate H modulo C. Then $\delta \longmapsto (k_1\delta, \ldots, k_r\delta, \delta|_C)$ is an embedding of Der(H,Z) into $Z^{(r)} \times \mathrm{Hom}(C,Z)$. But $\mathrm{Hom}(C,Z) \cong \mathrm{Hom}(C/C^{p^e}, Z)$ and $(C:C^{p^e})$ is (n,p^e)-bounded by Part a). Thus $|\mathrm{Hom}(C,Z)|$ is (n,p^e)-bounded, and the result follows. □

5.1.8. (Wehrfritz [19]). *Let* G *be a linear group of degree* n *and characteristic* $p \geq 0$, *and let* K *be a periodic normal subgroup of* G *such that* $|O_p(K)| = p^e$ *is finite. Then the index* $(G:KC_G(K))$ *is* (n,p^e)*-bounded.*

Proof: By 5.1.6 applied to $H = K/O_p(K)$ and $G/O_p(G)$ we have that $(G:KC_G(H))$ is n-bounded. Let $C = C_G(H) \cap C_G(O_p(K))$. Clearly $(C_G(H):C)$ divides $p^e!$. Let Z be the centre of $O_p(K)$. Then by stability theory $C/C_G(K)$ is isomorphic to a subgroup of Der(H,Z). The result follows from 5.1.7. □

We now come to the linear versions of the main results of this chapter. Not suprisingly, in the linear case we can derive stronger conclusions from weaker hypotheses. If G is a linear group with $u(G) = \langle 1 \rangle$ one can assume that G is completely reducible by 1.1.2, and hence absolutely completely reducible by Wehrfritz [2], 1.22. This sort of reduction is not available in the skew linear case.

Recall from Section 3.1 that $\underline{\bar{S}}$ denotes the class of groups with a local system, of finitely generated subgroups each of whose finite images is soluble. Also, if p is a prime then $\underline{E}_p$ denotes the class of p-groups of finite exponent, and $L_1\underline{F}$ the class of periodic groups. Note that while periodic linear groups are locally finite, periodic sections of linear groups need not be.

5.1.9. THEOREM. (Wehrfritz [18]). *Let* G *be a* $\acute{P}(\underline{\bar{S}} \cup L_1\underline{F})$ *subgroup of* GL(n,F), *where* F *is a field of characteristic* $p \geq 0$.

a) *If* $p = 0$ *then* G *is soluble-by-finite. More precisely,*

$$G \in (\underline{N} \cap \underline{F}^{-S})\underline{AF}.$$

b) *If* $p \neq 0$ *then* G *is soluble by locally-finite and moreover*

$$G \in (L\underline{F})\underline{AF} \cap (\underline{N} \cap \underline{E}_p)\underline{A}.L\underline{F} \subseteq \underline{S}.L\underline{F}.$$

Proof: Recall that by 3.1.4 the class $\bar{\underline{S}}$ is $\langle Q,S_f \rangle$ closed and that linear $\bar{\underline{S}}$-groups are soluble. By hypothesis the group G has an ascending series

$$\langle 1 \rangle = G_0 \trianglelefteq G_1 \trianglelefteq \dots \trianglelefteq G_\alpha \trianglelefteq \dots \trianglelefteq G_\gamma = G,$$

where each factor $G_{\alpha+1}/G_\alpha$ lies in $\bar{\underline{S}}$ or $L_1\underline{F}$. We induct on γ, proving first the case where γ is finite. Note that G contains a unique maximal soluble normal subgroup S (Wehrfritz [2], 3.8) and a unique maximal periodic normal subgroup T, and that G_1 lies in S or T. (This does not assume that G_1 is normal in G.)

a) Assume that γ is finite and greater than zero. Now S is (Zariski) closed in G, so G/S is isomorphic to a linear group (ibid. 6.4). If $G_1 \subseteq S$ then G/S is soluble-by-finite (and so finite) by induction. Suppose $G_1 \subseteq T$. By the Jordan-Schur Theorem (ibid. 4.9 & 9.4) the group T contains a diagonalizable characteristic subgroup A of finite index. Then $C = C_G(T/A)$ has finite index in G, and induction applied to $C/C \cap S$ shows that $C/C \cap S$ is soluble-by-finite. It follows that G is soluble-by-finite. The rest of a) follows from the structure of soluble linear groups.

b) Assume first that $G_1 \subseteq S$. Then by induction G/S, and hence G, is soluble by locally-finite. Now $O_p(G)$ is locally finite, so we may pass to $G/O_p(G)$ and assume that G is absolutely completely reducible. Hence S contains a diagonalizable characteristic subgroup A of finite index. Then $C = C_G(A)$ has finite index in G by Wehrfritz [2], 1.12. Also C is centre by locally-finite and hence C' is locally finite. Thus C is locally-finite by abelian; whence $G \in (L\underline{F})\underline{AF}$.

Now assume that $G_1 \subseteq T$. Let N be the normalizer in G of a Sylow 2-subgroup of T. Then $NT = G$ by the Frattini argument and the conjugacy of the Sylow subgroups of T (Wehrfritz [2], 9.10), and $N \cap T$ is soluble by the Odd Order Theorem (in fact there is no need to quote the Odd Order Theorem provided we choose N such that $NT = G$ and every Sylow p-subgroup of $N \cap T$, for $p \leq n$, is normal in N, ibid. proof of 9.10). Then

$$\langle 1 \rangle \trianglelefteq N \cap T \trianglelefteq N \cap G_2T \trianglelefteq \dots \trianglelefteq N \cap G_\gamma T = N$$

is an ascending series for N of the given type (since $\bar{\underline{S}}$ is Q-closed) and $N \cap T$ is soluble. By the first case N, and hence G, is locally-finite by abelian by finite. Also $G/C_G(T/O_p(G))$ is locally finite by 5.1.6. Therefore G is soluble by locally-finite and the rest of b) follows from the structure of soluble linear groups.

We have now completed the proof of the case where γ is finite. Suppose γ is infinite. If γ is a limit ordinal then by induction $G \in L(\underline{S}.L\underline{F}) = L(\underline{S}.\underline{F})$. But G is also linear, so $G \in \underline{S}.L\underline{F}$ by Wehrfritz [2], 10.8. If $\gamma - 1$ exists then $G_{\gamma-1} \in \underline{S}.L\underline{F}$ by induction, and $G \in \underline{S}.L\underline{S}.\overline{\underline{S}}.L_1\underline{F}$. Thus in either case we are back in the situation where γ is finite. The proof is complete. □

5.1.10. THEOREM. (Wehrfritz [20]). *Let* G *be a subgroup of* GL(n,F) *where* F *is a field, and let* H *be a normal* $\acute{P}(\overline{\underline{S}} \cup L_1\underline{F})$*-subgroup of* G *with* u(H) = ⟨1⟩.

a) H *contains an abelian normal subgroup* A *of* G *such that* H/A *and* $G/AC_G(H)$ *are locally finite. Moreover* $G/HC_G(H)$ *has an abelian normal subgroup of* n*-bounded index, and exponent dividing* n!.

b) *If the centre* Z *of* H *has finite index* m *in* H *then* $G/C_G(H)$ *is finite and of order dividing* $m!n!m^n$.

c) *If* H *is nilpotent of class* c *then* $G/C_G(H)$ *is finite with* (n,c)*-bounded order.*

d) *If* H *is periodic then* $G/HC_G(H)$ *is finite of* n*-bounded order.*

Part d) we have added for completeness; it is simply a restatement of 5.1.6. In Part a) we cannot, in general, choose A so that H/A has finite exponent, for H could be an infinite simple group. However if H is soluble or if charF = 0 then A can be chosen with (H:A) finite and n-bounded. This is essentially Mal'cev's and Jordan's Theorems (Wehrfritz [2], 3.5 and 9.4).

Let q be a prime, and let G be the wreath product of a Prüfer q^∞-group by a cyclic group of order q. Then G has a faithful absolutely irreducible linear representation of degree q and of any characteristic except q. Set H = G. Then H is abelian-by-finite, locally nilpotent, metabelian, and locally finite, and yet $G/C_G(H)$ is infinite. For later reference note that H is not contained in Δ(G) (cf. 3.3.1). It is possible for $G/HC_G(H)$ to be infinite. For such to happen H can be neither nilpotent nor periodic (by Parts c) and d)). However H can be abelian-by-finite. Let A be any non-trivial torsion-free abelian group and let i be the inversion automorphism of A. Let G denote the split extension ⟨i⟩ [A and set H = $\langle i\rangle A^2$. Since $i^a = ia^2$ for all $a \in A$ it follows that H is normal in G. Also $C_G(A^2) = A$ and $C_A(i) = \langle 1\rangle$, so $C_G(H) = \langle 1\rangle$ and $G/HC_G(H) \cong A/A^2$. Finally G has a faithfully absolutely irreducible linear representation of degree 2 in every characteristic.

Proof of 5.1.10: a) H has a unique maximal soluble normal subgroup S, and by 5.1.9 the group H/S is locally finite. Now S contains a triangularizable normal subgroup A of G such that (S:A) is n-bounded (cf. Wehrfritz [2], Exercise 3.1), and since $u(A) \leq u(H) = \langle 1 \rangle$ the group A is abelian.

Since $u(G) \cap H = u(H) = \langle 1 \rangle$ and $u(G) \leq C_G(H)$ we may pass to G/u(G) and assume that G is absolutely completely reducible. Set $K = C_H(A)$. Now $(G:C_G(A))$ divides n!, for example by Wehrfritz [2], 1.12 and 1.8. Hence (H:K) also divides n!. By enlarging A we may assume that A is the centre of K. Then K/A acts faithfully by conjugation on $F[K] \leq F^{n\times n}$. If charF = 0 then (K:A) is n-bounded by the choice of A and Jordan's Theorem applied to K/A and its action on F[K]. Consequently 5.1.8 applied to $G/C_G(K)$ in its action on F[K] yields that $(G:KC_G(K/A))$ is n-bounded in all cases. Therefore so is $(G:K(C_G(A) \cap C_G(K/A)))$. By stability theory $C_G(A) \cap C_G(K/A)$ modulo $C_G(K)$ is isomorphic to a subgroup of Hom(K/A,A), and the torsion subgroup of A has rank at most n.

If charF = 0 then Hom(K/A,A) has order at most $(K:A)^n$ and it follows that $(G:KC_G(K))$ is n-bounded. Suppose charF > 0. By 5.1.5 and Part c) of the proof of 5.1.6 there exist characteristic subgroups $A \leq B \leq C$ of K with (B:A) and (K:C) n-bounded and C/B a direct product of perfect simple groups. Thus $|\mathrm{Hom}(K/A,A)| \leq (K:C)^n(B:A)^n$, and so in this case too $(G:KC_G(K))$ is n-bounded. Since (H:K) divides n! the group

$$(C_G(K) \cap C_G(H/K))/AC_G(H)$$

is abelian of exponent dividing n! by 5.1.2, and trivially $(G:C_G(H/K))$ is n-bounded. Part a) follows.

b) Here $(G:C_G(Z))$ divides n!, the index $(G:C_G(H/Z))$ divides m!, and $C_G(Z) \cap C_G(H/A)$ modulo $C_G(H)$ embeds into Hom(H/Z,Z). The latter has order dividing m^n since the torsion subgroup of Z has rank at most n.

c) Again by passing to G/u(G) we may assume that G, and hence H, is completely reducible. Thus if Z is the centre of H then (H:Z) is (n,c)-bounded by Wehrfritz [2], 3.13. Now apply Part b).

d) This is 5.1.6. □

5.2 AUTOMORPHISMS OF CERTAIN SEMISIMPLE ALGEBRAS

Throughout this section P is a prime field of characteristic $p \geq 0$ and $S = P[H]$ is a semisimple Artinian P-algebra, generated as such by the locally finite subgroup H of its group of units. Let G be the group of all automorphisms of S normalizing H, and let $\bar{H}$ be the natural image of H in G. Denote the maximal number of orthogonal idempotents in S by n; that is, n is the composition length of S as a left (or right) S-module.

We intend to apply the results of this section in the situation where H is a locally finite normal subgroup of some skew linear group J and $G = J/C_J(H)$. Recall that a group G is *super residually an* $\underline{X}$*-group* if for every finite subset X of G there is a homomorphism Φ of G *onto* an $\underline{X}$-group such that $|X| = |X\Phi|$.

5.2.1. (Wehrfritz [23]) *Suppose* $p \geq 0$. *Then there are normal subgroups* $\bar{H} \leq X \leq Y$ *of G such that*

(X:$\bar{H}$) *is* n*-bounded, and is* 1 *if* n = 1;

Y/X *is super-residually a finite* n*-generator abelian group; in particular* Y/X *is abelian, residually finite, and its torsion subgroup has rank at most* n;

and G/Y *is isomorphic to a subgroup of* Sym(n).

Further G/$\bar{H}$ *is residually finite and has an abelian normal subgroup with* n*-bounded index, whose torsion subgroup has rank at most* n.

Proof: $S = \oplus_{1 \leq i \leq r} S_i$ where S_i is a matrix ring of degree n_i over a locally finite field k_i, and $\Sigma\, n_i = n$. Clearly G permutes the S_i. Let $Y = \bigcap_i N_G(S_i)$. Then $r \leq n$ and G/Y is isomorphic to a subgroup of Sym(n).

By the Skolem-Noether Theorem (Cohn [2], Vol. 2, p. 364) we have $\mathrm{Aut}(S_i) = \mathrm{Inn}(S_i) \rtimes \mathrm{Aut}(k_i)$. Let U be the group of units of S, and set $\bar{U} = \mathrm{Inn}(S) \geq \bar{H}$. Put $X = G \cap \bar{U} \leq Y$. Since k_i is locally finite, $\mathrm{Aut}(k_i)$ is procyclic. Hence Y/X is a subdirect product of r super-residually cyclic groups. As such Y/X is super-residually a finite r-generator abelian group.

Clearly U is isomorphic to the direct product of the $GL(n_i,k_i)$. Let k be an algebraic closure of P. Then k contains a copy of each k_i and U is isomorphic to a subgroup of GL(n,k). Let V be the inverse image of X in U. Then H is a normal subgroup of V. If H_i is the natural projection of H in S_i then H_i is an irreducible linear group of characteristic p. Therefore each $O_p(H_i) = \langle 1\rangle$, and so $O_p(H) = \langle 1\rangle$. By 5.1.6 the group $V/HC_V(H)$ has n-bounded order. Hence so does $X/\bar{H}$. If $n = 1$ then S is a field and $X = \bar{H} = \langle 1\rangle$.

Let $C = C_Y(X/\bar{H})$ and $T = C \cap X$. Then (G:C) is n-bounded and $C/\bar{H}$ is nilpotent of class at most 2. There is a finite subgroup T_0 of U covering $T/\bar{H}$. Clearly $U \cong \times_{i=1}^{r} GL(n_i,k_i)$. If we identify U with the latter direct product then there are finite subfields F_i of the k_i such that $T_0 \leq T_1 = \times_{i=1}^{r} GL(n_i,F_i)$. Set $A = C_C(T_1)$. Since $C_Y(T_1) \leq \times_{i=1}^{r} Aut_{F_i}(k_i) \leq Aut(S)$ it follows that $A \cap T = \langle 1\rangle$. Thus $AT/\bar{H}$ is an abelian residually finite normal subgroup of $C/\bar{H}$ with C/AT finite abelian of rank at most r. In particular $G/\bar{H}$ is residually finite.

Let $C = \langle c_1,\dots,c_r\rangle AT$. Then $[c_i,A] \leq T$ and so $(A:C_A(c_i\bar{H}/\bar{H})) \leq (T:\bar{H})$. Set $Z = C_A(C/\bar{H})$. Then $(A:Z) \leq (T:\bar{H})^r$, and hence C/Z can be generated by an n-bounded number, s say, of generators. It follows that the central quotient of $C/\bar{H}$ is abelian, has exponent dividing $(T:\bar{H})$, and is generated by s elements. Thus the centre B of $C/\bar{H}$ is an abelian normal subgroup of $G/\bar{H}$ with n-bounded index. Possibly the rank of the torsion subgroup of B exceeds n. However the rank of the torsion subgroup of $B/(X \cap B)$ is at most n. Thus the following lemma completes the proof.

5.2.2. *Let* B *be an abelian group with a finite subgroup* T *such that the torsion subgroup of* B/T *has rank at most* n. *Then* B *has a subgroup* A *with* (B:A) *dividing* |T| *such that the torsion subgroup of* A *has rank at most* n.

Proof: By an elementary induction on |T| we may assume that $|T| = p$, a prime. Suppose first that B/T is a p-group. If $\text{rank}B \leq n$, let $A = B$. If not then $\text{rank}B = n + 1$ and $B = B_0 \oplus B_1 \oplus\dots\oplus B_n$ where each B_i is locally cyclic (Fuchs [1], Vol. 1, Corollary 27.4). Let $b = b_0 +\dots+ b_n$ generate T, where each $b_i \in B_i$. If $\langle b_i\rangle \neq B_i$ for each i then

$$n = \text{rank}(B/T) \geq \text{rank}(B/\langle b_0,\dots,b_n\rangle) = n + 1.$$

Thus $\langle b_i\rangle = B_i$ for some i. Set $A = \oplus_{j\neq i} B_j$. Then $\text{rank}A = n$ and $A \oplus T = B$. This completes the proof in this case.

In general let $P \geq T$ be the p-primary component of B. By the above there is a subgroup Q of P with rank at most n and (P:Q) dividing |T|. By Fuchs [1], Vol. 1, Theorem 27.5, the group B/Q splits over P/Q.

Thus there is a subgroup A of B with $A + P = B$ and $A \cap P = Q$. Clearly $(B:A)$ divides $|T|$. The maximal p'-subgroup of A is ismorphic to that of B/T, and so has rank at most n. Also Q is the p-primary component of A. Thus the torsion subgroup of A has rank at most n. □

5.2.3. (Wehrfritz [23]) *Suppose* $p = 0$ *and assume the centre* Z *of* H *has finite index* m *in* H. *Then there are normal subgroups* $\overline{H} \leq X \leq Y$ *of* G *such that*

$|X|$ *is* (m,n)-*bounded,*

Y/X *is abelian and residually finite,*

and G/Y *is ismorphic to a subgroup of* Sym(n).

Further G *and* $G/\overline{H}$ *are residually finite and have abelian normal subgroups with* (m,n)-*bounded index.*

Proof: We have $S = S_1 \oplus \ldots \oplus S_r$ where S_i is a matrix ring of degree n_i over some division $\mathbb{Q}$-algebra D_i and $\Sigma n_i = n$. Let F_i denote the centre of D_i and P_i the subfield of F_i generated by the projection of Z into S_i. By the Skolem-Noether Theorem (Cohn [2], Vol. 2, p. 364) $\mathrm{Inn}(S_i)$ is the centralizer of F_i in $\mathrm{Aut}(S_i)$. Thus restriction yields the following exact sequences:

$$\langle 1\rangle \longrightarrow \mathrm{Inn}(S_i) \longrightarrow \mathrm{Aut}(S_i) \longrightarrow \mathrm{Aut}(F_i),$$

$$\langle 1\rangle \longrightarrow \mathrm{Aut}_{P_i}(F_i) \longrightarrow \mathrm{Aut}(F_i) \longrightarrow \mathrm{Aut}(P_i).$$

Note that P_i is cyclotomic (and hence normal) and $\mathrm{Aut}(P_i)$ is abelian and residually finite. Further by Galois theory

$$|\mathrm{Aut}_{P_i}(F_i)| \leq (F_i:P_i) \leq (S_i:P_i) \leq (H:Z) = m.$$

As before G permutes the S_i and we set $Y = \bigcap_i N_G(S_i)$. Let $X = C_Y(Z)$. Then $X = C_G(\bigcup_i P_i)$ and Y/X is abelian and residually finite. Let U be the group of units of S and set $\overline{U} = \mathrm{Inn}(S)$. Then $\overline{H} \leq G \cap \overline{U} \leq X$ and $X/(G \cap \overline{U})$ has order at most $\Pi_i\ |\mathrm{Aut}_{P_i}(F_i)| \leq m^r$. Let F be an algebraic closure of $\mathbb{Q}$. Then F contains a copy of each F_i. Also S_i is isomorphic to a subalgebra of the matrix algebra over F_i of degree $(D_i:F_i)n_i \leq m.n_i$. Thus U is isomorphic to a subgroup of GL(mn,F). Let V be the inverse image of $G \cap \overline{U}$ in U. Then $H \leq V \leq N_U(H)$ and $(V:HC_V(H))$ is (m,n)-bounded by 5.1.6. Therefore $(G \cap \overline{U}:\overline{H})$ is (m,n)-bounded. Trivially $|\overline{H}| = m$. Thus $|X|$ is (m,n)-bounded.

Now Z has rank at most r since the image of Z in P_i is locally cyclic. Then Z has an m-divisible (i.e. every element is an m-th power) subgroup of finite index dividing a power of m. Let K be the kernel of the transfer homomorphism of H into the maximal m-divisible subgroup of Z. Then K is a finite characteristic subgroup of H with $H = KZ$ and $|K|$ involving only primes dividing m (cf. Kegel & Wehrfritz [1], 3.9). Put $t = |X|$ and $C = C_Y(X)$. Then $(G:C)$ is (m,n)-bounded, C is nilpotent of class at most 2, and $|C'|$ divides t. Thus C^t is central in C. Also

$$C_Y(K) \cap X = C_Y(K) \cap C_Y(Z) = C_Y(H) = \langle 1 \rangle.$$

Therefore $C_Y(K)$ is abelian. Set $A = C^t C_C(K)$. Then A is an abelian normal subgroup of G.

Let s be the number of prime divisors of m. If L_i is the projection of $L = K \cap Z$ into S_i then L_i is cyclic. The automorphism group of a cyclic group of prime power order is abelian of rank at most 2, and is cyclic if the prime is odd. Also $\bigcap_i C_Y(L_i) = C_Y(L)$. Hence $C/C_C(L)$ is finite abelian of rank at most $(s + 1)r \leq mr$. Trivially $|C/C_C(K/L)|$ divides $m!$. Finally $C_C(L) \cap C_C(K/L)$ modulo $C_C(K)$ is isomorphic to a subgroup of $\mathrm{Hom}(K/L,L)$, and the latter has order dividing m^r. Therefore $(C:A)$ divides $t^{mr}.m!.m^r$.

We have now shown that A is an abelian normal subgroup of G with (m,n)-bounded index. Trivially then $G/\bar{H}$ also has such a subgroup, namely the image of A. Let $Z_i = \{ a \in Z : a^i = 1 \}$. Then each Z_i is finite and normalized by G, and

$$C_G(K) \cap \bigcap_i C_G(Z_i) = C_G(K) \cap C_G(Z) = \langle 1 \rangle.$$

Consequently G is residually finite, and since $\bar{H}$ is finite, so is $G/\bar{H}$. □

5.2.4. (Wehrfritz [23]). *Suppose* $p = 0$. *Then there are normal subgroups* $W \leq X \leq Y$ *of* G *such that*

W is abelian, and its torsion subgroup has n-bounded rank,

$(X:W)$ is n-bounded and G/W is residually finite,

Y/X is abelian and residually finite,

and *G/Y is isomorphic to a subgroup of $\mathrm{Sym}(n)$.*

Further G has a metabelian normal subgroup containing W with n-bounded index.

Proof: As before $S = \bigoplus_{1 \leq i \leq r} S_i$ where each S_i is simple. Apply 2.5.14 to the projections of H in the S_i. Thus H contains a metabelian subgroup H_o normalized by G such that $m = (H:H_o)$ is n-bounded, and every primary component of H_o is abelian and (by 2.5.1) of rank at most n. Let A be the Fitting subgroup of H_o. Then $A = C_{H_o}(A)$. Set $N = C_H(A)$ and let Z denote the centre of N. Then $N \cap H_o = A \le Z$ and so $(N:Z) \le m$.

By 1.2.6 the projection of $\mathbb{Q}[N]$ into S_i is semisimple Artinian. Therefore $\mathbb{Q}[N]$ is semisimple Artinian. Let $W = C_G(N)$. Then 5.2.3 applied to $\mathbb{Q}[N]$ yields normal subgroups $W \le X \le Y$ of G such that X/W, G/W, Y/X, and G/Y are as in the statement of 5.2.4. Moreover G/W has an abelian normal subgroup with n-bounded index. It remains only to show that W is abelian with its torsion subgroup of n-bounded rank.

Now $[\bar{H},W] \le \bar{H} \cap W = C_{\bar{H}}(N)$, and $C_H(N) = Z$. Hence $[H,W] \le Z$. By stability theory W can be embedded into Der(H/N,Z). In particular W is abelian. Since $(H:H_o) = m$ there are elements $x_1,\dots,x_m$ of H/N that generate H/N modulo $M = H_oN/N$. Now H_o/A is abelian of rank at most n by 3.13 of Kegel & Wehrfritz [1]. Hence M is too. Now $\delta \mapsto (x_1\delta,\dots,x_m\delta,\delta|_M)$ is an embedding of Der(H/N,Z) into $Z^{(m)} \times \mathrm{Der}(M,Z)$. If the torsion subgroup of Der(M,Z) has n-bounded rank then so too does the torsion subgroup of W. Thus the following lemma completes the proof.

5.2.5. *Let* M *and* Z *be periodic abelian groups of rank at most* n *and suppose that* Z *is also an* M*-module. Then the torsion subgroup of* Der(M,Z) *has rank at most* $2n^2$.

If M and Z are elementary abelian of rank n and Z is M-trivial then Der(M,Z) is elementary abelian of rank n^2. If Z is cyclic of order 3, M is cyclic of order 6, and M acts on Z by inversion, then Der(M,Z) has exponent 3, order 9, and rank 2.

Proof: For each prime p let Z_p be the p-primary component of Z. Then Der(M,Z) is contained in the cartesian product of the $\mathrm{Der}(M,Z_p)$. Moreover the torsion subgroup of $\mathrm{Der}(M,Z_p)$ is a p-group. Thus the torsion subgroup of Der(M,Z) lies in the direct product of the torsion subgroups of the $\mathrm{Der}(M,Z_p)$. We may therefore assume that Z is a p-group.

Periodic groups of automorphisms of Z are finite (Kegel & Wehrfritz [1], 1.F.3). Set $C = C_M(Z)$. Then $M = \langle x_1,\dots,x_n\rangle C$ for some $x_i \in M$. Hence $\delta \mapsto (x_1\delta,\dots,x_n\delta, \delta|_C)$ is an embedding of Der(M,Z) into $Z^{(n)} \times \mathrm{Hom}(C,Z)$. The torsion subgroup of Hom(C,Z) has rank at most n^2. Thus Der(M,Z) has rank at most $2n^2$. □

5.2.6. **REMARKS.** In the above we can regard G as a subgroup of Aut(H), and then G is usually very much smaller than Aut(H). For example let H be the direct product of n Prüfer p^∞-groups and S the direct product of n isomorphic fields (of characteristic not equal to p). Then $Aut(H) \cong GL(n,\mathbb{Z}_p)$, while $G = Aut(S)$ corresponds to a monomial subgroup of $GL(n,\mathbb{Z}_p)$

In Section 5.4 below we give examples which show that even in the special case of skew linear groups there are no material improvements possible to the conclusions of 5.2.1, 5.2.3, and 5.2.4.

5.3 RECOGNIZING CROSSED PRODUCTS

This section contains a number of criteria for certain rings to be crossed products. Ultimately we wish to prove that certain sections of some absolutely irreducible skew linear groups are periodic. These criteria enable us to use the following simple result.

5.3.1. *Let* (R,S,G,H) *be a crossed product such that* R *is either left or right Artinian. Then* H *is periodic.*

Proof: Set $N = S \cap G$ and suppose that g is an element of G such that $gN \in H = G/N$ has infinite order. By 1.4.1 the element $1 + g$ is regular in $T = S[\langle g\rangle] \subseteq R$. Now R is free as say right T-module, so $1 + g$ is left regular in R. Since R is left (or right) Artinian it follows that $1 + g$ is a unit of R. Thus $1 + g$ has a left inverse, u say, in T. Then $u = u_r g^r + ... + u_s g^s$ for some $r \leq s$ and $u_r, ..., u_s \in S$ with $u_r \neq 0 \neq u_s$. Clearly $u(1 + g) = u_r g^r + ... + u_s g^{s+1} \neq 1$. This contradiction proves that H is periodic. (The final part of the argument is a very special case of the Higman Units Theorem.) □

We now come to the crossed product recognition lemmas. We prove only what we need.

5.3.2. *Let* R *be a ring with* C *a subring of* R *and* k *a central subring of* C. *Suppose* C *is* k*-torsion-free and* G *is a group of units of* R *normalizing* C *and* k *such that* $C \cap G = C_G(k)$ *and* $R = C[G]$. *Then* $(R, C, G, G/C_G(k))$ *is a crossed product.*

In our applications k will usually be a field, so then C is automatically k-torsion-free, and $C = C_R(k)$, so $C \cap G = C_G(k)$ too is a triviality. Also we will have $R = F[G]$ where F is central, so $F \subseteq C$ and the condition $R = C[G]$ follows.

Proof: Clearly $C \cap G = C_G(k)$ is normal in G. If R is not a crossed product of C and $G/C_G(k)$ as given, there is a relation $g_1c_1 +\dots+ g_rc_r = 0$ with the $c_i \in C\backslash\{0\}$, the g_i in distinct cosets of $C_G(k)$ in G, and $r \geq 1$ minimal. By pre-multipying by g_1^{-1} we may assume that $g_1 = 1$. Clearly $r > 1$ and $g_2 \notin C_G(k)$. Then $s^{g_2} \neq s$ for some $s \in k$. Since C is k-torsion-free $c_2(s^{g_2} - s) \neq 0$. Further k centralizes C, so

$$0 = s(\Sigma\, g_ic_i) - (\Sigma\, g_ic_i)s = \Sigma_{i>1}\, g_ic_i(s^{g_i} - s),$$

which contradicts the choice of r. Therefore R is a crossed product of C by $G/C_G(k)$. □

EXERCISE. If in 5.3.2 the ring C is simple, prove that R is simple. Is the converse true ?

5.3.3. *Let* R *be a ring with* k *a subfield of* R, H *a group of automorphisms of* R *normalizing* k, *and* G *a group of units of* R *normalizing* k. *Suppose* R *is generated as a ring by* k, G *and* $C_R(H \cup k) = C_R(H) \cap C_R(k)$. *Also assume that* $[C_G(k), H] \subseteq k$. *Then*

$$(R,\ C_R(k),\ G,\ G/C_G(k))$$

and

$$(k[C_G(k), C_R(H \cup k)],\ k[C_R(H \cup k)],\ C_G(k),\ C_G(k)/(C_G(k) \cap k^*C_{k^*G}(H \cup k)))$$

are crossed products. If also $C_R(H \cup k)$ *is normalized by* G *then*

$$(R,\ k[C_R(H \cup k)],\ G,\ G/(G \cap k^*C_{k^*G}(H \cup k)))$$

is a crossed product and $C_R(k) = k[C_G(k), C_R(H \cup k)]$.

Proof: By enlarging G and H we may assume that G contains k^* and that H contains the image of k^* in Aut(R). Note that $C_G(k)$ is normal in G and since $[C_G(k), H] \subseteq k$ we have that $C_G(k)$ normalizes $k^*C_G(H)$. Further, if G normalizes $C_R(H)$ then $G \cap k^*C_G(H)$ is normal in G. Thus the above crossed products make sense.

Let $C \subseteq C_R(H)$ be a basis of $k[C_R(H)]$ over k, and let T be a transversal of $k^*C_G(H)$ to $C_G(k)$. We prove first that $TC = \{tc : t \in T,\ c \in C\}$ is linearly independent over k. Suppose that $r = \Sigma_{i=1}^{m} t_ic_i\alpha_i = 0$, where

$t_i \in T$, $c_i \in C$, the $\alpha_i \in k^*$, the pairs (t_i,c_i) are distinct, and $m \geq 1$ is minimal. By multipliying by α_1^{-1} we may assume that $\alpha_1 = 1$. If the t_i are all equal then $\Sigma\, c_i\alpha_i = 0$, contradicting the choice of C. Thus we may assume that $t_1 \neq t_2$. Let $h \in H$. Then $t_1^{-1}t_i \in C_G(k)$ and $(t_1^{-1}t_i)^h = t_1^{-1}t_i\beta_i$ for some $\beta_i \in k^*$. Hence

$$0 = t_1(t_1^{-1}r)^h - r = \Sigma_{i\geq 1}\, t_ic_i(\beta_i\alpha_i^{\,h} - \alpha_i).$$

By the choice of m we have that $\beta_2\alpha_2^{\,h} = \alpha_2$. It follows that $t_1^{-1}t_2\alpha_2 \in C_G(h)$ and this is true for every $h \in H$. Thus $t_2 \in t_1k^*C_G(H)$. This contradiction of the choice of T completes the proof of the claim.

We are given that $R = k[G, C_R(H)]$. By 5.3.2 we have that

$$(R,\ C_R(k),\ G,\ G/C_G(k))$$

is a crossed product. The preceding paragraph shows that

$$(k[C_G(k),C_R(H)],\ k[C_R(H)],\ C_G(k),\ C_G(k)/k^*C_G(H))$$

is also a crossed product. Suppose now that G normalizes $C_R(H)$. Then

$$(k[G,C_R(H)],\ k[C_R(H)],\ G,\ G/k^*C_G(H))$$

is a crossed product and $k[G,C_R(H)] = R$. Clearly $C_R(k) \supseteq k[C_G(k),C_R(H)]$ and $R = \oplus\, tC_R(k) = \oplus\, tk[C_G(k),C_R(H)]$ where t ranges over a transversal of $C_G(k)$ to G. Thus $C_R(k) = k[C_G(k),C_R(H)]$. In view of the first paragraph of the proof, 5.3.3 follows. □

5.3.4. *Let* R *be a ring,* E *a division subring of* R *and* G *a group of units of* R *normalizing* E *such that* $E^* \leq G$ *and* $R = E[G]$. *Set* $C = C_R(E)$. *Then*

$$(R,\ E[C],\ G,\ G/E^*(C \cap G))$$

is a crossed product.

Proof: Let T be a transversal of $E^*(C \cap G)$ to G and let B be a basis of C over the centre F of E. Clearly $E[C] = \Sigma_{b\in B}\, bE$. If the conclusion of 5.3.4 is false we can choose a relation $\Sigma\, t_ib_ie_i = 0$ of minimal length r, where the (t_i,b_i) are distinct elements of $T \times B$ and the $e_i \in E^*$. By right multiplication by e_1^{-1} we may assume that $e_1 = 1$. Since t_1 is a unit and $b_1 \neq 0$ we have $r > 1$.

Let $x \in E^*$. Then

$$0 = t_1(t_1^{-1}\Sigma\, t_ib_ie_i)^x - \Sigma\, t_ib_ie_i = \Sigma_{i\geq 2}\, t_ib_i([t_1^{-1}t_i,x]e_i^{\,x} - e_i)$$

where we have used that $[G,E^*] \leq E^*$, which commutes with B. By the minimal choice of r we have $[t_1^{-1}t_i,\ x]e_i^x = e_i$ for all $x \in E^*$ and $i > 1$. Then $(t_1^{-1}t_ie_i)^x = t_1^{-1}t_ie_i$ for all such x and $t_1^{-1}t_ie_i \in C$ for all $i > 1$. Thus

$t_i \in t_1E^*C \cap G = t_1E^*(C \cap G)$ and so $t_i = t_1$ for all $i \geq 1$. But then $e_i^x = e_i$ for all $i \geq 1$ and $e_i \in C \cap E = F$ for all $i \geq 1$. However we also have $\Sigma\, b_ie_i = 0$ and this contradicts the choice of B. The proof is complete. □

We continue this section with a somewhat different type of condition for a ring to be a crossed product. It depends upon a mild generalization of Zalesskiĭ's theory of annihilator-free ideals, as expounded for example in Passman [2], pp. 336-344 and 516-523. We need a little preparation before we can give the crossed product criterion (5.3.10) itself.

Recall that if G is a group acting on a ring R, a *G-ideal* of R is an ideal of R normalized by G.

5.3.5. *Let* J *be a ring,* G *a group,* H *a group acting on* G *and* N *a subgroup of* G *for which there is an ascending series*

$$N = G_0 \triangleleft G_1 \triangleleft \dots \triangleleft G_\lambda \triangleleft G_{\lambda+1} \triangleleft \dots G_\rho = G$$

of H*-invariant subgroups of* G*. Suppose that:*

a) $(H:C_H(G_{\lambda+1}/G_\lambda)) < \infty$ *for all* $\lambda < \rho$*, and*

b) *for any* H*-ideals* $\underline{b} \subset \underline{a}$ *of* JN *there exists* $\alpha \in \underline{a}\backslash\underline{b}$ *with the index* $(H:C_H(\alpha + \underline{b}/\underline{b})$ *finite.*

Then for any H*-ideals* $\underline{b} \subset \underline{a}$ *of* JG *there exists* $\alpha \in \underline{a}\backslash\underline{b}$ *with the index* $(H:C_H(\alpha + \underline{b}/\underline{b}) < \infty$.

Proof: (After J. E. Roseblade and P. F. Smith). We induct on ρ. If $\rho = 0$ there is nothing to prove. If ρ is a limit ordinal then $G_\rho = \bigcup_{\tau<\rho} G_\tau$ and there exists $\tau < \rho$ with $\underline{b} \cap JG_\tau \subset \underline{a} \cap JG_\tau$. The result then follows by induction. Hence assume that $\rho = \tau + 1$.

Set $S = JG_\tau$. Since $JG = \Sigma_{g\in G}\, Sg$ there exists a finite set $g_1,\dots,g_n$ of elements of G with

$$\underline{b} \cap (Sg_1 +\dots+ Sg_n) \subset \underline{a} \cap (Sg_1 +\dots+ Sg_n).$$

Choose $g_1,\dots,g_n$ such that n is minimal. By multiplying through by g_1^{-1} we may then assume that $g_1 = 1$. Set $\underline{a}_1 = S \cap (\underline{a} + Sg_2 +\dots+ Sg_n)$ and $\underline{b}_1 = S \cap (\underline{b} + Sg_2 +\dots+Sg_n)$. Then $\underline{a}_1$ and $\underline{b}_1$ are ideals of S since $G_\tau \triangleleft G$, and $\underline{a}_1 \supset \underline{b}_1$ by construction.

Clearly $\underline{a}_1$ and $\underline{b}_1$ are $C_H(G/G_\tau)$-invariant, so by induction there exists $\alpha_1 \in \underline{a}_1\backslash\underline{b}_1$ and a subgroup $K \leq C_H(G/G_\tau)$ of finite index in H such that $\alpha_1^k - \alpha_1 \in \underline{b}_1$ for all $k \in K$. There exists $\sigma_2,\dots,\sigma_n \in S$ such that

$\alpha = \alpha_1 + \sigma_2 g_2 + \ldots + \sigma_n g_n$ lies in $\underline{a}$. Now if $\alpha \in \underline{b}$ then $\alpha_1 \in \underline{b}_1$ by the definition of $\underline{b}_1$, so in fact $\alpha \in \underline{a} \backslash \underline{b}$. Also for $k \in K$ we have

$$\begin{aligned} \alpha^k - \alpha &\in \underline{a} \cap (\alpha_1{}^k - \alpha_1 + Sg_2 + \ldots + Sg_n) && \text{(using } [G,K] \subseteq G_T \subseteq S) \\ &\subseteq \underline{a} \cap (\underline{b}_1 + Sg_2 + \ldots + Sg_n) && \text{(by the choice of } \alpha_1) \\ &= \underline{b}_1 + (\underline{a} \cap (Sg_2 + \ldots + Sg_n)) && \text{(by the modular law)} \\ &= \underline{b}_1 + (\underline{b} \cap (Sg_2 + \ldots + Sg_n)) && \text{(by the choice of } n) \\ &\subseteq \underline{b}. \end{aligned}$$

The proof is complete. □

Let J be a ring, G a group, H a subgroup of G and $\underline{a}$ an ideal of JG. Then $\underline{a}$ is ***left annihilator-free in*** JG ***over*** H if for every infinite subset X of H, whenever $\alpha \in JG$ satisfies $\alpha(x - 1) \in \underline{a}$ for all $x \in X$, then $\alpha \in \underline{a}$. Equivalently $\underline{a}$ is left annihilator-free in JG over H if for all X as above, the left annihilator in $JG/\underline{a}$ of the image of $X - 1$ is $\{0\}$. There is a corresponding notion of ***right annihilator-free in*** JG ***over*** H. If $H = G$ we omit the phrase "over H"; this is the case studied by Zalesskiĭ and Passman. Notice that in the definition the phrase "infinite **subset** X of H" can be replaced by "infinite ***subgroup*** X of H", since $\alpha(x^{-1} - 1) = -\alpha(x - 1)x^{-1}$ and $\alpha(xy - 1) = \alpha(x - 1)y + \alpha(y - 1)$, for $\alpha \in JG$ and $x, y \in H$.

5.3.6. *Let* J *be a ring,* G *a group,* N *a normal subgroup of* G, $\underline{b}$ *a* G*-ideal of* JN, *and* H *a group acting on* G *that normalizes* N *and* $\underline{b}$. *Suppose* α *is an element of* JG *such that* $(\alpha^H + \underline{b}G)/\underline{b}G$ *is finite (i.e.* $(H:C_H(\alpha \bmod \underline{b}G)) < \infty$).

a) *If* $M/N = \Delta_{G/N}(H)$ *then* $\alpha \in JM + \underline{b}G$.

b) *If* $\underline{b}$ *is left annihilator-free in* JN *over* $N \cap [G,H]$ *and if* $N \leq \Delta_G(H)$ *then* $\alpha \in J\Delta_G(H) + \underline{b}G$.

Proof: a) Let $K = C_H(\alpha \bmod \underline{b}G)$ and let

$$\alpha = \alpha_1 x_1 + \ldots + \alpha_r x_r + \beta_1 y_1 + \ldots + \beta_s y_s$$

where $x_1, \ldots, x_r, y_1, \ldots, y_s$ lie in distinct cosets of N in G, the $\alpha_i \in JN \backslash \underline{b}$ and the $\beta_i \in \underline{b}$. If $k \in K$ then $\alpha^k - \alpha \in \underline{b}G$, so

$$(\Sigma\, \alpha_i{}^k x_i{}^k) - \Sigma\, \alpha_i x_i \in \underline{b}G = (\oplus_i\, \underline{b}x_i) \oplus \underline{b}Y,$$

for some $Y \subseteq G$. But $\alpha_i \notin \underline{b}$, so $x_i \equiv x_j{}^k \bmod N$ for some j. It follows that each $x_i N \in \Delta_{G/N}(K) = \Delta_{G/N}(H) = M/N$. Therefore $\alpha \in JM + \underline{b}G$.

b) By the proof of a) and the hypothesis that $N \leq \Delta_G(H)$, there exists a subgroup L of K of finite index centralizing each x_i modulo N and centralizing each α_i. Hence for $t \in L$ we have

$$\Sigma\, \alpha_i(x_i^t\, x_i^{-1} - 1)x_i = (\Sigma\, \alpha_i x_i)^t - \Sigma\alpha_i x_i \in \underline{b}G.$$

Consequently $\underline{b}$ contains $\alpha_i(x_i^t\, x_i^{-1} - 1)$ for each i and every $t \in L$. Clearly $L_i = \{ x_i^t\, x_i^{-1} : t \in L \} \subseteq N \cap [G,H]$. Since the $\alpha_i \in JN\backslash\underline{b}$ the hypothesis on $\underline{b}$ yields that L_i is finite for each i. But then so is $x_i^L = L_i x_i$. Therefore each $x_i \in \Delta_G(L) = \Delta_G(H)$ and $\alpha \in J\Delta_G(H) + \underline{b}G$. □

5.3.7. *Let* J *be a ring,* G *a group,* H *a normal subgroup of* G *that is hyper* FC*-central,* N *a normal subgroup of* G *with* $N \leq \Delta_G(H)$, $\underline{a}$ *an ideal of* JG *and* $\underline{b}$ *a* G*-ideal of* JN *with* $\underline{b}G \subset \underline{a}$. *Define* M *by* $M/N = \Delta_{G/N}(H)$. *Then:*

a) $\underline{b}M \subset \underline{a} \cap JM$;

b) *if* $\underline{b}$ *is left annihilator-free in* JN *over* $N \cap [G,H]$ *then* $\underline{b}\Delta_G(H) \subset \underline{a} \cap J\Delta_G(H)$;

c) *if* $\underline{a} \cap JN$ *is left annihilator-free in* JN *over* $N \cap [G,H]$ *and* $N = \Delta_G(H)$ *then* $\underline{a} = (\underline{a} \cap JN)G$.

Proof: a) H has an ascending normal series $\{H_\lambda\}_{\lambda\leq\rho}$ say, with its factors generated by finite H-sets. Clearly then $(H:C_H(H_{\lambda+1}/H_\lambda))$ is finite for each $\lambda < \rho$. Also $H = C_H(G/H)$ trivially. Set $G_\lambda = H_\lambda N$ for $\lambda \leq \rho$, and $G_{\rho+1} = G$. Since $N \leq \Delta_G(H)$ we have that α^H is finite for every $\alpha \in JN$, and since $\underline{b}$ is a G-ideal, $\underline{b}G$ is an ideal of JG. Hence by 5.3.5 there is an $\alpha \in \underline{a}\backslash\underline{b}G$ with $(\alpha^H + \underline{b}G)/\underline{b}G$ finite. Then by 5.3.6 a) we have $\alpha \in \underline{a} \cap (JM + \underline{b}G) = (\underline{a} \cap JM) + \underline{b}G$, by the modular law. But $\alpha \notin \underline{b}G$. Therefore $\underline{a} \cap JM \supset \underline{b}G \cap JM \supseteq \underline{b}M$.

b) We repeat the proof of Part a). Since $\underline{b}$ is left annihilator-free over $N \cap [G,H]$ Part b) of 5.3.6 yields that $\alpha \in \underline{a} \cap (J\Delta_G(H) + \underline{b}G) = (\underline{a} \cap J\Delta_G(H)) + \underline{b}G$. Again $\alpha \notin \underline{b}G$ and therefore $\underline{a} \cap J\Delta_G(H) \supset \underline{b}\Delta_G(H)$.

c) Set $\underline{b} = \underline{a} \cap JN$. If $\underline{a} \neq (\underline{a} \cap JN)G$ then by Part b) we have $\underline{b} \subseteq \underline{b}\Delta_G(H) \subset \underline{a} \cap J\Delta_G(H) = \underline{a} \cap JN = \underline{b}$. This contradiction proves the result. □

5.3.8. *Let* J *be a ring and* K *and* N *normal subgroups of the group* G.

a) *If* $\underline{a}$ *is an ideal of* JG *which is left annihilator-free in* JG *over* K *then* $\underline{a} \cap JN$ *is a left annihilator-free* G*-ideal in* JN *over* $K \cap N$.

b) *If* $\underline{b}$ *is a* G*-ideal of* JN *which is left annihilator-free in* JN *over* $K \cap N$, *then* $\underline{b}G$ *is a left annihilator-free ideal in* JG *over* K.

Proof: a) Trivially $\underline{a} \cap JN$ is a G-ideal of JN. Let X be an infinite subset of $K \cap N$. If $\alpha \in JN$ is such that $\alpha(X - 1) \subseteq \underline{a} \cap JN$ then $\alpha \in \underline{a}$ by assumption. Part a) follows.

b) Certainly $\underline{b}G$ is an ideal of JG. Let X be a subgroup of K and let $\alpha \in JG\backslash\underline{b}G$ with $\alpha(X - 1) \subseteq \underline{b}G$. We have to prove that X is finite. Let $\alpha = \Sigma\, \alpha_i x_i + \Sigma\, \beta_j y_j$ where the x_i and y_j are in distinct cosets of N in G, the $\alpha_i \in JN\backslash\underline{b}$, and the $\beta_j \in \underline{b}$. Let $x \in X$. Then $\alpha(x - 1) \in \underline{b}G$, so

$$\Sigma\, \alpha_i x_i x - \Sigma\, \alpha_i x_i \in \underline{b}G. \qquad (*)$$

Since $Nx_1, \ldots, Nx_r$ are distinct cosets of N, for each i there is a unique j with $Nx_j x = Nx_i$. Let $Y = \{ x \in X : Nx_i x = Nx_i \text{ for each } i \}$. Then Y is a subgroup of X of finite index. Also for $y \in Y$, (*) yields

$$\Sigma\, \alpha_i(x_i y x_i^{-1} - 1)x_i \in \underline{b}G$$

and each $x_i y x_i^{-1} \in K^G \cap N = K \cap N$. Consequently each $x_i Y x_i^{-1}$ is finite by the hypothesis on $\underline{b}$. Therefore X is finite. □

5.3.9. THEOREM. *Let J be a ring, H a normal subgroup of the group G that is hyper FC-central, and set* $K = [G,H]$ *and* $N = \Delta_G(H)$. *Then the maps*

$$\underline{a} \longmapsto \underline{a} \cap JN \quad , \quad \underline{b} \longmapsto \underline{b}G$$

are inverse bijections between the sets of

ideals $\underline{a}$ of JG that are left annihilator-free in JG over K, and

G-ideals $\underline{b}$ of JN that are left annihilator-free in JN over $K \cap N$.

An ideal $\underline{a}$ of JG is left annihilator-free in JG over K if and only if $\underline{a} \cap JN$ is left annihilator-free in JN over $K \cap N$.

Proof: Always $\underline{a} \cap JN$ is a G-ideal, $\underline{b}G$ is an ideal, and $\underline{b}G \cap JN = \underline{b}$. If $\underline{a}$ is left annihilator-free in JG over K then $\underline{a} \cap JN$ is left annihilator-free in JN over $K \cap N$, by 5.3.8 a) and $\underline{a} = (\underline{a} \cap JN)G$ by 5.3.7 c). Conversely if $\underline{b}$ is left annihilator-free in JN over $K \cap N$ then $\underline{b}G$ is left annihilator-free in JG over K by 5.3.8 b). The final statement of the theorem follows from 5.3.8 and 5.3.7 c). □

5.3.10. COROLLARY. *Assume the notation of 5.3.9 and let $\underline{a}$ be an ideal of JG that is left annihilator-free over K. Set $\underline{b} = \underline{a} \cap JN$. Then*

$$(JG/\underline{a},\ JN/\underline{b},\ (G + \underline{a})/\underline{a},\ G/N)$$

is a crossed product.

Proof: For any G-ideal $\underline{b}$ of JN, $(JG/\underline{b}G,\ JN/\underline{b},\ G + \underline{b}G/\underline{b}G,\ G/N)$

is a crossed product, almost by definition. The corollary follows from 5.3.9. □

In order to make use of 5.3.10 we need to use Zalesskiĭ's subgroup. Our approach follows Passman [2]. Let G be a soluble group. Define E(G) inductively by $E(\langle 1\rangle) = \langle 1\rangle$ and $E(G) = E(G').\Delta_G(E(G'))$. Clearly E(G) is a characteristic subgroup of G. Then the ***Zalesskiĭ subgroup*** Z(G) of G is defined to be $\Delta_G(E(G))$.

5.3.11. *With the above notation,*

a) E(G) *is hyper* FC*-central,*

b) $\Delta_G(E(G)) \subseteq E(G)$,

c) $Z(G) = \Delta(E(G))$.

Proof: We induct on the derived length of G, there being nothing to prove if $G = \langle 1\rangle$. Since $E(G) \supseteq E(G')$, so $\Delta_G(E(G)) \subseteq \Delta_G(E(G')) \subseteq E(G)$. This proves b), and c) is a trivial consequence of b). It remains to prove a).

Let $F = E(G')$ and $D = \Delta_G(F)$, so $E = E(G) = FD$. Clearly $E/D \cong F/(F \cap D)$, which is hyper FC-central by induction. If D is hyper FC-central then since d^F is finite by definition of D for all $d \in D$, it will follow that D is hyper FC-central in E, proving that E is hyper FC-central. Now $D' \subseteq G' \cap \Delta_G(F) = \Delta_{G'}(F) \subseteq F$ by induction. Hence $D = \Delta_D(D')$, and the following completes the proof:

Let G *be a soluble group with* $G = \Delta_G(G')$. *Then* G *is hyper* FC*-central.*

If $G' = \langle 1\rangle$ the result is trivial, so assume $G' \neq \langle 1\rangle$. We induct on the derived length of G. By the usual transfinite induction argument it is sufficient to prove that $\Delta(G) \neq \langle 1\rangle$. Let A be the penultimate term of the derived series of G, and set $C = C_G(A)$. Then G/C has derived length less than G. Hence by induction there exists $g \in G\backslash C$ such that $K = C_G(gC/C)$ has finite index in G. Now $A \subseteq G'$ and $G = \Delta_G(G')$, so the set

$$S = \{ g^{-1}g^a : a \in A \}$$

is finite, and $S \neq \langle 1\rangle$ since $g \notin C$. If $k \in K$ then $g^k = gc$ for some $c \in C$, and hence

$$(g^{-1}g^{a})^{k} = [g^{k},a^{k}] = [gc,a^{k}] = [g,a^{k}] \in S.$$

Thus K normalizes the finite set S. Since also $(G:K) < \infty$ it follows that $S \subseteq \Delta(G)$. □

EXERCISE. Let G be a group with a normal subgroup S such that G/S is locally finite. Let E be a characteristic subgroup of S such that $\Delta_S(E) \subseteq E$. Set $Z = \Delta_G(E)$. Prove that Z' is locally finite and that G has an abelian normal subgroup A with $A \subseteq Z$ and Z/A locally finite.

If S is soluble then E could be E(S) for example. This exercise describes an actual situation where we will want to apply 5.3.10 and 5.3.11. (We shall not, however, assume that the reader has done this exercise !)

In applications of 5.3.10 with G soluble we can by 5.3.11 take H = E(G) and N = Z(G). If G is actually nilpotent we can do better since trivially then G and Δ(G) satisfy the properties given in 5.3.11 for E(G) and Z(G) respectively. The following will be useful.

5.3.12. *Let G be a finitely generated nilpotent group, Z the centre of G and T the torsion subgroup of G. Then $ZT \leq \Delta(G)$, $\Delta(G)/T$ is the centre of G/T and $(\Delta(G):Z)$ is finite.*

Proof: Δ(G/T) is just the centre S/T of G/T; for example this is a trivial consequence of the fact that G/T is isomorphic to a unipotent subgroup of $GL(n,\mathbb{Z})$ for some n (see Wehrfritz [2], p. 23, Point 3, and 5.5). Thus $Z \leq \Delta(G) \leq S$. Now T is finite, so $T \leq \Delta(G)$, and G is residually finite, so $H \cap T = \langle 1 \rangle$ for some normal subgroup H of G of finite index. Then $[G,H \cap S] \leq H \cap T = \langle 1 \rangle$ and $H \cap S \leq Z$. Therefore (S:Z) is finite. Also $S \leq \Delta(G)$. The lemma is proved. □

5.3.13. THEOREM. (Zalesskiĭ). *Let J be a ring, G a finitely generated nilpotent group and $\underline{a}$ a prime ideal of JG with $G \cap (1 + \underline{a}) = \langle 1 \rangle$. Set N = Δ(G) and $\underline{b} = \underline{a} \cap JN$. Then*

$$(JG/\underline{a},\ JN/\underline{b},\ (G + \underline{a})/\underline{a},\ G/N)$$

is a crossed product.

Proof: Let Z be the centre of G and suppose X is a subgroup of N and α is an element of $JN \setminus \underline{a}$ such that $\alpha(X - 1) \subseteq \underline{a}$. Now Z is central in JG and $\underline{a}$ is prime. Thus $JG/\underline{a}$ is torsion-free over the subring generated by the image of Z. Therefore $X \cap Z \subseteq G \cap (1 + \underline{a}) = \langle 1 \rangle$, and 5.3.12 yields that X is finite. This proves that $\underline{b}$ is annihilator-free in JN. Then $\underline{a} = \underline{b}G$ by 5.3.7 c) and the conclusion follows. □

5.3.14. COROLLARY. *Let $R = J[G]$ be a ring where J is a subring of R and G is a locally nilpotent subgroup of the group of units of R centralizing J such that for every finite subset X of R there is a finitely generated subgroup Y of G with $J[Y] \subseteq R$ a prime ring containing X. Let T be the torsion subgroup of G and Z/T the centre of G/T. Then*

$$(R, J[Z], G, G/Z)$$

is a crossed product.

Proof: Let $\Sigma_{i=1}^{r} t_i\alpha_i = 0$ where the t_i are distinct elements of a transversal of Z to G and the α_i are non-zero elements of J[Z]. If $i \neq j$ then $t_i^{-1}t_j \notin Z$ and so there is some $g_{ij} \in G$ with $t_{ij} = [t_i^{-1}t_j, g_{ij}]$ of infinite order. By hypothesis there exists a finitely generated subgroup Y of G such that J[Y] is a prime subring of R containing the t_i, t_i^{-1} and g_{ij} and with $J[Y \cap Z]$ containing the α_i.

Now J[Y] is a crossed product of $J[\Delta(Y)]$ by $Y/\Delta(Y)$ by 5.3.13. Also $Y \cap Z \leq \Delta(Y)$ by 5.3.12. Therefore $t_i^{-1}t_j \in \Delta(Y)$ for some $i \neq j$. Then again by 5.3.12, since $g_{ij} \in Y$, we have t_{ij} of finite order. This contradiction completes the proof. □

EXERCISE. Let $R = J[G]$ be a ring, where J is a subring of R and G is a locally soluble subgroup of the group of units of R centralizing J, such that for every infinite subgroup X of G the left annihilator of X - 1 in R is {0}. Let T be the maximal periodic normal subgroup of G and B/T the Hirsch-Plotkin radical of G/T. Then

$$(R, J[B], G, G/B)$$

is a crossed product. Hint: try and copy the proof of 5.3.14, using 5.3.10 (with N the Zalesskiǐ subgroup) in place of 5.3.13, cf. Wehrfritz [25], Point 7.

We have now proved all that we require of annihilator-free ideal theory. The reader who has not met these ideas before may well find the whole concept, and 5.3.9 in particular, rather mysterious. We conclude this section with some simple remarks, which although we do not use, may help to set the above into context.

5.3.15. *Let* J *be a ring and* G *a group. Then* {0} *is left and right annihilator-free in* JG.

Proof: Let $X \leq G$ and $\alpha \in JG$ be such that $\alpha(X - 1) = \{0\}$. Then $\alpha x = x$ for every $x \in X$, so X normalizes the support of α by right multiplication. Thus $|X| \leq |\text{supp } \alpha| < \infty$. This proves the left version. The right version is similar. □

5.3.16. *In the situation of* 5.3.9, *for every non-zero ideal* $\underline{a}$ *of* JG *we have* $\underline{a} \cap JN \neq \{0\}$.

Proof: Suppose $\underline{a} \cap JN = \{0\}$. Then by 5.3.15 and 5.3.9 we have $\underline{a} = (\underline{a} \cap JN)G = \{0\}$. □

5.3.17. *Assume the notation of* 5.3.9 *and let* $\underline{a}$ *be an ideal of* JG *that is left annihialtor-free in* JG *over* K.

a) $\underline{a}$ *is a prime ideal of* JG *if and only if* $\underline{a} \cap JN$ *is a* G*-prime ideal of* JN.

b) $\underline{a}$ *is a semiprime ideal of* JG *if and only if* $\underline{a} \cap JN$ *is a* G*-semiprime ideal of* JN.

A G-ideal P of JN is *G-prime* if whenever $\underline{a}$, $\underline{b}$ are G-ideals of JN with $\underline{ab} \subseteq P$ then $\underline{a} \subseteq P$ or $\underline{b} \subseteq P$, and is *G-semiprime* if whenever $\underline{a}$ is a G-ideal of JN with $\underline{a}^2 \subseteq P$ then $\underline{a} \subseteq P$.

Proof: a) Suppose $\underline{a}$ is prime, and let $\underline{b}$, $\underline{c}$ be G-ideals of JN with $\underline{bc} \subseteq \underline{a} \cap JN$. Then $(\underline{b}G)(\underline{c}G) \subseteq \underline{a}$, so $\underline{b} \subseteq \underline{a}$ or $\underline{c} \subseteq \underline{a}$. Conversely suppose that $\underline{b}$ is a left annihilator-free G-prime ideal of JN over $K \cap N$ and that $\underline{a}$, $\underline{c}$ are ideals of JG with $\underline{ac} \subseteq \underline{b}G \subseteq \underline{a} \cap \underline{c}$. Then $(\underline{a} \cap JN)(\underline{c} \cap JN) \subseteq \underline{b}$, so $\underline{a} \cap JN = \underline{b}$ or $\underline{c} \cap JN = \underline{b}$. If $\underline{a} \cap JN = \underline{b}$ then $\underline{a} = \underline{b}G$ by 5.3.9. Thus $\underline{b}G$ is prime.

b) The proof here is an obvious modification of that of Part a). □

5.4 LOCALLY FINITE NORMAL SUBGROUPS OF SKEW LINEAR GROUPS

Most of the work for this section has been done in Section 5.2. We begin with the positive characteristic case.

5.4.1. THEOREM (Wehrfritz [23]). ***Let G be a subgroup of GL(n,D), where D is any division ring of positive characteristic p and let H be a locally finite normal subgroup of G with $O_p(H) = \langle 1 \rangle$. Then $G/HC_G(H)$ is residually finite and has an abelian normal subgroup with n-bounded index and with its torsion subgroup of rank at most n. Further, there are normal subgroups $HC_G(H) \subseteq X \subseteq Y \subseteq G$ such that***

$X/HC_G(H)$ is of n-bounded order and is trivial if $n = 1$,

Y/X is abelian, residually finite, and its torsion subgroup has rank at most n,

and G/Y is isomorphic to a subgroup of Sym(n).

In fact $O_p(H) = s(H) = u(H)$ by 1.3.9.

Proof: Clearly $H \cap s(G) \subseteq O_p(H) = \langle 1 \rangle$, and therefore $C_G(H) = C_G(s(G)H/s(G))$. Hence we may pass to G/s(G) via 1.1.2 and assume that G is completely reducible. By Clifford's Theorem (1.1.7) H is completely reducible. Let P be the prime subfield of D. Then S = P[H] is semisimple Artinian by 1.1.14 c), and the number of primitive orthogonal idempotents of S is at most n by 1.1.9. The result now follows from 5.2.1. □

5.4.2. COROLLARY (Wehrfritz [19]). ***Let G be an absolutely irreducible subgroup of GL(n,D), where D is any division ring of positive characteristic, and let H be a locally finite normal subgroup of G. Then $G/C_G(H)$ is locally finite and $G/HC_G(H)$ has a reduced abelian normal subgroup of rank at most n and finite n-bounded index. Moreover there are normal subgroups $HC_G(H) \subseteq X \subseteq Y$ of G such that $X/HC_G(H)$ and G/Y are as in 5.4.1 and Y/X is a reduced periodic abelian group of rank at most n.***

An abelian group is *reduced* if it has no non-trivial divisible subgroups. A periodic abelian group of finite rank is reduced if and only if it is residually finite.

Proof: Here $O_p(H)$ is a stability subgroup by 1.3.11 c) and is completely reducible by Clifford's Theorem. Thus $O_p(H) = \langle 1 \rangle$ and 5.4.1 applies. If we can show that $G/C_G(H)$ is periodic then each factor $HC_G(H)/C_G(H)$, $X/HC_G(H)$, Y/X and G/Y is locally finite and hence $G/C_G(H)$ is locally finite. Then 5.4.2 follows from 5.4.1.

We prove that $G/C_G(H)$ is periodic by induction on n. Let P be the prime subfield of D and set $S = P[H]$. By 1.1.14 c) we have $S = S_1 \oplus \ldots \oplus S_r$ where the S_i are simple Artinian. Suppose $r > 1$ and set $Y = \bigcap_i N_G(S_i)$. Let F be the centre of D and set $T = F[Y]$. Then T is Artinian by 1.2.8 and semiprime by 1.2.5. Thus T is semisimple Artinian. The central idempotents of S are central in T, so T is not simple. Therefore $T = \oplus_i T_i$ where T_i are simple Artinian and of degree less than n. Let Y_i and H_i be the images of Y and H in T_i. By induction $Y_i/C_{Y_i}(H_i)$ is periodic. Thus $Y/C_Y(H_i)$ is periodic. But $\bigcap_i C_Y(H_i) = C_Y(H)$. Therefore $G/C_G(H)$ is also periodic.

Now assume that $r = 1$. Then S is a matrix ring, $k^{m \times m}$ say, for some locally finite field k. Let $R = F[G] = D^{n \times n}$ and $C = C_R(k)$. Then R is a crossed product of C by $G/C_G(k)$ by 5.3.2, and hence $G/C_G(k)$ is periodic by 5.3.1. But if $g \in C_G(k)$ then g induces an inner automorphism of S by the Skolem-Noether Theorem, and S is a locally finite ring. Consequently $C_G(k)/C_G(H)$ is locally finite, and the proof of the claim is complete. □

5.4.3. EXAMPLES. We give examples to show that there are no obvious improvements to 5.4.1, 5.4.2 and 5.2.1 (apart from possible reductions of the bounds).

Let C be any subgroup of a procyclic group, and let p be any prime. Then there exists a locally finite field E of characteristic p such that C is isomorphic to a group of automorphisms of E. Let $A_1 \rightarrowtail Q_1 \twoheadrightarrow C$ be a group extension with Q_1 free abelian. By 1.4.4 the skew group ring E^*Q_1 is an Ore domain. Let D be its division ring of quotients, set $A = E(A_1)^*$ and $Q = Q_1A$. Then A is an abelian normal subgroup of Q and $Q/A \cong C$. Also if P is the prime subfield of E then $D = P(Q)$, and if C is periodic then $D = P[Q]$.

Let m and r be positive integers and put $n = mr$. Let $S \cong \mathrm{Sym}(r)$ be the subgroup of $GL(n,P)$ of block $r \times r$ matrices whose r by r entries are zero or the $m \times m$ identity matrix and which have exactly one 1 in each row and column. Set

$$K = \mathrm{diag}(Q1_m,\ldots,Q1_m),$$
$$L = \mathrm{diag}(A1_m,\ldots,A1_m),$$
$$M = \mathrm{diag}(GL(m,E),\ldots,GL(m,E)),$$
$$N = \mathrm{diag}(SL(m,E),\ldots,SL(m,E)),$$
$$Z = \mathrm{diag}(E^*1_m,\ldots,E^*1_m),$$

as subgroups of $GL(n,D)$. Let $G = SKM \le GL(n,D)$. If C is periodic then G generates $D^{n\times n}$ as a ring and so in this case G is absolutely irreducible.

If $m \ge 1$ set $H = N$; otherwise set $H = M$. Then H is a locally finite normal subgroup of G with $O_p(H) = \langle 1\rangle$. Clearly $C_G(H) = LZ$. Let $X = LM$ and $Y = KM$. Then $HC_G(H) \le X \le Y$ are normal subgroups of G such that $X/HC_G(H)$ is a direct product of r copies of the cyclic group $E^*/(E^*)^m$ and Y/X is the direct product of r copies of C, so the torsion subgroup of Y/X has rank r or 0, and G/Y is isomorphic to $\mathrm{Sym}(r)$. If $\Phi(m)$ is the Euler function and $B_1 = XC_K(X/HC_G(H))$ then $B_1/HC_G(H)$ is an abelian normal subgroup of $G/HC_G(H)$ of finite index dividing $\Phi(m)^r.r!$. Also its torsion subgroup has rank at most $2r$. If $B_2 = C_K(X/HC_G(H))$ then $B_2 \subseteq B_1$, $(G:B_2)$ divides $m^r.\Phi(m)^r.r!$ and the torsion subgroup of $B_2/HC_G(H)$ has rank r if C is not torsion-free, and is 0 otherwise.

C could be a free abelian group of uncountable rank, or it could be the additive group of any proper subring of the rationals. For each prime p let K_p be a cyclic p-group (possibly trivial). Then C could be the direct product, over all p, of the K_p. In this case of course C is periodic.

The condition $O_p(H) = \langle 1\rangle$ cannot be removed from 5.4.1; indeed without it $G/HC_G(H)$ can be any group. For let C be any group and let E be the field of rational functions over a finite field, with the elements of C as the indeterminates. Then C acts as a group of automorphisms of E via right multiplication in the obvious way. Let Q be any free group mapping onto C, and form the skew group ring $E{}^{\bullet}Q$ via the induced action of Q. By 1.4.14 there is a division ring D containing $E{}^{\bullet}Q$. Let $H = Tr_1(2,E)$ and $G = \langle H, Q1_2\rangle$, regarded as subgroups of $GL(2,D)$. Then H is a periodic abelian normal subgroup of G and $G/C_G(H) = G/HC_G(H) \cong C$, which is arbitrary.

EXERCISE. If in the above C is soluble show that G can be chosen to be soluble. Hint: repeat the above proof with Q a free soluble group, and use 1.4.4, cf. the proof of 5.7.1 below.

However, as in the linear case, something can be salvaged if $O_p(H)$ in 5.4.1 is finite.

5.4.4. COROLLARY (Wehrfritz [23]). *Let* D *be a division ring of characteristic* $p > 0$, *let* G *be a subgroup of* GL(n,D), *and let* K *be a locally finite normal subgroup of* G *with* $O_p(K)$ *finite of order* p^e. *Then there are normal subgroups* $KC_G(K) \leq X \leq Y$ *of* G *such that*

$(X:KC_G(K))$ *is* (n,p^e)*-bounded, and is* 1 *if* $n = 1$,

Y/X *is abelian and residually finite, with its torsion subgroup of rank at most* n,

and G/Y *is isomorphic to a subgroup of* Sym(n).

Proof: Set $H = K/O_p(K)$. Clearly $O_p(K) = K \cap s(G)$. We can apply 5.4.1 to G/s(G) via 1.1.2. Hence there exist normal subgroups $KC_G(H) \leq X \leq Y$ of G with $(X:KC_G(H))$ n-bounded, and Y/X and G/Y as required.

Clearly the index $(C_G(H):C_G(H) \cap C_G(O_p(K)))$ divides $p^e!$. Also $C_G(H) \cap C_G(O_p(K))$ modulo $C_G(K)$ embeds into Der(H,Z), for Z the centre of $O_p(K)$. Now H is linear of degree n and characteristic p by 2.3.1. Thus $|Der(H,Z)|$ is (n,p^e)-bounded by 5.1.7 b), and the corollary follows. □

We now come to the more involved case of characteristic zero. Stronger conclusions hold in this case but it takes considerably more work to prove them. We therefore do not prove all that is known.

5.4.5. THEOREM (Wehrfritz [23]). *Let* G *be a subgroup of* GL(n,D), *where* D *is a division ring of characteristic zero, and let* H *be a locally finite normal subgroup of* G. *Then* $G/C_G(H)$ *has a metabelian normal subgroup with* n*-bounded index, and there are normal subgroups* $C_G(H) \leq W \leq X \leq Y$ *of* G *such that*

$W/C_G(H)$ *is abelian with its torsion subgroup of finite* n*-bounded rank,*

X/W *has* n*-bounded order, and* G/W *is residually finite,*

Y/X *is abelian and residually finite,*

and G/Y *is isomorphic to a subgroup of* Sym(n).

Proof: $H \cap s(G) = \langle 1 \rangle$ since stability subgroups of GL(n,D) are torsion-free (1.3.1). By 1.1.2 we may assume that G is completely reducible. Then $S = \mathbb{Q}[H]$ is semisimple Artinian by 1.1.14 c) and the result follows from 5.2.4 and 1.1.9. □

5.4.6. COROLLARY (Wehrfritz [19]). *Let* G *be an absolutely irreducible subgroup of* GL(n,D), *where* charD = 0, *and let* H *be a locally finite normal subgroup of* G. *Then* $G/C_G(H)$ *is locally finite and has a metabelian normal subgroup with* n*-bounded index. Moreover there exist normal subgroups* $W \leq X \leq Y$ *of* G *as in* 5.4.5.

Proof: In view of 5.4.5 we merely have to prove that $G/C_G(H)$ is periodic. This we do by induction on n.

By 2.5.13 there is a metabelian characteristic subgroup H_o of H such that every primary subgroup of H_o is abelian and $(H:H_o)$ is finite. Suppose $G/C_G(H_o)$ is periodic. Clearly $G/C_G(H/H_o)$ is finite and by stability theory $C_G(H_o) \cap C_G(H/H_o)$ modulo $C_G(H)$ embeds into the direct product of $(H:H_o)$ copies of H_o. As such it is periodic, and therefore $G/C_G(H)$ is periodic. We may therefore assume that $H = H_o$.

Let A be the Fitting subgroup of H, so $A = C_H(A) \supseteq H'$ by the above reduction. By 1.1.14 c) the subring $S = \mathbb{Q}[A]$ is semisimple Artinian. Let Y be the normalizer in G of the simple components of S, so (G:Y) divides n!, and let F be the centre of D. As in the proof of 5.4.2 the subalgebra $T = F[Y]$ is semisimple Artinian, and if T is not simple, induction on n yields that $Y/C_Y(H \cap Y)$ is periodic. Stability theory, as in the previous paragraph, implies that $Y/C_Y(H)$ is periodic, and the result follows in this case.

We are left with the possiblity that T is simple. The idempotents of S are central in T by the definition of Y. Hence S too is simple. Thus S is a field and $T = F[G]$. Now $[C_G(S),H] \leq C_G(S) \cap H = A \leq S$. Also $C_T(H) \subseteq C_T(S)$ and G normalizes H and $C_T(H)$. Hence the group $G/(G \cap S^*C_{S^*G}(H))$ is periodic by 5.3.3 and 5.3.1. Let $g = bx \in G$ where $b \in S^*$ and $x \in C_{S^*G}(H)$. Then $[H,b] = [H,g] \subseteq S \cap H = A$. Also S is algebraic over $\mathbb{Q}$, since A is periodic, and therefore b has only finitely many conjugates in S. Hence [H,b] is a finitely generated, and hence finite, subgroup of A. Stability theory now yields that some power, say b^m, of b centralizes H. But then $g^m = b^m x^m \in C_G(H)$ so $G/C_G(H)$ is periodic, as required. □

5.4.7. REMARK Suppose in 5.4.5 and 5.4.6 that the centre of H has finite index m in H. Then 5.2.3 shows that $G/C_G(H)$ and $G/HC_G(H)$ are residually finite with abelian normal subgroups with (m,n)-bounded index, and are moreover ((m,n)-boundedly finite)-by-abelian-by-(embeddable in Sym(n)).

5.4.8. EXAMPLES In 5.4.5 and 5.4.6 the group $G/C_G(H)$ is metabelian-by-finite. It need not be nilpotent-by-finite, nor even locally-nilpotent by finite, even if G = H. For by the proof of 2.5.10 there exists a division ring $D = \mathbb{Q}[H]$ where $H = H_1 \times H_2 \times \ldots$ with each H_i a non-trivial split extension of a cyclic p_i-group by a cyclic q_i-group, where the p_i and q_i for i = 1, 2,... are distinct primes. Then H = G is absolutely irreducible and the Hirsch-Plotkin radical of $G/C_G(H)$ is a $\{p_i : i = 1, 2,\ldots\}$-group and has infinite index.

We now show that $G/HC_G(H)$ need not be countable or metabelian, or even periodic if G is not absolutely irreducible. Let E be an infinite cyclotomic extension of $\mathbb{Q}$ and let C be a group of automorphisms of E. Note that for a suitable choice of E, C could be isomorphic to the torsion subgroup of the cartesian product of cyclic groups of order p-1, one for each prime p. This group is periodic but uncountable. C could also be a free abelian group of uncountable rank.

Let $A_1 \rightarrowtail Q_1 \twoheadrightarrow C$ be a group extension with Q_1 free abelian and let D be the division ring of quotients of $E^{\circ}Q_1$. Set $A = E(A_1)^*$ and $Q = Q_1A$. Then $Q/A \cong C$. Further $D = \mathbb{Q}[Q]$ if C is periodic. Let n be any positive integer and let S be the standard copy of Sym(n) in GL(n,D). Let Z be the subgroup of E^* of roots of unity and set

$$G = \langle \operatorname{diag}(g,1,\ldots,1),\ S : g \in QZ \rangle,$$

and

$$H = \langle \operatorname{diag}(h_1,\ldots,h_n) : h_1,\ldots,h_n \in Z \rangle.$$

Then H is periodic abelian normal subgroup of the subgroup G of GL(n,D), and $G/C_G(H) = G/HC_G(H)$ is isomorphic to the permutational wreath product $C \wr \mathrm{Sym}(n)$. If C is periodic then G is an absolutely irreducible subgroup of GL(n,D).

The positive characteristic results above are about $G/HC_G(H)$, while the characteristic-zero results are really about $G/C_G(H)$. If the characteristic is positive then H could be infinite and simple for example, so there is no possibility of proving in general that $G/C_G(H)$ has a soluble normal subgroup of finite index. Also by 2.3.1 the group H is isomorphic to a linear group of degree n and the same characteristic. Thus there seems to be little more to say in the positive-characteristic case.

Suppose now that G and H are as in 5.4.6. Then certainly $G/HC_G(H)$ is locally finite with a metabelian normal subgroup of n-bounded index. In the examples of 5.4.8 this group is always abelian by n-bounded. Whether the latter is always true we do not know, but certainly much more can be proved about this group. For a proof of the following we refer the reader to the original paper.

5.4.9. THEOREM (Wehrfritz [20]). *Let* G *and* H *be as in* 5.4.6. *Then* $G/HC_G(H)$ *is residually finite and abelian-by-finite, and* G *contains normal subgroups* $HC_G(H) \leq X \leq Y \leq G$ *such that* $X/HC_G(H)$ *and* G/Y *are finite with* n*-bounded order, and* Y/X *is abelian. In particular* $G/HC_G(H)$ *has a normal subgroup with finite* n*-bounded index that is nilpotent of class at most* 2*, and an abelian normal subgroup modulo which* $G/HC_G(H)$ *has finite* n*-bounded exponent.* □

None of these claims are true if we no longer assume that G is absolutely irreducible, as the following example shows.

5.4.10. (Wehrfritz [23]). *There exists a division ring* D *of characteristic zero, a subgroup* G *of* D^**, and a locally finite metabelian normal subgroup* H *of* G *such that* $G/HC_G(H)$ *is torsion-free, uncountable, and neither residually finite nor locally-nilpotent by periodic. Moreover* $G/C_G(H)$ *is metabelian.*

Proof: By the proof of 2.5.10 there is a division $\mathbb{Q}$-algebra D_0 generated by its multiplicative subgroup H, where H is the direct product $\times_{i=1}^{\infty} H_i$, $H_i = \langle a_i, b_i \rangle$, $|a_i| = p_i$, $|b_i|$ is a power of q_i, b_i normalizes $\langle a_i \rangle$ and acts on it as an automorphism of order q_i, and the p_i, q_i for $i = 1,2,\ldots$ are distinct primes. Of course H is locally finite and metabelian.

Let $\bar{H}$ be the cartesian product of the H_i. Then H is a normal subgroup of $\bar{H}$ and the action of $\bar{H}$ on H extends to one of $\bar{H}$ on D_0. For if $X \subseteq D_0$ is finite and $k \in \bar{H}$ then there exists an $h \in H$ such that $k|_X$ is conjugation by h. Pick a surjection of a free metabelian group Q onto $\bar{H}$. Then Q acts on D_0 and the skew group ring $D_0^{*}Q$ has a division ring D of quotients by 1.4.3. Let $G = QH \leq D^*$. Then H is a normal subgroup of G.

The centralizer of H in the split extension of H by $\bar{H}$ is the set of all pairs (k,h) with $k \in \bar{H}$, $h \in H$, such that $kh \in \bar{H}$ lies in the cartesian product of all the $\langle b_i^{q_i}\rangle$. Let $c_i = b_i\langle b_i^{q_i}\rangle$ and let A and $\bar{A}$ (respectively C and $\bar{C}$) be the direct and cartesian products of the $\langle a_i\rangle$ (resp. the $\langle c_i\rangle$). Then $\bar{C}$ acts on $\bar{A}$ componentwise, and $G/C_G(H) \cong \bar{C} \wr \bar{A}$. Hence $G/HC_G(H) \cong (\bar{C}/C) \wr (\bar{A}/A)$. Clearly $G/C_G(H)$ is metabelian.

Now $\bar{A}/A$ and $\bar{C}/C$ are uncountable, torsion-free, and divisible. Thus $G/HC_G(H)$ is uncountable, torsion-free, and not residually finite. Let $a = (a_i) \in \bar{A}$ and $c = (c_i) \in \bar{C}$. If $G/HC_G(H)$ were locally-nilpotent by periodic then $[a^e,_m c^e] \in A$ for some positive integers e and m. This implies that H_i is nilpotent for all large i, which is false. Therefore $G/HC_G(H)$ is not locally-nilpotent by periodic.

5.5 LOCALLY FINITE-DIMENSIONAL DIVISION ALGEBRAS

One part of the general soluble case reduces to the study of normal subgroups of skew linear groups over locally finite-dimensional division algebras. Working over the latter one can derive stronger conclusions from weaker hypotheses. We therefore treat this special case separately.

5.5.1. (Wehrfritz [21]). *Let* D *be a locally finite-dimensional division algebra,* G *a subgroup of* GL(n,D) *and* H *a normal* $\acute{P}(\underline{S} \cup L_1\underline{F})$*-subgroup of* G *with* $u(H) = \langle 1\rangle$.

a) H *contains an abelian normal subgroup* A *of* G *such that* H/A *and* $G/AC_G(H)$ *are locally finite. In particular* $G/C_G(H)$ *is abelian by locally-finite.*
b) *If* H *modulo its centre* Z *is periodic then* $G/C_G(H)$ *is locally finite.*
c) *If* H *is locally nilpotent then* H *is centre by locally-finite.*
d) *If* H *is a Baer group (in particular if* H *is a Fitting group) then* $H \le \Delta(G)$.
e) *If* H *is periodic then* $G/C_G(H)$ *is locally finite and, if* charD $= 0$, *is metabelian by* n*-bounded.*

In the proof of e) below we prove substantially more than stated above, but the conclusions are difficult to state concisely, especially if charD $\neq 0$. Not suprisingly, unlike the rest of this chapter, the proof of 5.5.1 depends upon the results of Chapter 3. We also need the following.

5.5.2. *Let* G *be a subgroup of* GL(n,R), *where* R *is a finitely generated integral domain, such that* u(G) = ⟨1⟩, *and suppose that* G *has an abelian normal subgroup* A *with* G/A *periodic. Then* G *is abelian by finite. If* A *is central in* G *then the centre* Z *of* G *has finite index in* G.

Proof: To begin with G/A is locally finite by Wehrfritz [2], 4.9 (and 6.4, 5.9 and 5.11). Let $C = C_G(A)$. Then C' is locally finite by Schur's Theorem (e.g. Wehrfritz [2], p.213, Lemma). Also G has a normal subgroup N of finite index whose only torsion elements are unipotent (ibid. 4.8). Let $B = C \cap N$. Then $[B,C] \leq C' \cap N \leq u(G) = \langle 1 \rangle$, and so B is abelian. It follows from Wehrfritz [2], 1.12 that (G:C) divides n!. Therefore (G:B) is finite. Finally if $A \leq Z$ then $C = G$ and $[B,G] = \langle 1 \rangle$; that is, $B \leq Z$.□

Proof of 5.5.1: a) If K is any $\acute{P}(\overline{\underline{S}} \cup L_1\underline{F})$-subgroup of GL(n,D) then by the linear case (5.1.9) and an elementary transfinite induction the group K is locally soluble-by-finite and hence K^+ is locally soluble by 3.3.9. Thus $A = H^+$ is abelian by 3.1.5 and H/A is locally finite.

Let $a \in A$, $h \in H$ and $g \in G^+$. There is a finitely generated subgroup Y of GL(n,D) such that $a, h \in X = G \cap Y$ and $g \in X^o$. Now $u(A \cap X) \leq u(A) = \langle 1 \rangle$. Hence by 5.1.10 b) we have $[A \cap X, X^o] = \langle 1 \rangle$. Thus $[a,g] = 1$ and so $[A,G^+] = \langle 1 \rangle$. Also $(H \cap X)^o \leq A \cap X$ by the definition of A and consequently $A \cap X$ is closed in $H \cap X$. Therefore $C_X(H \cap X/A \cap X)$ is closed of finite index in X and $[H \cap X, X^o] \leq A$. Consequently $[h,g] \in A$ and we have shown that $[H,G^+] \leq A$.

Again let $g \in G^+$. Then $[H,g] \leq A$ and $A\langle g \rangle$ is abelian and normal in $H\langle g \rangle$. Let $U = u(A\langle g \rangle)$. By 1.1.2 and 3.3.1 we have that $gU \in \Delta(H\langle g \rangle/U)$, and so $[H,g]U/U$ is finitely generated. But $H \cap U = \langle 1 \rangle$ and therefore $A_g = [H,g]$ is a finitely generated subgroup of A, and it is normal in H. Now $H^1(H/A, A_g)$ is periodic by 5.1.3. Also $hA \longmapsto [h,g]$ is a derivation of H/A into A_g. Therefore for some positive integer r, its r-th power is inner. Thus for some $a \in A_g$ we have that $[h,g^r] = [h,g]^r = [h,a]$ for every $h \in H$. Hence $g^r \in aC_G(H)$ and so $G^+/AC_{G^+}(H)$ is periodic. Since G^+ stabilizes the series $\langle 1 \rangle \leq A \leq H$, the group $G^+/C_{G^+}(H)$ is abelian. It follows that $G/AC_{G^+}(H)$ is locally finite.

b) We prove that $A \leq Z$, where $A = H^+$ is as above. It will then follow from Part a) that $G/C_G(H)$ is locally finite. Let $a \in A$ and $h \in H$. There is a finitely generated subgroup Y of GL(n,D) with $a \in (H \cap Y)^o$ and

$h \in Y$. Now $(H \cap Y)/u(H \cap Y)$ is centre-by-finite by 5.5.2 and so $[a,h]$ is unipotent. Therefore $[A,H] \leq u(H) = \langle 1 \rangle$. Consequently $A \leq Z$, as claimed.

c) By 3.2.3 we have $[A,H] \leq u(H) = \langle 1 \rangle$. Part c) follows.

d) If p is a prime and Z is the centre of H let H_p/Z be the p-primary component of the periodic locally nilpotent group H/Z. Clearly H_p is normal in G and $u(H_p) = \langle 1 \rangle$. Then H_p is abelian by 3.2.9 if p = char F; otherwise H_p is nilpotent-by-finite by 3.2.10 and a Fitting group by 3.5.3 b). Thus H_p is nilpotent in all cases. But then $H_p \leq \Delta(G)$ by 3.3.1 for all p, and hence $H \leq \Delta(G)$, as claimed.

e) We may pass to G/u(G) via 1.1.2 and assume that G is completely reducible as a subgroup of GL(n,D). Let F be the centre of D. Then R = F[G] is semisimple Artinian by 1.1.12. Let $R = \oplus_i R_i$ where the R_i are simple and let $\pi_i : R \twoheadrightarrow R_i$ be the natural projection. Then $G\pi_i$ is absolutely irreducible and the structure of $G/C_G(H\pi_i)$ is given by 5.4.2 and 5.4.6. Also $\bigcap_i C_G(H\pi_i) = C_G(H)$. Consequently $G/C_G(H)$ is locally finite. Also if charF = 0 then $G/C_G(H)$ is metabelian by n-bounded and has an abelian normal subgroup of n-bounded rank modulo which it is residually finite and abelian by n-bounded. If charF > 0 then $G/H_iC_G(H\pi_i)$ is abelian by n-bounded for each i, where $H_i = \{ g \in G : g\pi_i \in H\pi_i\}$. □

Note that in view of 1.2.4 and 1.1.2, the difference in 5.5.1 between the hypothesis "$u(H) = \langle 1 \rangle$", and the hypothesis "G is absolutely irreducible" is marginal. This certainly was not the case in Section 5.4, and is very far from being the case in Sections 5.6 and 5.7 below.

We have already remarked (see the comments after the statement of 5.1.10 that in 5.5.1 we need not have $H \leq \Delta(G)$, even if H is abelian-by-finite, locally nilpotent, and metabelian. The examples of 3.2.8 d) are metabelian groups H as in 5.5.1, with n = 1 and H^m non-abelian for every positive integer m. A modification of that construction gives rise to a group H of the following form: H is a direct product of groups H_p, where p ranges over an infinite set of primes and H_p is the split extension of a cyclic group of order p by an infinite cycle acting as an automorphism of order p-1. Clearly H is metabelian but neither abelian by finite-exponent nor locally-nilpotent by finite. Further, it is possible to have H abelian in 5.5.1 and yet $G/HC_G(H)$ not of finite exponent; simply let H be the torsion subgroup of the previous example. Alternatively this also follows from

1.4.12, since if H is a maximal abelian normal subgroup of a locally poly-C_∞ group G such that G/H is locally finite then $H = HC_G(H)$. In particular the structure of $G/HC_G(H)$ can be very complicated indeed.

As in the linear case, if u(H) is infinite then nothing comparable to 5.5.1 is possible, while if u(H) is finite then stability theory enables much to be salvaged; cf. 5.1.8 and 5.4.4.

5.6 SOLUBLE NORMAL SUBGROUPS OF SKEW LINEAR GROUPS

When we considered locally finite normal subgroups of skew linear groups in Section 5.4, although our strongest conclusions were reserved for absolutely irreducible groups, nevertheless we were able to draw quite strong conclusions in general. The situation with soluble normal subgroups is quite different. In the absolutely irreducible case we obtain results analogous to those of Section 5.4, while in general the following example shows that nothing positive can be said, even for abelian normal subgroups of irreducible groups.

5.6.1. *Let* p *be zero or a prime and let* C *be any group. Then there exists a division ring* D *of characteristic* p, *a subgroup* G *of* D^* *and an abelian normal subgroup* H *of* G *such that* $G/C_G(H) = G/HC_G(H)$ *is isomorphic to* C. *If* C *is soluble then* G *can be chosen to be soluble.*

Proof: Let F be any field and let $E = F(x_c : c \in C)$ be the field of rational functions in indeterminates x_c in one-to-one correspondence with the elements of C. The group C permutes the x_c by right multiplication on the suffixes, and hence acts faithfully on E. If C is soluble pick a surjection of a free soluble group Q onto C. Otherwise pick a surjection of a free group Q onto C. Form the skew group ring $E^{\cdot}Q$, where Q acts via the projection of Q onto C. By either 1.4.4 or 1.4.14 we can embed $E^{\cdot}Q$ into a division ring, D say. Let $H = \langle x_c : c \in C\rangle$ and $G = QH \le D^*$. Then H is a free abelian normal subgroup of G and $G/C_G(H)$ is isomorphic to C. (In fact $G/C_Q(H) \cong C_\infty \wr C$). □

We have already seen in Section 1.4 that soluble irreducible skew linear groups need not be abelian-by-periodic. For example every poly torsion-free abelian group is isomorphic to an irreducible skew linear group of degree 1 by 1.4.4, as is every torsion-free locally nilpotent group by the same result. Further examples arise from 1.4.14 and 1.4.23.

We can now concentrate on the absolutely irreducible case. For the remainder of this section D denotes a division ring with centre F and characteristic $p \geq 0$.

5.6.2. (Wehrfritz [21]). *Let A be an abelian normal subgroup of the absolutely irreducible subgroup G of GL(n,D). Then $G/C_G(A)$ is periodic.*

Whether $G/C_G(A)$ in 5.6.2 is actually locally finite we do not know (and this same comment applies to other results below). It can certainly be any locally finite group, as is shown by Snider's example, see the Remark after 1.4.10. Before we prove 5.6.2 we point out the following.

5.6.3. a) *Let R = S[X] be a ring where X is a subgroup of the group of units of R normalizing the subring S. If C is a right divisor subset of S normalized by X then C is a right divisor subset of R.*

b) *If C is a right Ore subset of the ring R then $T = \bigcup_{c \in C} \ell_R(c)$ is an ideal of R.*

Proof: a) Let $a \in R$ and $c \in C$. Then $a = \Sigma\, x a_x$ where the $x \in X$, the $a_x \in S$, and the sum is finite. Now each $c^x = x^{-1}cx \in C$ by hypothesis, so $a_x d_x = c^x b_x$ for some $b_x \in S$ and $d_x \in C$, since C is right Ore in S. The sum is finite, so we can choose the d_x all equal, to d say. Then

$$ad = \Sigma\, x a_x d = \Sigma\, x c^x b_x = c(\Sigma x b_x).$$

Consequently C is right Ore in R.

b) Trivially T is a left ideal of R. Let $a \in R$ and $x \in \ell_R(c)$ for some $c \in C$. Then $ad = cb$ for some $b \in R$ and $d \in C$. Hence $(xa)d = xcb = 0$ and thus $xa \in \ell_R(d) \subseteq T$. Therefore T is also a right ideal of R. □

Proof of 5.6.2: Choose, if possible, a counter-example (D,G,A) with n minimal. Set $J = F[A] \leq D^{n \times n}$. By 1.2.5 the ring J is semiprime and certainly J has the maximal condition on annihilators. Thus J has only a finite number r of minimal prime ideals $\mathfrak{P}_i$ and $\bigcap \mathfrak{P}_i = \{0\}$, see Chatters and Hajarnavis [1], 1.16. Also each $\mathfrak{P}_i$ is an annihilator ideal of J. Clearly G permutes the $\mathfrak{P}_i$; let $Y = \bigcap N_G(\mathfrak{P}_i)$. Then $(G{:}Y)$ is finite and $T = F[Y]$ is Artinian by 1.2.8 and semiprime by 1.2.5. Thus $T = T_1 \oplus \ldots \oplus T_s$ where each T_i is simple Artinian, of degree n_i say. Since $D^{n \times n}$ can contain at most n pariwise orthogonal idempotents $\Sigma n_i \le n$. Let $\pi_i : T \twoheadrightarrow T_i$ be the natural projection.

Suppose $s > 1$. Then each $n_i < n$ and by the choice of n each $Y/C_Y(A\pi_i)$ is periodic. But $C_Y(A) = \bigcap C_Y(A\pi_i)$. Therefore $Y/C_Y(A)$ and hence $G/C_G(A)$ is periodic. This contradiction shows that $s = 1$ and that T is simple. Now $\mathfrak{P}_iY$ is an ideal of T with $\ell_J(\mathfrak{P}_i).\mathfrak{P}_iY = \{0\}$ and $\mathfrak{P}_i = r_J\ell_J(\mathfrak{P}_i)$. Hence $r = 1$, J is a domain and $Y = G$. Then $C = J\backslash\{0\}$ is right Ore in J and hence in T, and $\bigcup_{c \in C} \ell_T(c)$ is an ideal of T, see 5.6.3. As T is simple this ideal is $\{0\}$; that is, T is J-torsion-free as right J-module. By 5.3.2 the ring T is a crossed product of $C_T(A)$ by $G/C_G(A)$ and consequently $G/C_G(A)$ is periodic by 5.3.1. This contradiction proves the result. □

We now consider soluble normal subgroups. In fact we can combine this with the results of Section 5.4 and consider $P(\underline{A} \cup L\underline{F})$ normal subgroups. First we prove some preliminary lemmas.

5.6.4. *Let G be a $P(\underline{A} \cup L\underline{F})$-group. Then G has a characteristic series of finite length whose factors are abelian or locally finite.*

Proof: Certainly G has a characteristic series of finite length whose factors are either torsion-free locally nilpotent or locally finite. Hence we may assume that G is torsion-free and locally nilpotent.

G has a series $\langle 1 \rangle = G_0 \le G_1 \le \ldots \le G_{2r-1} \le G_{2r} = G$ where the G_{2i+1}/G_{2i} are abelian and the G_{2i}/G_{2i-1} are locally finite. We prove by induction on r that G is soluble of derived length at most r. The result will then follow.

By induction applied to G_{2r-2} the subgroup $H = G_{2r-1}$ is soluble of derived length at most r. Let X be any finitely generated subgroup of G and regard X as a unipotent linear group over $\mathbb{C}$. The Zariski closure Y of $H \cap X$ in X is soluble of derived length at most r and X/Y is finite and isomorphic to a unipotent linear group over $\mathbb{C}$, by Wehrfritz [2], 5.11, 5.9, and 6.6. Therefore $X = Y$ and G is soluble of derived length at most r, as claimed. □

5.6.5. (Wehrfritz [18]). *Let* N *be a normal subgroup of the absolutely irreducible subgroup* G *of* GL(n,D). *Assume that*

a) $F[A] \leq D^{n\times n}$ *is a prime ring for every abelian normal subgroup* A *of* G *contained in* N,

b) $N \in L(\underline{S} \cap \underline{F}).\ \underline{F}\ \underline{F}^{-S}$ *and there is an abelian normal subgroup* $Z \subseteq N$ *of* G *with* N/Z *periodic.*

Then the kernel of the natural map of FG *onto* F[G] *is (left and right) annihilator-free in* FG *over* N.

Proof: Suppose first that A is an abelian normal subgroup of G contained in N. We claim that $F[A]^* = F[A]\backslash\{0\}$ is contained in $C = \underline{C}_{F[G]}(0)$. For F[A] is prime by hypothesis and so is a domain. Consequently $F[A]^*$ is Ore in F[A] and hence in F[G] by 5.6.3 a). Therefore $\bigcup_{x\in F[A]^*} \ell_{F[G]}(x)$ is an ideal of the simple ring F[G] by 5.6.3 b), and hence is zero. But in the Artinian ring $D^{n\times n}$ left regular elements are regular. (Alternatively interchange left and right.) Thus $F[A]^* \subseteq C$ as claimed.

Let X be any subgroup of N with $\ell_{F[G]}(X-1) \neq \{0\}$. Then $(X-1) \cap C = \emptyset$. We are given an abelian normal subgroup $Z \subseteq N$ of G with N/Z periodic. If $x \in X$ then $x^m \in Z$ for some positive integer m, and $x^m - 1 \in (X-1) \cap F[Z] \subseteq (X-1) \cap (C \cup \{0\}) = \{0\}$. Consequently X is periodic. But N contains a locally finite characteristic subgroup T with N/T torsion-free. Therefore $X \subseteq T$.

Suppose charF = 0. By 2.5.14 there is a characteristic metabelian subgroup M of T of finite index such that the Hirsch-Plotkin radical B of M is abelian. Since $(X-1) \cap C = \emptyset$ and $F[B]^* \subseteq C$ we have $X \cap B = \langle 1\rangle$. Also $B = C_M(B)$, so $C_{X\cap M}(B) \leq X \cap B = \langle 1\rangle$. Since B is periodic E = F[B] is a field, being by assumption prime. Thus if Y is any finite subgroup of $X \cap M$ then E[Y] is a classical crossed product (with trivial factor set) and is therefore a matrix algebra of degree |Y| over $C_E(Y)$ (Cohn [2], Vol. 2, p. 377, but this follows easily from 5.3.2 and the exercise following it). But $E[Y] \subseteq D^{n\times n}$, so $|Y| \leq n$, and $X \cap M$ is finite. Therefore in this case X is finite.

Now assume that charF = p > 0. Then $u(T) = \langle 1\rangle$ since G is irreducible. By 2.3.1 the group T is isomorphic to a linear group of characteristic p. Also $T \in (L\underline{S})\underline{F}$. Hence T has an abelian characteristic subgroup B of finite index. Since $F[B]^* \subseteq C$ we have that $X \cap B = \langle 1\rangle$. Consequently in this case too X is finite. The proof is complete. □

5.6.6. (Wehrfritz [18]). *Let* G *be an absolutely irreducible subgroup of* GL(n,D) *and let* S *be a normal soluble subgroup of* G. *Set* $M = \Delta_G(E(S))$, *where* E(S) *is as in* 5.3.11, *and* $N = M \cap S$. *Assume that* G *and* N *satisfy Condition* a) *of* 5.6.5. *Then* G/M *is periodic.*

Proof: By 5.3.11 we have that N is an FC-group. In particular N' is locally finite-soluble and the centre Z of N is an abelian normal subgroup of G with N/Z locally finite. By 5.6.5 the kernel $\underline{a}$ of the natural map of FG onto F[G] is annihilator-free in FG over N. Hence $\underline{a} \cap FM$ is a G-ideal of FM, which is annihilator-free over N by 5.3.8. Therefore F[G] is a crossed product of F[M] by G/M by 5.3.10 (and 5.3.9), and consequently G/M is periodic by 5.3.1. □

5.6.7. THEOREM (Wehrfritz [21]). *Let* H *be a normal* $P(\underline{A} \cup L\underline{F})$*-subgroup of the absolutely irreducible subgroup* G *of* GL(n,D). *Then* H *contains an abelian normal subgroup* A *of* G *with* H/A *locally finite.*

Again Snider's example (1.4.10) shows that in general nothing further can be said about H/A.

Proof: We induct on the length of a characteristic series of H with abelian or locally finite factors; such a series exists by 5.6.4. Then by induction H contains normal subgroups $C \leq K \leq B \leq H$ of G with C and B/K abelian, and K/C and H/B locally finite. By 5.6.2 the group $G/C_G(C)$ is periodic. Replacing K and B by $C_K(C)$ and $C_B(C)$ we may assume that C is central in B. By Schur's Theorem (e.g. Wehrfritz [2], p. 213, Lemma) K' is locally finite. Then $G/C_G(K')$ is locally finite by 5.4.2 and 5.4.6. Replace B by $C_B(K')$. We have now reduced the problem to the case where B is soluble.

By 1.2.13 we may assume that G is persistent, and hence prime by 1.2.15. Thus condition a) of 5.6.5 holds for any normal subgroup of G. Therefore by 5.6.6 there is a characteristic subgroup E of B such that $G/\Delta_G(E)$ is periodic and $\Delta_B(E) \leq E$. In particular $H/\Delta_B(E)$ is locally finite and $\Delta_B(E)$ is an FC-group. The latter implies that $\Delta_B(E)$ is locally finite modulo its centre A (Scott [1], 15.1.16). Thus A is an abelian normal subgroup of G with $A \leq H$ and H/A locally finite. □

As an immediate consequence of 5.6.7 we have the following, which historically was the starting point for much of this chapter.

5.6.8. COROLLARY (Snider [2]). *A soluble absolutely irreducible skew linear group is abelian by locally-finite.* □

5.6.9. THEOREM (Wehrfritz [21]). *Let H be a normal subgroup of the absolutely irreducible subgroup G of GL(n,D), and assume that H is locally finite modulo its centre A. Then $G/C_G(H)$ is periodic.*

Proof: By Schur's Theorem H' is locally finite. Let T be the maximal locally finite subgroup of H. Then H/T is torsion-free abelian and H/AT is periodic. Thus $C_G(A) \subseteq C_G(AT/T) = C_G(H/T)$. Note that $G/C_G(A)$ is periodic by 5.6.2. Let L be a normal subgroup of G of finite index and set $K = H \cap L$. If $L/C_L(K)$ is periodic then so is G modulo $C_o = C_G(K) \cap C_G(H/K) \cap C_G(H/T)$. By stability theory $C_o/C_G(H)$ is isomorphic to a subgroup of the direct product of (H:K) copies of $K \cap T$ and as such is locally finite. Consequently $G/C_G(H)$ is also periodic. It now follows from 1.2.13 and 1.2.15 that we may assume that G is persistent and prime.

Let Z be the centre of T. Then F[AZ] is prime and commutative, and hence a domain. The F[AZ]-torsion right (resp. left) submodule of the simple ring R = F[G] is an ideal (by 5.6.3), and so R is F[AZ]-torsion- free. Let $C = C_G(T)$ and $C_1 = C_C(A)$. By 5.4.2 and 5.4.6 the group G/C is locally finite and therefore F[C] is Artinian by 1.2.12. As such its regular elements are units and hence the quotient field K of F[AZ] is naturally embedded in F[C]. Since C_1 centralizes H/T we have $[H,C_1] \leq Z \leq K$ by stability theory. Consequently 5.3.3 and 5.3.1 applied to F[C] yield that $C/(C \cap K^*C_{K^*C}(H))$ is periodic, and hence so is $G/(C \cap K^*C_{K^*C}(H))$. Also $C \cap K^*C_{K^*C}(H) \leq N_{K^*C}(H) \cap K^*C_{K^*C}(H) = N_{K^*}(H).C_{K^*C}(H)$.

Let J be the quotient field of F[A] in K. Clearly H centralizes J. Also F[H] is a prime ring since G is prime. Any ideal $\underline{a}$ of S = J[H] satisfies $\underline{a} = J(\underline{a} \cap F[H])$. Thus S is prime. Also J is a central subfield of S and, since H/A is locally finite, S is locally finite-dimensional over J. Since Z is periodic, $K = J[Z] \leq S$. By 1.2.6 the ring S is simple Artinian, so we can apply 5.5.1 b) to the normal subgroup H of $M = N_{K^*}(H).H \subseteq S$. Thus $M/C_M(H)$ is locally finite. But then so too is $N_{K^*}(H)/C_{K^*}(H)$. Hence $G/C_G(H)$ is periodic. □

5.6.10. COROLLARY. (Wehrfritz [21]). *Let* $A \leq H$ *be normal subgroups of the absolutely irreducible subgroup* G *of* $GL(n,D)$ *with* A *abelian and* H/A *locally finite. Then* $G/C_G(H)$ *is abelian-by-periodic.*

Proof: Let $K = C_H(A)$ and let Z denote the centre of K. Then $G/C_G(K)$ is periodic by 5.6.9 and $[H,C_G(K)] \leq C_H(K) = Z$. Thus $C_G(K)/C_G(H)$ is isomorphic to a subgroup of the abelian group $Der(H/K,Z)$ by stability theory. The result follows. □

The following is now an immediate consequence of 5.6.7 and 5.6.10. Note that in 5.6.11 there is no need for $G/HC_G(H)$ to be periodic, even if H is metabelian (cf. 5.6.2), see Wehrfritz [21].

5.6.11. COROLLARY (Wehrfritz [21]). *Let* H *be a normal* $P(\underline{A} \cup L\underline{F})$-*subgroup of the absolutely irreducible subgroup* G *of* $GL(n,D)$. *Then* $G/C_G(H)$ *is abelian-by-periodic.*

EXERCISE. (Wehrfritz [21]). Let G and H be as in 5.6.11. Prove that $S = F[H]$ is a semiprime Goldie subring of $R = D^{n\times n}$ whose ring of quotients embeds naturally into R.

5.7 GENERALIZED SOLUBLE GROUPS

In fact very little is known about generalized soluble skew linear groups over arbitrary division rings. We have given many examples in Section 1.4. The few positive results we describe here. Most of the section concerns locally nilpotent normal subgroups of absolutely irreducible groups. We start with an elementary lemma.

5.7.1. *Let* G *be a localy nilpotent group and suppose* G *is a finite extension of a centre-by-periodic group. Then* G *is centre-by-periodic.*

Proof: By hypothesis there are subgroups $A \leq C$ of G with A central in C, C/A periodic, C normal in G and G/C finite. We can replace A by one of its free abelian subgroups of maximal rank. Now A has at most $(G{:}C)$ conjugates in G and therefore $A_G = \bigcap_{x\in G} A^x$ is normal in G with G/A_G

periodic. Thus assume also that A is normal in G.

Since G is locally nilpotent it has a periodic normal subgroup T with G/T torsion-free. Clearly G/T is the isolator of AT/T in G/T. Since centralizers in G/T are isolated (P. Hall [1], 4.8) it follows that G/T is abelian. Consequently $[A,G] \subseteq A \cap T = \langle 1 \rangle$, since A is torsion-free. Therefore A lies in the centre of G. □

5.7.2. THEOREM. (Wehrfritz [22]). *Let* H *be a normal subgroup of the absolutely irreducible subgroup* G *of* GL(n,D) *that is hypercentral. Then* H *is centre by locally-finite and* $G/C_G(H)$ *is periodic.*

Thus in view of the previous section the situation here is exactly what one would expect. Much of this section is devoted to extending this result to locally nilpotent normal subgroups H. Here and for the remainder of this section D is a division ring of characteristic $p \geq 0$ and with centre F.

Proof: In view of 5.6.9 it suffices to prove that H is centre-by-periodic. By 1.2.13, 1.2.15 and 5.7.1 we may assume that G is prime. Set $N = \Delta_G(H)$ and $K = [G,H] \leq H$. Now $\Delta(H)$ is locally nilpotent, and hence locally-finite by torsion-free, and an FC-group, and so centre-by-periodic (Scott [1], 15.1.16). Hence the kernel $\underline{a}$ of the natural map of FG onto $F[G] = D^{n \times n}$ is annihilator-free over $\Delta(H)$ by 5.6.5. Consequently $\underline{a} \cap FN$ is annihilator-free in FN over $K \cap N \leq \Delta(H)$. But then F[G] is a crossed product of F[N] by G/N by 5.3.9 and 5.3.10. Therefore 5.3.1 yields that G/N is periodic.

We now have that $H \cap N = \Delta(H)$ is an FC-group with $H/\Delta(H)$ periodic. Let A be the centre of $\Delta(H)$ and set $Z = C_A(H)$. Clearly A is normal in G and H/A is periodic. In particular (P. Hall [1], 4.8 again) $H' \subseteq T$, the torsion subgroup of H. Let $a \in A$. Then $\langle a^H \rangle$ is finitely generated abelian and its torsion subgroup has finite order m, say. Hence $\langle a^H \rangle^m$ is torsion-free and normal in H, and so $[\langle a^H \rangle^m, H] \leq \langle a^H \rangle^m \cap T = \langle 1 \rangle$. It follows that $a^m \in Z$ and consequently A/Z is periodic. The theorem follows. □

The reader may spot that in the proof of 5.7.2 we have not used the full force of the hypercentrality of H. In fact we only used that H is locally nilpotent and hyper FC-central. The following exercise shows why we have not made this explicit.

EXERCISE. Prove that a locally nilpotent, hyper FC-central group is hypercentral.

5.7.3. COROLLARY (Wehrfritz [22]). *Let* G *be a hyperabelian absolutely irreducible skew linear group. Then* G *is abelian by locally-finite.*

This gives yet another generalization of Snider's theorem 5.6.8. Consideration of hyperabelian normal subgroups of absolutely irreducible groups will have to wait until we have dealt with locally nilpotent normal subgroups of such groups; see 5.7.13.

Proof: G contains a nilpotent normal subgroup H such that $C_G(H) \leq H$, see Kegel & Wehrfritz [1], 1.A.8. Then $C_G(H)$ is abelian and $G/C_G(H)$ is locally finite by 5.7.2. □

We now start our serious study of locally nilpotent groups. It is convenient to isolate a special case first (5.7.5), to which the general case is then reduced. We need a simple lemma.

5.7.4. *Let* G *be a locally nilpotent skew linear group, such that* $u(G) = \langle 1 \rangle$ *if the characteristic is positive.*

a) *If* T *is the torsion subgroup of* G *then* T *lies in the hypercentre of* G.

b) *If* G *is abelian-by-periodic then* G *is hypercentral.*

Proof: a) T satisfies the minimal condition on q-subgroups for every prime q by 2.3.1 and 2.5.2. Thus T has an ascending series of normal subgroups of G with finite factors. Each of these finite factors is G-hypercentral, so T is too.

b) With T as in Part a) we have $G' \leq T$ by P. Hall [1], 4.8. Thus Part b) follows from Part a). □

5.7.5. *Let* H *be a locally nilpotent normal subgroup of the absolutely irreducible skew linear group* G. *If* $H \in P(\underline{A} \cup L\underline{F})$ *then* H *is centre by locally-finite and* $G/C_G(H)$ *is periodic.*

Proof: By 5.6.7 the group H is abelian by periodic. Also $u(H) = s(H) \leq s(G) = \langle 1 \rangle$ by 1.3.4. Hence H is hypercentral by 5.7.4 and the result follows from 5.7.2. □

We now come to our main result concerning locally nilpotent skew linear groups in general. It is a sort of weak analogue of the proposition that every subgroup of a completely reducible locally nilpotent linear group is completely reducible (reduce to the finitely generated case and use Clifford's Theorem; cf. the proof of Wehrfritz [2], 7.7).

5.7.6. THEOREM. (Wehrfritz [22]). *Let* H *be a locally nilpotent normal subgroup of the completely reducible subgroup* G *of* GL(n,D) *and let* S *be a right Noetherian subring of* $D^{n\times n}$ *normalized by* G. *Suppose that either* G *is irreducible or* $S \subseteq F$. *Then for every subgroup* K *of* H *the subring* S[K] *of* $D^{n\times n}$ *is semiprime and right Goldie of dimension at most* n.

The dimension here is the uniform or Goldie dimension of S[K] as right module over itself. Trivially the complete reducibility condition cannot be removed even for linear groups. Note that although a subring of a division ring is trivially semiprime with the maximal condition on annililators, it need not be right Goldie; it could be a free ring for example.

Proof: If G is irreducible then S[H] is semiprime by 1.1.15. If G is not irreducible then $S \subseteq F$ and S[H] is semiprime by applying the same result to the irreducible constiuents of G. Note that any nil subring of $D^{n\times n}$ is nilpotent by 1.3.9. If K is any subgroup of H let $\underline{n}(K)$ denote the nilpotent radical of S[K]. We claim that

$$\underline{n}(H) = \cup_X \underline{n}(X) = \cup_K \underline{n}(K),$$

where X ranges over the finitely generated subgroups of H and K ranges over all the subgroups of H. Once this is shown, since S[H] is semiprime, we will have $\underline{n}(H) = \{0\}$ and each S[K] is semiprime.

Clearly $\underline{n}(K) \cap S[X] \subseteq \underline{n}(X)$ for X any finitely generated subgroup of K. If Y is also a finitely generated subgroup of K, set $Z = \langle X,Y \rangle$. Then X is subnormal in Z since Z is nilpotent; say

$$X \lhd X_1 \lhd \dots \lhd X_r = Z.$$

Now $\underline{n}(X)$ is nilpotent and normalized by X_1, so $\underline{n}(X)X_1$ is a nilpotent ideal of $S[X_1]$. Thus $\underline{n}(X) \subseteq \underline{n}(X_1)$ and an elementary induction shows that $\underline{n}(X) \cup \underline{n}(Y) \subseteq \underline{n}(Z)$. It follows that $\bigcup_X \underline{n}(X)$ is a nilpotent ideal of $S[K]$ and therefore $\bigcup_X \underline{n}(X) = \underline{n}(K)$. Consequently $\bigcup_{K \leq H} \underline{n}(K) = \bigcup_{K \leq H} \bigcup_{X \leq K} \underline{n}(X) = \bigcup_{X \leq H} \underline{n}(X) = \underline{n}(H)$, where X is always finitely generated. We have now proved that $S[K]$ is semiprime.

Trivially $S[K]$ has the maximal condition on right annihilators. Suppose $\oplus_{i=1}^{m} U_i \subseteq S[K]$, where the U_i are non-zero right ideals of $S[K]$. There is a finitely generated subgroup X of K such that each $U_i \cap S[X] \neq \{0\}$. Now $S[X]$ is right Noetherian (Passman [2], 10.2.7) and hence right Goldie. By the above it is semiprime. If $\dim S[X]_{S[X]} \leq n$, dimension here being the (right) uniform dimension, then $m \leq n$. It will follow that $\dim S[K]_{S[K]} \leq n$ and thus that $S[K]$ is right Goldie. The following therefore completes the proof.

5.7.7. *Let S be a semiprime right Ore subring of* $D^{n\times n}$. *Then for some integer* $m \leq n$ *the ring Q of right quotients of S is isomorphic to a subring of* $D^{n\times n}$. *If S is also right Goldie then* $\dim S_S \leq n$.

This lemma is a generalization from S prime to semiprime of an unpublished lemma of J. C. Robson and J. T. Stafford. In 5.7.7 we cannot in general extend the embedding of S into $D^{n\times n}$ to one of Q, even if Q is isomorphic to some subring of $D^{n\times n}$; for consider $S = F[\,\mathrm{diag}(x,0)] \subseteq D^{2\times 2}$, where $D = F(x)$ and x is transcendental over the field F.

Proof: Regard $V = D^n$ as a D-S bimodule in the usual way and let $V_1, V_2, \dots, V_r$ be a full set of non-isomorphic D-S composition factors of V. Let P_i be the annihilator of V_i in S. If $\underline{a}$ is an ideal of S then $V_i\underline{a}$ is a D-S submodule of the irreducible module V_i. It follows that each P_i is prime. Also $\bigcap P_i$ acts nilpotently on V, and S is semiprime and faithful on V. Therefore $\bigcap_i P_i = \{0\}$. It follows that every prime ideal of S contains at least one of the P_i. Suppose $P_1, \dots, P_s$ are the distinct minimal prime ideals of S. Then $\bigcap_{i=1}^{s} P_i = \{0\}$ and we have ring embeddings

$$S \hookrightarrow \oplus_{i\leq s} S/P_i \hookrightarrow \oplus_{i\leq s} \mathrm{End}_D V_i \hookrightarrow \mathrm{End}_D(\oplus_{i\leq s} V_i) \cong D^{m\times m},$$

where $m = \Sigma_{i\leq s} \dim_D V_i \leq n$.

Let $C = \underline{C}_S(0)$; by hypothesis C is a right Ore subset of S. Trivially $C_i = C + P_i/P_i$ is a multiplicative submonoid of $S_i = S/P_i$ and C_i is a right Ore subset of S_i. Let $i \leq s$. Then $P_i = \ell_S r_S(P_i)$ since $D^{n\times n}$ and hence S satisfies the maximal condition on annihilators (Chatters & Hajarnavis [1], 1.16). Let $c \in C$ and $a \in S$, and suppose $ca \in P_i$. Then $ca.r_S(P_i) = \{0\}$ and yet $c \in \underline{C}_S(0)$. Thus $a.r_S(P_i) = \{0\}$ and $a \in \ell_S r_S(P_i) = P_i$. Also $P_i = r_S\ell_S(P_i)$ and a similar argument shows that $a \in P_i$ if $ac \in P_i$. Consequently $C_i \subseteq \underline{C}_{S_i}(0)$ and we can form the ring $S_iC_i^{-1}$ of right quotients. Clearly the natural embedding of S into $\oplus_{i\le s} S_i$ extends to one of $Q = SC^{-1}$ into $\oplus_{i\le s} S_iC_i^{-1}$.

By definition V_i is an S_i-faithful D-S_i bimodule. Let

$$Z_i = \{ v \in V_i : vc = 0 \text{ for some } c \in C_i \}.$$

By the Ore condition Z_i is a D-S_i submodule of V_i. If $Z_i = V$ then some element of C_i annihilates a left D-basis of V_i and hence also annihilates V_i itself. This impossibility shows that $Z_i = \{0\}$. Thus the elements of C_i are one-to-one on V_i. Consequently $C_i \subseteq \mathrm{Aut}_D V_i$ and the subring $S_i[C_i^{-1}]$ of $\mathrm{End}_D V_i$ is naturally isomorphic to $S_iC_i^{-1}$. Therefore we have embeddings

$$Q \hookrightarrow \oplus_{i\le s} S_iC_i^{-1} \hookrightarrow \oplus_{i\le s} \mathrm{End}_D V_i \hookrightarrow D^{m\times m}$$

Always $\dim S_S = \dim Q_Q$. Suppose S is right Goldie. Then Q is semisimple Artinian, so $\dim Q_Q$ is the maximum number of orthogonal idempotents in Q. The maximum number of orthogonal idempotents in $D^{m\times m}$ is m. Therefore $\dim S_S = \dim Q_Q \le m \le n$. □

5.7.8. *Let G be an irreducible subgroup of $GL(n,D)$ and let S be a right Ore subring of $D^{n\times n}$ normalized by G. Then the ring Q of right quotients of S is naturally embedded in $D^{n\times n}$.*

Note that the exercise at the end of Section 5.6 is effectively a special case of 5.7.8. In 5.7.8. it is not sufficient to assume that G is completely reducible (try $G = \langle 1\rangle$ and D a field). Suppose in 5.7.6 that G is irreducible. Then the ring of right quotients of $S[H]$ embeds naturally in $D^{n\times n}$ by 5.7.8. Does the same apply to that of $S[K]$? It certainly does in many cases, as we shall see during the proof of 5.7.11.

Proof: Let $V = D^n$, regarded as D-G bimodule in the obvious way. By hypothesis $C = \underline{C}_S(0)$ is right Ore in S and therefore

$$Z = \{ v \in V: vc = 0 \text{ for some } c \in C \}$$

is a D-S submodule of V. But G normalizes C, so Z is in fact a D-G submodule of V, and the irreducibility hypothesis yields that either $Z = V$ or $Z = \{0\}$.

If $Z = V$ then by the Ore condition there exists $c \in C$ annihilating a left D-basis of V. But then $Vc = \{0\}$, and yet $c \neq 0$. Therefore $Z = \{0\}$ and each element of C is one-to-one on V. Consequently C consists of units of $D^{n \times n}$ and the subring $S[C^{-1}]$ of $D^{n \times n}$ is S-isomorphic to Q. □

We need just two more results before we can prove the main theorem of this section. We have to return to a construction we have carefully avoided since Chapter 1. Let K be a normal subgroup of a group H with H/K an ordered group, and suppose that $R = E[H]$ is a crossed product of the division ring E by H/K, so $K = E \cap H$. Let D be the set of all formal sums $x = \Sigma_{t \in T} t\varepsilon_t$, where T ranges over all well-ordered subsets of some fixed transversal of K to H, the order being that induced by the given order on H/K, and the $\varepsilon_t \in E^*$. Then the obvious addition and multiplication on D are well-defined and make D into a division ring containing R as a subring, see 1.4.14. and its proof.

5.7.9. *With the above notation assume also that* H *centralizes* E. *Then* $N_{D^*}(H) = H.C_{D^*}(H)$.

Proof: Trivially $N_{D^*}(H) \supseteq H.C_{D^*}(H)$. Let $x = \Sigma_{t \in T} t\varepsilon_t \in N_{D^*}(H)$ and $h \in H$. Then $k = h^x \in H$ and $hx = xk$. Thus

$$\Sigma_T ht\varepsilon_t = \Sigma_T tk\varepsilon_t.$$

Now left and right multiplication in H/K preserve the order and of course the supports of these two sums are equal. Consequently $htK = tkK$ for all $t \in T$. But then $h^tK = kK = h^sK$ and $st^{-1} \in C_H(hK/K)$ for all s and t in T. This is for all $h \in H$ and so $st^{-1} \in C_H(H/K)$ for all such s and t. Thus choosing a fixed t in T we have $x = (\Sigma_{c \in C} c\varepsilon_{ct})t$, where $C = Tt^{-1} \subseteq C_H(H/K)$.

Trivially t normalizes H and hence so does $y = xt^{-1}$. But then for $h \in H$ and $\ell = h^y \in H$ we have $hy = y\ell$ and

$$\Sigma_C ch[h,c]\varepsilon_{ct} = \Sigma_C c\ell\varepsilon_{ct}.$$

Consequently $chK = c\ell K$ for any $c \in C$ and $h^{-1}\ell \in K$. Comparing coefficients we obtain $[h,c]\varepsilon_{ct} = h^{-1}\ell\varepsilon_{ct}$ and so $h^c = \ell$ for all $c \in C$. Hence $c'c^{-1}$ centralizes h for all $c, c' \in C$ and all $h \in H$. Therefore if we pick any particular c in C, we have $yc^{-1} \in C_{D^*}(H)$ and so $x \in C_{D^*}(H)ct \subseteq H.C_{D^*}(H)$. The lemma is proved. □

5.7.10. *Let F be a field, R = F[G] an F-algebra, generated as such by the subgroup G of its group of units, K a normal subgroup of G and E = F(K) a division F-subalgebra of R generated as such by K. Let C denote the centre of E and set A = K ∩ C. Assume that K/A is an ordered group and that C[K] ⊆ E is both an Ore domain and a crossed product of C by K/A. Then K = A.*

Proof: Set $X = G \cap E^*C_{E^*G}(K)$. Then R is a crossed product of $E[C_R(K)]$ by G/X, see 5.3.4. But R = F[G], so $E[C_R(K)] = F[X]$. Trivially $X = G \cap N_{E^*}(K)C_{E^*G}(K)$, so $E[C_R(K)] = C[N_{E^*}(K), C_R(K)]$. Now E is a central simple C-algebra. Hence

$$E \otimes_C C_R(K) \cong E[C_R(K)] \leq R$$

by Theorem 2, p. 363 of Cohn [2], Vol. 2. Therefore $E = C[N_{E^*}(K)]$.

By hypothesis C[K] is a crossed product of C by the ordered group K/A. By 5.7.9 there is a division ring D containing C[K] as a subring such that $N_{D^*}(K) = K.C_{D^*}(K)$. But C[K] is an Ore domain. Consequently E is embedded naturally in D and $N_{E^*}(K) = KC^*$. Then E = C[K], and so E is a crossed product of C by K/A. Therefore by 5.3.1 the ordered group K/A is periodic. It follows that K/A = ⟨1⟩ and K = A as required. □

5.7.11. THEOREM. (Wehrfritz [25]). *Let H be a locally nilpotent normal subgroup of the absolutely irreducible subgroup G of GL(n,D). Then H is centre by locally-finite and $G/C_G(H)$ is periodic.*

Proof: Set $S = F[H] \leq R = F[G] = D^{n\times n}$. By 5.7.5 we have only to prove that $H \in P(\underline{A} \cup L\underline{F})$. In particular by 1.2.13 and 1.2.15 we may assume that G is prime. Thus S is now a prime ring, as well as a Goldie ring by 5.7.6. Then S has a simple Artinian ring Q of quotients (Goldie's Theorem) and Q can be taken to be a subring of R by 5.7.8. Now Q = EU for some division ring E and set U of matrix units centralizing E. There exists a regular element s of S with Us ⊆ S.

Let Y be a finitely generated subgroup of H with Us ⊆ F[Y]. Now F[Y] is Noetherian and, by 5.7.6 again, semiprime. Let Q_Y denote its ring of quotients. Since Us ⊆ F[Y] and s is a unit of Q it follows easily that Y is an irreducible subgroup of the group of units of Q. Therefore by 5.7.8 we may assume that Q_Y is contained in Q. Since 1 is a sum of elements of U we have s ∈ F[Y] and clearly s is regular in F[Y]. Hence

$s^{-1} \in Q_Y$, $U \subseteq Q_Y$ and Q_Y is a matrix ring over $Q_Y \cap E$ with U as a set of matrix units. But $Q_Y \cap E$ is a domain, so Q_Y is prime, and also semisimple Artinian. Therefore Q_Y is simple and $F[Y]$ is a prime ring, by Goldie's Theorem yet again.

Let T be the torsion subgroup of H and Z/T the centre of H/T. Then S is a crossed product of $F[Z]$ by H/Z, see 5.3.14. Let $K = C_H(Z)$ and $A = Z \cap K$. Then H/K is periodic by 5.7.5. Also H/T is torsion-free and locally nilpotent, so its centralizers are isolated and $(Z \cap KT)/T$ is the centre of KT/T. But $Z \cap KT = (Z \cap K)T = AT$, so AT/T is the centre of KT/T. Also $AT \cap K = A(T \cap K) = A$. Therefore $A/(K \cap T)$ is the centre of $K/(K \cap T)$. From the structure of S, the algebra $F[K]$ is a crossed product of $F[A]$ by K/A. Since G is prime, $F[A]$ is a domain. But K/A is torsion-free and locally nilpotent since H/Z is. Therefore $F[K]$ is a domain by 1.4.1. It is also Ore by 1.4.2 and K/A is orderable (Passman [2], 13.1.6 and 13.2.2). The division ring of quotients of $F[K]$ is naturally embedded in R by 5.7.8 and if C is its centre then $C[K]$ is Ore (1.4.2) and a crossed product of C by K/A by 5.3.14 and $A = K \cap C$. Thus all the conditions of 5.7.10 are satisfied and $K = A \leq Z$. We have already seen that H/K is periodic, so $H \in (L\underline{F})\underline{A}(L\underline{F})$ and the theorem follows from 5.7.5. □

Suppose G and H are as in 5.7.11. Then further properties of H can be read off from the results of Chapter 3. For a start 3.2.2 is directly applicable to H. The subring $F[H]$ of $D^{n \times n}$ is semiprime and Goldie by 5.7.6 (actually this special case is very easy - see the exercise at the end of Section 5.6). Let Q be its ring of quotients. Then Q is semisimple Artinian, say $Q = \oplus_i Q_i$ where each Q_i is simple. Let F_i be the (necessarily) central subfield of Q_i generated by F and the natural image in Q_i of the centre Z of H. Since H/Z is locally finite, Q_i is a locally finite-dimensional F_i-algebra (use 1.2.6 a) to show that $Q = (\oplus_i F_i)[H]$). Thus H is a subdirect product of a finite number of locally nilpotent skew linear groups of characteristic p over locally finite-dimensional division algebras. We can therefore apply the results of Section 3.2.

Finally in this section we consider the wider classes of generalized soluble groups. Here our information is much more limited. Since free groups are residually finite-nilpotent there is no hope of comparable results if one works with series that are not ascending. In view of the results of Sections 3.1 and 5.5 it would be nice to work with the all-embracing class $\bar{\underline{S}}$, but we are unable to do this. Thus the widest classes that are worth considering are the classes of radical (i.e. $\acute{P}L\underline{N}$) groups and of locally soluble groups (see the diagram on p. 80 of Robinson [1], Vol. 2). The first case is easily dealt with, using 5.7.11. In fact we can be a little more general.

5.7.12. (Wehrfritz [25]). *Let* G *be an absolutely irreducible skew linear group and let* H *be a normal* $\acute{P}L(\underline{N} \cup \underline{F})$*-subgroup of* G. *Then there is an abelian normal subgroup* $A \leq H$ *of* G *with* H/A *locally finite, and* $G/C_G(H)$ *is abelian by periodic.*

Proof: Suppose X is a $L(\underline{N} \cup \underline{F})$-group that is not locally nilpotent. Then X has a finitely generated subgroup Y that is not nilpotent. Hence every finitely generated subgroup of X containing Y is finite and X is locally finite. It now follows from the Hirsch-Plotkin Theorem (Robinson [1], 2.31) that

$$\acute{P}L(\underline{N} \cup \underline{F}) = \acute{P}_n(L\underline{N} \cup L\underline{F}).$$

Let K be the Hirsch-Plotkin radical of H and T/K the maximal locally finite normal subgroup of H/K. Now K is centre by locally-finite by 5.7.11 and therefore T is abelian by locally-finite and 5.6.11 yields that $G/C_G(T)$ is abelian by periodic. Let C be the centre of T. We claim that $C = C_H(T)$. If so then $H \in P(\underline{A} \cup L\underline{F})$ and the conclusion will follow from 5.6.7 and 5.6.11.

Suppose $C \neq C_H(T)$. Then there is a non-trivial normal subgroup L/C of H/C in $C_H(T)/C$ that is either locally nilpotent or locally finite. In the first case L too is locally nilpotent and $L \leq K \cap C_H(T) \leq C$. In the second case, as $C \subseteq K$, we have $L \subseteq T \cap C_H(T) = C$. We chose $L > C$. This contradiction completes the proof. □

The following result, which we promised some time back, is now immediate.

5.7.13. COROLLARY. *Let* H *be a hyperabelian normal subgroup of the absolutely irreducible skew linear group* G. *Then* H *and* $G/C_G(H)$ *are both abelian by periodic.* □

Now consider locally soluble groups. Here very little is known and what little is known we state without proof. Using the techniques of this section and the exercise at the end of Section 5.3, the following can be shown, see Point 3 of Wehrfritz [25].

5.7.14. *Let G be an absolutely irreducible skew linear group of degree 1.*

a) *If G is locally soluble then G is abelian by locally-finite.*
b) *Let H be a locally soluble normal subgroup of G, T the maximal periodic normal subgroup of H and B/T the Hirsch-Plotkin radical of H/T. Set* $K = C_H(B)$ *and* $A = B \cap K$, *and suppose that K/A is an ordered group. Then H and* $G/C_G(H)$ *are both abelian by periodic.*

We conclude the chapter with the following, as yet unpublished, result of I. A. Stewart, taken from his thesis.

5.7.15. (Stewart [1]). *Let G be an absolutely irreducible skew linear group and H a normal subgroup of G. Suppose that H has an ascending normal series whose factors are abelian of rank 1. Then H is hypercyclic and abelian by locally-finite, and* $G/C_G(H)$ *is abelian by periodic.*

Proof: Certainly H and $G/C_G(H)$ are abelian by periodic by 5.7.13. (This result was not available to Stewart we should point out.) Let A be an abelian normal subgroup of a group K with K/A periodic and B an abelian normal subgroup of K of rank 1. We claim that B is K-hypercyclic. If so, this and an elementary induction yields that H is hypercyclic and the proof will be complete. If B is periodic, since rankB = 1, B has an ascending characteristic series with finite cyclic factors. Hence we may assume that $B \leq A$ and B is torsion-free. Let $b \in B$ and $k \in K$. Then for some positive integers p, q, r with p and q coprime, we have $b^{pk} = b^q$ and $[b,k^r] = 1$. Consequently $b^{p^r} = b^{q^r}$ and $p/q = \pm 1$. Thus $\langle b \rangle$ is normal in K and the claim follows. □

6 AN APPLICATION TO GROUP RINGS

This brief chapter is not really part of the theory of skew linear groups. The latter is too recent for any real applications to have arisen. However we show here how it sheds a little clarification on some known results concerning group rings.

Throughout this chapter F is a field of charactersitic $p \geq 0$. A group G has ***finite endomorphism dimension over*** F if every irreducible FG-module has finite dimension over its endomorphism ring. (Since any left G-module V becomes a right G-module via $vg = g^{-1}v$, $v \in V$, $g \in G$, there is no need to specify left or right.) Let $\underline{Z}_F$ denote the class of all such groups G. An alternative formulation is given by the following: a group G is in $\underline{Z}_F$ if and only if every primitive image of FG is Artinian. $\underline{Z}_F$-groups seem first to have arisen in connection with injective modules, see Passman [2], Section 12.4, and they have been completely characterized group-theoretically in the following cases: locally finite $\underline{Z}_F$-groups (Hartley, see Hartley [2], [3] and Passman [2], Section 12.4), finitely genertaed soluble-by-finite $\underline{Z}_F$-groups (Snider [2]), soluble-by-finite $\underline{Z}_F$-groups for F not locally finite (also Snider [2]).

A smaller class of groups than $\underline{Z}_F$ is the class $\underline{Y}_F$ of groups G with ***finite central endomorphism dimension over*** F, meaning that every irreducible FG-module has finite dimension over the centre of its endomorphism ring. A group G is in $\underline{Y}_F$ if and only if every primitive image of FG satisfies a polynomial identity. The class $\underline{Y}_F$ is considered in Wehrfritz [24] and, not surprisingly, more is known about $\underline{Y}_F$ than about $\underline{Z}_F$. The classes $\underline{Z}_F$ are intimately connected with the theory of skew linear groups, whereas the study of $\underline{Y}_F$ really only depends upon linear groups. Now it so happens that for the classes considered by Hartley and Snider, namely $L\underline{F}$, $\underline{G} \cap \underline{SF}$ and, for F not locally finite, $\underline{SF}$, the structural conclusions for $\underline{Z}_F$ and $\underline{Y}_F$ are identical. Thus, for example, $\underline{Z}_F \cap \underline{G} \cap \underline{SF} = \underline{Y}_F \cap \underline{G} \cap \underline{SF}$. This and similar results we shall prove below, without needing to prove the full characterization first.

$\underline{Y}_F$ lies midway between the interesting class $\underline{Z}_F$ and the interesting and smaller class $\underline{X}_F$ of all groups G such that every irreducible FG-module has finite dimension over F. Such groups arose in works of P. Hall and of J. E. Roseblade on the residual finiteness of certain soluble groups, see Passman [2], Sections 12.2 and 12.3. They have been extensively studied, apart from Hall and Roseblade, by E. M. Levič, I. M. Musson, R. L. Snider, and Wehrfritz. Much of the information available about $\underline{X}_F$ can now be easily derived from our knowledge of $\underline{Y}_F$, see Wehrfritz [24]. As far as we know there is no connection between the classes $\underline{X}_F$ and the theory of skew linear groups. Therefore we do not mention the former again. Interested readers should consult Wehrfritz [24] for references.

6.1 PRELIMINARIES

6.1.1. Hall's Lemma. *Let* G *be a group,* H *a subgroup of* G *and* W *an irreducible* FH*-module. Then there is an irreducible* FG*-module* V *containing* W *as an* FH*-submodule.*

Proof: Let M be an FG-submodule of $\overline{V} = W \otimes_{FH} FG$ that is maximal subject to $M \cap (W \otimes 1) = \{0\}$. Since every non-zero element of $W \otimes 1$ generates $\overline{V}$ as FG-module, $V = \overline{V}/M$ is an irreducible FG-module. Moreover $V \supseteq (W \otimes 1) + M/M \cong_{FH} W$. □

6.1.2. *The class $\underline{Z}_F$ is ⟨S,Q⟩-closed.*

Proof: Trivially $\underline{Z}_F$ is Q-closed. Suppose H is a subgroup of the $\underline{Z}_F$-group G and let W be an irreducible FH-module with endomorphism ring D. By 6.1.1 there is an irreducible FG-module V containing W as FH-submodule. Set $E = \mathrm{End}_{FG}(V)$ and suppose we have $\Sigma_i e_i w_i = 0$ where the $e_i \in E$ and the w_i are elements of W that are (left) linearly independent over D. For each k there exists, by the Jacobson Density Theorem (Jacobson [3] p. 28 or [5] Vol. 2 p. 199), an element s_k of FH such that $w_j s_k = w_j$ if $j = k$ and 0 otherwise. Then $e_k w_k = (\Sigma\, e_i w_i) s_k = 0$, $w_k \neq 0$ and E is a division ring. Thus each $e_k = 0$. This shows that $\dim_D W$ is at most equal to $\dim_E V$, and the latter is finite by hypothesis. The proof is complete. □

EXERCISE. Use 1.2.6 and 1.2.12 to prove the following special case of 6.1.2. Let $H \lhd G \in \underline{Z}_F$ where H or G/H is locally finite. Then H is in $\underline{Z}_F$.

6.1.3. a) *$\underline{Y}_F$ is* $\langle S,Q\rangle$*-closed.*

b) *Let* H *be a subgroup of the group* G *of finite index. Then* $G \in \underline{Y}_F$ *if and only if* $H \in \underline{Y}_F$.

c) $\underline{AF} \subseteq \underline{Y}_F$.

(In fact $\underline{X}_F$ too is $\langle S,Q\rangle$-closed.)

Proof: a) Let H be a subgroup of the $\underline{Y}_F$-group G and let W be an irreducible FH-module. By 6.1.1 there is an irreducible FG-module V containing W as an FH-submodule. Now the ring $FH/\mathrm{Ann}_{FH}(W)$ is an image of $FH/\mathrm{Ann}_{FH}(V)$ and the latter is an F-subalgebra of $FG/\mathrm{Ann}_{FG}(V)$. From the hypothesis the latter is finite-dimensional over its centre and so it satisfies a standard polynomial identity (Passman [2], 5.1.6). Consequently so does the primitive ring $FH/\mathrm{Ann}_{FH}(W)$, and hence W is finite-dimensional over the centre of its endomorphism ring by Kaplansky's Theorem (Passman [2], 5.3.4). Thus $H \in \underline{Y}_F$ and so $\underline{Y}_F$ is S-closed. Trivially $\underline{Y}_F$ is Q-closed.

b) If $G \in \underline{Y}_F$ then $H \in \underline{Y}_F$ by Part a). Suppose $H \in \underline{Y}_F$. By Part a) we may assume that H is normal in G. Let V be an irreducible FG-module. By a version of Clifford's Theorem (Passman [2] 7.2.16 or use 1.2.8 c) V is a direct sum of a finite number of irreducible FH-modules. Let $R = FG/\mathrm{Ann}_{FG}(V)$ and let S be the natural image of FH in R. By hypothesis $H \in \underline{Y}_F$ so S is finitely generated as a module over its centre Z and Z is a direct sum of a finite number of fields. The finite group G/H acts on Z. By a well-known result from invariant theory, in fact an easy extension of a result from Galois theory, Z can be generated as $C_Z(G)$-module by (G:H) elements. Since $C_Z(G)$ is central in R the result follows.

c) Trivially $\underline{A} \subseteq \underline{Y}_F$. Thus $\underline{AF} \subseteq \underline{Y}_F$ by Part b). □

EXERCISE. Let H be a subgroup of the group G of finite index. Prove that $G \in \underline{Z}_F$ if and only if $H \in \underline{Z}_F$.

6.1.4. Snider's Lemma. *Let* H *be a normal subgroup of the* $\underline{Z}_F$*-group* G. *Then each irreducible* FH*-module has only finitely many conjugates under* G *(up to* FH*-isomorphism of course). If* H *is abelian and* $\underline{m}$ *is a maximal ideal of* FH *then* $N_G(\underline{m})$ *has finite index in* G.

If W is an FH-module its conjugate under the element $x \in G$ is the FH-module with underlying F-space W and H-action given by

$$w \cdot_x h = w h^{x^{-1}} \quad , w \in W,\ h \in H.$$

Proof: Let W be an irreducible FH-module and let V be an irreducible FG-module containing W as an FH-submodule. Then for all $x \in G$, the FH-submodule Wx of V is an isomorphic copy of the x-conjugate of W. Clearly $V = \Sigma_{x \in G} Wx$, so V is completely FH-reducible; let $\{ V_i : i \in I \}$ be its set of non-zero homogeneous components. Then

$$\Pi_{i \in I} \mathrm{End}_{FH}(V_i) = \mathrm{End}_{FH}(V) \geq \mathrm{End}_{FG}(V) = D$$

say. Since $\dim_D V$ is finite, I must be finite. Thus W has only finitely many conjugates under G. If H is abelian and $\underline{m}$ is a maximal ideal of FH then $W = FH/\underline{m}$ is irreducible, $FH/\underline{m}^x$ for $x \in G$ is the x-conjugate of W, and $\underline{m} = \mathrm{Ann}_{FH}(W)$. The lemma follows. □

EXERCISE. If $G \in \underline{Z}_F$ prove that $G/(G \cap (1 + J(FG)))$ is residually absolutely irreducible skew linear over division F-algebras. For any group G prove that

$$G \in \underline{Z}_F \text{ if and only if } G/(G \cap (1 + J(FG))) \in \underline{Z}_F.$$

(Here J(FG) is the Jacobson radical of FG.)

Note that for any group G

$$O_p(G) \cap \Lambda(G) \subseteq G \cap (1 + J(FG)) \subseteq O_p(G),$$

where $O_0(G) = \langle 1 \rangle$ and

$\Lambda(G) = \{ g \in G : (H:C_H(g)) < \infty$ for every finitely generated subgroup H of G $\}$.

The first containment follows since $(O_p(G) \cap \Lambda(G))-1$ generates a nil ideal of FG and the second follows from $F\langle g \rangle \cap J(FG) \subseteq J(F\langle g \rangle)$ for all $g \in G$, an immediate consequence of 6.1.1.

If $G \in \underline{Z}_F$ then probably $G \cap (1 + J(FG)) = O_p(G)$. This is of course related to the unipotence problem (Section 1.3).

6.2 SOLUBLE GROUPS

6.2.1. $\underline{Z}_F \cap \underline{G} \cap \acute{P}L(\underline{N} \cup \underline{F}) \subseteq \underline{Y}_F$.

Proof: Let G_0 be any finitely generated $\acute{P}L(\underline{N} \cup \underline{F})$-group in $\underline{Z}_F$ and let V be an irreducible FG_0-module of dimension n over its endomorphism ring D. Let G be the obvious image of G_0 in GL(n,D). Then G is absolutely irreducible and hence, by 5.7.12, is abelian by locally-finite. But G_0 is finitely generated, so G is in fact abelian-by-finite. But then $G \in \underline{Y}_F$ by 6.1.3 c) and so V is finite-dimensional over the centre of D. This is for all such V and therefore $G_0 \in \underline{Y}_F$. □

6.2.1 and Theorem 1 of Wehrfritz [24] immediately yield the first of the following results.

6.2.2. (Snider [2]). *Let F be any field. Then*

$$\underline{Z}_F \cap \underline{G} \cap \underline{SF} = \begin{cases} \underline{G} \cap \underline{AF} & \text{if F is not locally finite,} \\ (PC_\infty)\underline{F} & \text{if F is locally finite.} \end{cases} \quad \square$$

6.2.3. *Suppose F is not locally finite. Then* $\underline{Z}_F \cap \acute{P}L(\underline{N} \cup \underline{F}) \subseteq \underline{Y}_F$.

Before we give the proof, note that 6.2.3 and Wehrfritz [24] Theorem 2, immediately yield Snider's second theorem.

6.2.4. (Snider [2]). *Suppose F is not locally finite. Then* $\underline{Z}_F \cap \underline{SF}$ *is the class of all groups G satisfying*

$$G/O_p(G) \in \underline{AF}, \quad O_p(G) \subseteq \Lambda(G), \quad O_p(G) \in \underline{S}. \ \square$$

Proof of 6.2.3: Just as in the proof of 6.2.1 it suffices to consider an absolutely irreducible $\underline{Z}_F \cap \acute{P}L(\underline{N} \cup \underline{F})$-subgroup G of GL(n,D) say. By 5.7.12 there is an abelian normal subgroup A of G such that G/A is locally finite and, again as in the proof of 6.2.1, it suffices to prove that G is abelian-by-finite. By 1.2.13 and 1.2.15 (and 6.1.2) we may assume that G is prime and by a standard ladder argument we may also assume that G is countable. Let T be the maximal periodic normal subgroup of G. Since now F[A] is a domain, A ∩ T has rank at most 1.

Let B be a free abelian subgroup of A with A/B periodic and

let $\bar{F}$ be an algebraic closure of F. Then $A \cap T$ embeds into $\bar{F}^*$ and, since F is not locally finite, B embeds into F^*. Now $\bar{F}^*$ is divisible so these embeddings extend to a homomorphism Φ of A into $\bar{F}^*$. Since, by construction, Φ is one-to-one on $(A \cap T) \times B$, so Φ is itself one-to-one. Set W = $F[A\Phi]$. Then W is an irreducible FA-module via Φ. By 6.1.4 there is a normal subgroup $H \supseteq A$ of G of finite index normalizing W and then the action of H on A extends to one on W. But $B\Phi \subseteq F$, which is centralized by H, and Φ is one-to-one. Therefore H centralizes B.

The group H/B is locally finite. Set $P/B = O_p(H/B)$. Assuming the locally finite case, that is assuming Hartley's Theorem (6.3.2 below), H/P is abelian-by-finite. Thus with a slightly smaller choice of H and A we may assume that $H' \subseteq P$. Let $C = C_H(AT/T)$. Then $C/(H \cap T)$ is centre by locally finite, so $C' \subseteq T$ by Schur's Theorem. The argument of the previous paragraph shows that some normal subgroup K of H of finite index centralizes a basis of the torsion-free abelian group $C/(H \cap T)$. If $c \in C$ and $k \in K$ then for some $r > 0$ we have $c^{rk}T = c^rT$. Thus $[c,k]^r \in T$. Consequently $[c,k]$ belongs to T and K centralizes $C/(H \cap T)$. But clearly

$$C_H(C/(H \cap T)) \subseteq C_H(A/(A \cap T)) = C.$$

Therefore $K \leq C$ and so $K' \subseteq T$. But then $K' \subseteq H' \cap T \subseteq P \cap T$, $= \langle 1 \rangle$ since $B \cap T = \langle 1 \rangle$, P/B is a p-group, and $O_p(T) = \langle 1 \rangle$. Thus K is an abelian subgroup of G of finite index.

6.3 LOCALLY FINITE GROUPS

Here skew linear group theory has less impact than in the soluble case above. This is made even more unsatisfactory by our use of the locally finite case in the proof of 6.2.3. However let us say what we can.

6.3.1. $\underline{Z}_F \cap L\underline{F} \subseteq \underline{Y}_F$.

Proof: As in the proof of 6.2.1 we may assume that G is an absolutely irreducible locally finite $\underline{Z}_F$-subgroup of GL(n,D), where D is a division F-algebra and $F[G] = D^{n\times n}$. We need to prove that D is finite-dimensional over its centre. Consider first the case p = 0. We suppose that G is not abelian-by-finite and seek a contradiction. By 2.5.14 we may assume that G is metabelian with all its primary subgroups abelian and by

1.2.13 and 1.2.15 we may assume that G is prime. This implies that for any abelian normal subgroup A of G the subalgebra F[A] is a domain and so A has rank at most 1.

Let A be a maximal abelian subgroup of G containing G'. Then $A = C_G(A)$. Also E = F[A] is a field as A is periodic, and (F[G], E, G, G/A) is a crossed product by 5.3.2. Also G/A is residually finite, since Aut(A) is, and of finite rank (at most n in fact). Thus replacing G by a normal subgroup of finite index we may assume that G/A involves only primes greater than n. Let G_1/A be any finite subgroup of G/A. Then $(F[G_1], E, G_1, G_1/A)$ is a classical crossed product and as such is simple (e.g. see the exercise after 5.3.2). It is therefore a matrix ring over a division ring, and its degree is at most n and divides $(G_1:A)$. Thus $F[G_1]$ is a division ring, and consequently so too is F[G].

G is therefore the split extension of its Hall subgroup G' by an abelian Hall subgroup B say, see 2.5.9 (although this much also follows from the elementary result 2.1.3). Let $Z = C_B(G')$. Then Z is the centre of G and the centre of G/Z is trivial. Now let E be the field $F[G']$. Then G acts on E and we can form the crossed product $(F[G/Z], E, G/Z, G/ZG')$. Then R = F[G/Z] is simple (exercise after 5.3.2 again) and so primitive. Since $G \in \underline{Z}_F$ the ring R is Artinian and so is a matrix ring of degree m say over some division F-algebra. As in the preceding paragraph G/ZG' has a subgroup H/ZG' of finite index involving only primes greater than m and then $F[H/Z] \subseteq R$ is a division ring. By 2.5.9 every proper subgroup of H/Z has a non-trivial centre. Then the definition of Z yields that $H = G'$, a contradiction since G is not abelian-by-finite. Thus G is in fact abelian-by-finite and therefore a $\underline{Y}_F$-group, which completes the proof of this case. (Note that we have shown, for p = 0, that for a locally finite $\underline{Z}_F$-group G, in any primitive image of FG the image of G is abelian-by-finite, a fact which we use later.)

Now let $p > 0$. Here $R = \mathbf{F}_p[G]$ is locally finite and semisimple Artinian by 1.2.6 b) and its central idempotents are central in the simple ring F[G]. Hence R is a matrix ring of finite degree m say over a field K. Then $F[K] \leq F[G]$ is a commutative ring and $D^{n\times n} = F[G] \cong F[K]^{m\times m}$. It follows that F[K] is a field, namely the centre of D, and the Artin-Wedderburn Theorem yields that D = F[K]. The proof is complete. □

6.3.2. **(Hartley [2], [3]).** $\underline{Z}_F \cap L\underline{F}$ ***is the class of all locally finite groups*** G ***with*** $G/O_p(G) \in \underline{AF}$.

Proof: Suppose first that G is a locally finite group with $G/O_p(G)$ abelian-by-finite. If X is any finite subgroup of $O_p(G)$ then the augmentation ideal of FX is nilpotent. Thus $O_p(G) - 1$ generates a nil ideal of FG and therefore lies in its Jacobson radical J(FG). Hence an irreducible FG-module is actually an irreducible $F(G/O_p(G))$-module and $G/O_p(G) \in \underline{Y}_F$ by 6.1.3. Therefore $G \in \underline{Y}_F \subseteq \underline{Z}_F$.

Now assume that $G \in \underline{Z}_F \cap L\underline{F}$ and suppose first that $p = 0$. Then on any irreducible FG-module, G acts as an abelian-by-finite group by the proof of 6.3.1. Thus FG satisfies a polynomial identity by Passman [2], 12.4.1 and therefore G is abelian-by-finite (Passman [2], 5.3.7).

Now let $p > 0$. The first step in Hartley's proof (his Lemma 1.1) is, in our language, to prove that $G \in \underline{Y}_F$. This is 6.3.1. The remainder of the proof involves no skew linear groups and we therefore refer the reader to Hartley's original paper (Hartley [3]).

Perhaps we should at least point out why an image H of G, acting faithfully on an irreducible FH-module, is abelian-by-finite. If X is any infinite periodic simple linear group of characteristic p then X is countable (Wehrfritz [2], 9.5) and FX is prime (Passman [2] 4.2.10) with zero Jacobson radical (Passman [4] 2.5). Hence FX is primitive (Passman [2], 9.2.5). Consequently $X \notin \underline{Y}_F$ and therefore H is abelian-by-finite by 5.1.5 and 5.1.6. □

BIBLIOGRAPHY

A. A. ALBERT

[1] Structure of Algebras, (Amer. Math. Soc. Colloquium Publications, 24, 1961).

R. K. AMAYO & I. STEWART,

[1] Infinite-Dimensional Lie Algebras, (Noordhoff International, Leyden 1974).

S. A. AMITSUR

[1] Finite subgroups of division rings, Trans. Amer. Math. Soc. 80 (1955), 361-386.

E. ARTIN

[1] Geometric Algebra, (Interscience, New York 1957).

S. BACHMUTH & H. Y. MOCHIZUKI

[1] Automorphism groups and subgroups of SL_2 over division rings, Comm. Algebra 7 (1979), 1531-1558.

B. BANIEQBAL

[1] Ph.D. Thesis, (Manchester, 1982).

A. BOREL

[1] Linear Algebraic Groups, (Benjamin, New York, 1969).

K. A. BROWN & B. A. F. WEHRFRITZ

[1] Division rings associated with polycyclic groups, J. London Math. Soc. (2) 30 (1984), 465-467.

R. H. BRUCK

[1] Engel Conditions in Groups and Related Questions, (Lecture Notes, Australian National University, Canberra, January 1973).

R. W. CARTER

[1] Simple Groups of Lie Type, (Wiley & Sons, London 1972).

J. W. S. CASSELS & A. FRÖHLICH

[1] Algebraic Number Theory, (Academic Press, London, 1967).

A. W. CHATTERS & C. R. HAJARNAVIS

[1] Rings with Chain Conditions, (Pitman, London, 1980).

G. H. CLIFF

[1] Zero divisors and idempotents in group rings, Canad. J. Math. 32 (1980), 596-602.

P. M. COHN

[1] On the embedding of rings in skewfields, Proc. London Math. Soc.(3) 11 (1961), 511-530.

[2] Algebra, 2 vols. (Wiley & Sons, London, 1974 & 1977).

[3] Skew Field Constructions, (London Mathematical Society Lecture Note Series, Vol. 27, 1977).

C. W. CURTIS & I. REINER

[1] Representation Theory of Finite Groups and Associative Algebras, (Wiley & Sons, New York, 1962).

J. D. DIXON

[1] The Structure of Linear Groups, (Van Nostrand, London, 1971).

R. J. FAUDREE

[1] On locally finite and solvable subgroups of sfields, Proc. Amer. Math. Soc. 22 (1969), 407-413.

E. FORMANEK & A. V. JATEGAONKAR

[1] Subrings of Noetherian rings, Proc. Amer. Math. Soc. 46 (1974), 181-186.

L. FUCHS

[1] Infinite Abelian Groups, 2 vols. (Academic Press, New York, 1970 & 1973).

K. W. GRUENBERG

[1] The Engel elements of a soluble group, Illinois J. Math. 3 (1959), 151-168.

[2] The upper central series in soluble groups, Illinois J. Math. 5 (1961), 436-466.

M. HALL Jr.

[1] The Theory of Groups, (Macmillan, New York, 1959).

P. HALL

[1] The Edmonton Notes on Nilpotent Groups, (Queen Mary College Math. Notes, 1969).

B. HARTLEY

[1] Sylow subgroups of locally finite groups, Proceedings of 1973 Canberra Conference on Theory of Groups, Springer Lecture Notes 372, 1974.

[2] Injective modules over group rings, Quart. J. Math. (2) 28 (1977), 1-29.

[3] Locally finite groups whose irreducible modules are finite-dimensional, Rocky Mount. J. Math. 13 (1983), 255-263.

B. HARTLEY & P. MENAL

[1] Unipotent representations of torsion-free nilpotent groups over skew-fields, Bull. London Math. Soc. 15 (1983), 378-383.

B. HARTLEY & M. A. SHAHABI SHOJAEI

[1] Finite groups of matrices over divison rings, Math. Proc. Camb. Phil. Soc. 92 (1982), 55-64.

[2] Finite quasi-simple subgroups of 2×2 matrices over a division ring, Math. Proc. Camb. Phil. Soc. 97 (1985), 415-420.

H. HEINEKEN

[1] Endomorphismenringe und engelsche Elemente, Arch. Math. (Basel) 13 (1962), 29-37.

H. HEINEKEN & I. J. MOHAMED

[1] A group with trivial centre satisfying the normalizer condition, J. Algebra 10 (1968), 368-376.

I. N. HERSTEIN

[1] Finite multiplicative subgroups in division rings, Pacific J. Math. 1 (1953), 121-126.

G. HIGMAN

[1] The units of group rings, Proc. London Math. Soc. (2) 46 (1940), 231-248.

B. HUPPERT

[1] Endliche Gruppen, (Springer, Berlin, 1967).

B. HUPPERT & N. BLACKBURN

[1] Finite Groups III, (Springer, Berlin, 1982).

N. JACOBSON

[1] A note on two-dimensional division ring extensions, Amer. J. Math. 77 (1955), 593-599.

[2] Lie Algebras, (Interscience, New York, 1962).

[3] Structure of Rings, (American Mathematical Society Colloquium publications 37, 1964).

[4] Lectures in Abstract Algebra III, (Graduate Texts in Mathematics, Vol. 32, Springer, Berlin, 1975).

[5] Basic Algebra, 2 vols. (Freeman & Co., San Fransisco, 1980).

G. JANUSZ

[1] Simple components of Q[SL(2,q)], Comm. Algebra 1 (1974), 1-22.

A. V. JATEGAONKAR

[1] Integral group rings of polycyclic-by-finite groups, J. Pure Appl. Algebra 4 (1974), 337-343.

I. KAPLANSKY

[1] Fields and Rings, (Chicago University Press, 1969).

[2] Commutative Rings, (Allyn & Bacon, Boston, Mass., 1970).

O. H. KEGEL & B. A. F. WEHRFRITZ

[1] Locally Finite Groups, (North Holland, Amsterdam, 1973).

T. Y. LAM

[1] The Algebraic Theory of Quadratic Forms, (Benjamin, Reading, Mass., 1973).

S. LANG

[1] Algebraic Number Theory, Addison-Wesley, New York, 1970.

J. LEWIN & T. LEWIN

[1] An embedding of the group algebra of a torsion-free one-relator group in a field, J. Algebra 52 (1978), 39-74.

A. I. LICHTMAN

[1] Normal subgroups of the multiplicative group of a division ring, Soviet Math. 4 (1963), 1425-1429.

[2] On subgroups of the multiplicative group of a skew field, Soviet Math. 6 (1965), 915-917.

[3] On embeddings of group rings in division rings, Israel J. Math. 23 (1976), 288-297.

[4] On subgroups of the multiplicative group of skew fields, Proc. Amer. Math. Soc. 63 (1977), 15-16.

[5] On normal subgroups of the multiplicative group of skew fields generated by a polycyclic-by-finite group, J. Algebra 78 (1982), 548-577.

[6] On linear groups over the field of fractions of a polycyclic group ring, Israel J. Math. 42 (1982), 318-326.

[7] Localization in enveloping algebras, Math. Proc. Camb. Phil. Soc. 93 (1983), 467-475.

[8] Matrix rings over fields of fractions of group rings and enveloping algebras I, J. Algebra 88 (1984), 1-37.

[9] Matrix rings over fields of fractions of group rings and enveloping algebras II, J. Algebra 90 (1984), 516-527.

M. LORENZ

[1] Finite normalizing extensions of rings, Math. Zeit. 176 (1981), 447-484.

W. MAGNUS, A. KARRASS, & D. SOLITAR

[1] Combinatorial Group Theory, (Interscience, New York, 1966).

H. MATSUMURA

[1] Commutative Algebra, (Benjamin, New York, 1970).

H. Y. MOCHIZUKI

[1] Unipotent matrix groups over division rings, Canad. Math. Bull. 21 (1978), 249-250.

M. NAGATA

[1] Local Rings, (Interscience, New York, 1962).

H. NEUMANN

[1] Varieties of Groups, (Springer, Berlin, 1967).

D. S. PASSMAN

[1] Permutation Groups, (Benjamin, New York, 1968).

[2] The Algebraic Structure of Group Rings, (Wiley & Sons, New York, 1977).

[3] Group Rings of Polycyclic Groups, Essays for Philip Hall, Ed. K. W. Gruenberg & J. E. Roseblade, (Academic Press, London, 1984).

[4] On the semisimplicity of group rings of linear groups II, Pacif. J. Math. 48 (1973), 215-234.

V. P. PLATONOV

[1] The theory of algebraic linear groups and periodic groups (Russian), Izv. Akad. Nauk SSSR, Ser. Mat. 30 (1966), 573-620 = Amer. Math. Soc. Transl. (2) 69 (1968), 61-110.

I. REINER

[1] Maximal Orders, (Academic Press, London, 1975).

A. H. RHEMTULLA & B. A. F. WEHRFRITZ

[1] Isolators in soluble groups of finite rank, Rocky Mount. J. Math. 14 (1984), 415-421.

D. J. S. ROBINSON

[1] Finiteness Conditions and Generalized Soluble Groups, 2 vols.(Springer, Berlin, 1972).

J. E. ROSEBLADE

[1] Group rings of polycyclic groups, J. Pure Appl. Algebra 3 (1973), 307-328.

[2] Applications of the Artin-Rees lemma to group rings, Symp. Math. 17 (1976), 471-478.

[3] Prime ideals in group rings of polycyclic groups, Proc. London Math. Soc. (3) 36 (1978), 385-447.

J. E. ROSEBLADE & P. F. SMITH

[1] A note on the Artin-Rees property of certain polycyclic group algebras, Bull. London Math. Soc. 11 (1979), 184-185.

P. SAMUEL

[1] A propos du théorème des unités, Bull. Sci. Math. 90 (1966), 89-96.

W. R. SCOTT

[1] Group Theory, (Prentice Hall, Englewood Cliffs, N. J., 1964).

D. SEGAL

[1] On abelian-by-polycyclic groups, J. London Math. Soc. (2) 11 (1975), 445-452.

[2] Unipotent groups of module automorphisms over polycyclic group rings, Bull. London Math. Soc. 8 (1976), 174-178.

[3] Polycyclic Groups, (Cambridge University Press, 1982).

J.-P. SERRE

[1] Corps locaux, (Hermann, Paris, 1962).

[2] Cours d'arithmetique, (Presses Univirsit. d. France, Paris, 1970).

[3] Trees, (Springer, Berlin, 1980).

M. SHIRVANI

[1] On finite and locally finite subgroups of division rings, preprint.

W. S. SIZER

[1] Triangularizing solvable groups of unipotent matrices over a skewfield, Canad. Math. Bull. 20 (1977), 121-123.

R. L. SNIDER

[1] The zero divisor conjecture for some solvable groups, Pacific J. Math. 90 (1980), 191-196.

[2] Group rings with finite endomorphism dimension, Arch. Math. (Basel), 41 (1983), 219-225.

[3] Private Communication.

[4] Solvable linear groups over division rings, Proc. Amer. Math. Soc. 91 (1984), 341-344.

L. SOLOMON

[1] The representation of finite groups in algebraic number fields, J. Math. Soc. Japan 13 (1961), 144-164.

I. A. STEWART

[1] Locally Supersoluble Skew Linear Groups, (Ph.D. Thesis, London, 1986).

V. P. ŠUNKOV

[1] On locally finite groups of finite rank (Russian), Algebra i. Logika 10 (1971), 199-225.

D. A. SUPRUNENKO

[1] On nilpotent matrices and unipotent matrix groups over a division ring (Russian), Vessti. Akad. Nauk BSSR Ser. Fiz.-Mat. 5, 5 (1979), 5-10.

R. STEINBERG

[1] Lectures on Chevalley Groups, (Yale University Lecture Notes, 1967).

S. THOMAS

[1] The classification of the simple periodic linear groups, Arch. Math. (Basel) 41 (1983), 103-116.

J. TITS

[1] Quaternions over $\mathbb{Q}(\sqrt{5})$, Leech's lattice, and the sporadic group of Hall-Janko, J. Algebra 63 (1980), 56-75.

B. A. F. WEHRFRITZ

[1] Remarks on centrality and cyclicity in linear groups, J. Algebra 18 (1971), 229-236.

[2] Infinite Linear Groups, (Springer, Berlin, 1973).

[3] Finite central height in linear groups, Linear Algebra Appl. 17 (1977), 59-64.

[4] Hypercentral unipotent subgroups of linear groups, Bull. London Math. Soc. 10 (1978), 310-313.

[5] Lectures around Complete Local Rings, (Queen Mary College Math. Notes, 1979).

[6] Invariant maximal ideals of commutative rings, J. Algebra 56 (1979), 472-480.

[7] Finitely generated modules over polycyclic groups, Quart. J. Math. (2) 31 (1980), 109-127.

[8] On finitely generated soluble linear groups, Math. Zeit. 170 (1980), 155-167.

[9] The rank of a linear p-group, J. London Math. Soc. (2) 21 (1980), 237-243.

[10] Nilpotence in certain linear groups over division rings, Proc. London Math. Soc. (3) 46 (1983), 334-346.

[11] The upper central factors of certain skew-linear groups, Arch. Math. (Basel) 40 (1983), 481-494.

[12] The upper central factors of certain skew-linear groups; an addendum, Arch. Math. (Basel) 44 (1985), 229.

[13] Complete reducibility in skew-linear groups, J. London Math. Soc. (2) 28 (1983), 301-309.

[14] On certain soluble skew-linear groups, J. Pure Appl. Algebra 29 (1983), 209-218.

[15] Hyperabelian skew-linear groups, J. Algebra 90 (1984), 220-229.

[16] On division rings generated by polycyclic groups, Israel J. Math. 47 (1984), 154-164.

[17] Faithful representations of finitely generated abelian-by-polycyclic groups over division rings, Quart. J. Math. (2) 35 (1984), 361-372.

[18] Soluble-by-periodic skew-linear groups, Math. Proc. Camb. Phil. Soc. 96 (1984), 379-389.

[19] Locally finite normal subgroups of absolutely irreducible skew-linear groups, J. London Math. Soc. (2) 32 (1985), 88-102.

[20] On absolutely irreducible skew-linear groups in characteristic zero, Arch. Math. (Basel) 45 (1985), 193-199.

[21] Soluble normal subgroups of skew-linear groups, J. Pure & Appl. Algebra, to appear.

[22] Locally nilpotent skew linear groups, Proc. Edinburgh Math. Soc. 29 (1986), 101-113.

[23] Normal subgroups of skew linear groups, Mathematika 33 (1986), 122-130.

[24] Group rings with finite central endomorphism dimension, Glasgow Math. J. 24 (1983), 169-176.

[25] Locally nilpotent skew linear groups II, Proc. Edinburgh Math. Soc., to appear.

E. WEISS

[1] Algebraic Number Theory, (McGraw-Hill, New York, 1963).

A. E. ZALESSKIĬ

[1] The structure of several classes of matrix groups over a division ring (Russian), Sibirsk. Mat. Ž. 8 (1967), 1284-1298 = Siberian Math. J. 8 (1967), 978-988.

[2] Sylow p-subgroups of the full linear group over a field (Russian), Izv. Akad. Nauk SSSR Ser. Mat. 31 (1967), 1149-1158.

O. ZARISKI & P. SAMUEL

[1] Commutative Algebra, 2 vols. (Van Nostrand, Princeton, 1958 & 1960).

NOTATION INDEX

CONTENTS

ABSTRACT GROUPS

GROUP INVARIANTS

Let G be a group.

h(G) : the Hirsch number of the polycyclic-by-finite group G.

hah(G) : see page 99.

hph(G) : see page 99.

rankG : the (Prüfer) rank of G, viz. the maximum of the minimum number of generators of the finitely generated subgroups of G.

$\pi(G)$: the set of prime divisors of the orders of the elements of G of finite order.

$|G|$: the order of G.

CANONICAL SUBGROUPS & SUBSETS

Let G be a group, α an ordinal number and i a positive integer.

$G^{(\alpha)}$: the α-th term of the derived series of G.

E(G) : see page 71.

F(G) : see page 71.

$F^*(G)$: see page 71.

L(G) : the set of left Engel elements of G (see page 116).

$\bar{L}(G)$: the set of bounded left Engel elements of G (see page 116).

$O_\pi(G)$: the unique maxiaml normal π-subgroup of G.

$O_{\pi,\tau}(G)$: $O_{\pi,\tau}(G)/O_\pi(G) = O_\tau(G/O_\pi(G))$, where π and τ are sets of primes.

R(G) : the set of right Engel elements of G (see page 116).

$\bar{R}(G)$: the set of bounded right Engel elements of G (see page 116).

Z(G) : the Zalesskiĭ subgroup of the soluble group G (see page 188).

$\gamma^i(G)$: the i-th term of the lower central series of G.

$\Delta(G)$: the FC-centre of G (see page 95).

$\zeta_\alpha(G)$: the α-th term of the upper central series of G.

$\zeta(G)$: the hypercentre of G, viz. $\bigcup_\alpha \zeta_\alpha(G)$.

$\eta(G)$: the Hirsch-Plotkin radical of G.

$\eta^{(\alpha)}(G)$: the α-th term of the upper locally nilpotent series of G (see page 99).

$\eta_1(G)$: the Fitting subgroup of G.

$\Lambda(G)$: see page 222.

$\rho(G)$: see page 116.

$\bar{\rho}(G)$: see page 116.

$\sigma(G)$: the Gruenberg radical of G (see page 116).

$\bar{\sigma}(G)$: the Baer radical of G (see page 116).

SPECIAL GROUPS

Let n be a positive integer.

Alt(n) : the alternating group of degree n.

C_n : the cyclic group of order n.

C_∞ : the infinite cyclic group.

GL(n,q) : the general linear group of degree n over the field GF(q).

$O(n,\mathbb{R})$: the n-dimensional real orthogonal group.

Q_{2^n} : the generalized quaternion group of order 2^n.

$Sym(n)$: the symmetric group of degree n.

${}^*X_n(q)$: a group of Lie type (see page 70).

COMMUTATOR NOTATION

Let x, y be elements of the group G.

$[x,_1y] = [x,y] = x^{-1}y^{-1}xy$.

$[x,_ny] = [[x,_{n-1}y],y]$ for $n \geq 2$.

S e T, S |e T : see page 103.

GROUP CONSTRUCTIONS

Let $\{ G_i : i \in I \}$ be a family of groups.

$G_1] G_2$: the split extension of G_1 by G_2, where G_2 acts on G_1.

$G_1 \wr G_2$: a wreath product of G_1 by G_2, being the standard product unless a permutation representation is specified.

$\prod G_i$: the cartesian product of the G_i.

$\times G_i$: the direct product of the G_i.

GROUP ACTIONS

Let V be a set and G a group acting on V. Write v^g for the image of $v \in V$ under $g \in G$. Let X be a subset of V and let $v \in V$.

$C_G(X) = \{ g \in G : x^g = x \text{ for all } x \in X \}$.

$C_X(G) = \{ x \in X : x^g = x \text{ for all } g \in G \}$, the set of fixed points of X under G.

$N_G(X) = \{ g \in G : X^g = X \}$.

$v^G = \{ v^g : g \in G \}$.

$\Delta_X(G) = \{ x \in X : x^G \text{ is finite} \}$.

Assume further that V is actually a G-module.

$[V,_1G] = [V,G] = \langle v - v^g : \text{all } v \in V, \text{ all } g \in G \rangle$.

$[V,_nG] = [[V,_{n-1}G],G]$ for $n \geq 2$.

CALCULUS OF CLASSES

SPECIAL CLASSES

$\underline{A}$: the class of abelian groups.

$\underline{E}_p$: the class of p-groups of finite exponent.

$\underline{F}$: the class of finite groups.

$\underline{F}_p$: the class of finite p-groups.

$\underline{G}$: the class of finitely generated groups.

$\underline{G}_1$: the class of cyclic groups.

$\underline{I}$: the class of locally indicable groups (see page 35).

$\underline{N}$: the class of nilpotent groups.

$\underline{N}_1$: the class of groups all of whose subgroups are subnormal.

$\underline{O}$: see page 26.

$\underline{O}_F$: see page 162.

$\underline{O}_1$: see page 34.

$\underline{O}_2$: see page 37.

$\underline{P}$: the class of polycyclic groups.

$\underline{S}$: the class of soluble groups.

$\overline{\underline{S}}$: see page 82.

$\underline{S}_g$: see page 82.

$\underline{U}$: see page 18.

$\underline{X}$: see page 147.

$\underline{X}_F$: see page 220.

$\underline{Y}_F$: see page 219.

$\underline{Z}$: see page 46.

$\underline{Z}_F$: see page 219.

$\underline{Z}_0$: see page 46.

CLASS OPERATORS

Let $\underline{X}$ be a class of groups.

D : the direct product operator.

L : the local operator.

L_1 : $G \in L_1\underline{X}$ if and only if every cyclic subgroup of G is contained in an $\underline{X}$-subgroup of G.

P : the poly (or extension) operator.

$\acute{P}$: $\acute{P}\underline{X}$ is the class of groups with an ascending series whose factors are $\underline{X}$-groups.

$\grave{P}$: as in $\acute{P}$, with "descending" in place of "ascending".

$\acute{P}_n$: as in $\acute{P}$, but where the series is normal.

P_n : as in $\acute{P}$, but where the series is finite and normal.

Q : the quotient (or homomorphic image) operator.

R : the residual operator.

R_o : the finite-subcartesian product operator.

S : the subgroup operator.

S_f : the subgroup-of-finite-index operator.

Let A be a unary closure operator.

$\underline{X}^A$: the largest A-closed subclass of $\underline{X}$.

$\underline{X}^{-A}$: the largest A-closed class of groups disjoint from $\underline{X}$.

$\langle A,B \rangle$: the closure operator generated by the operators A and B.

RINGS & FIELDS

Rings except nilrings have a multiplicative identity, which is preserved by homomorphisms.

SUBSETS & INVARIANTS

Let R be a ring, X a subset of R, and $\underline{a}$ an ideal of R.

$\underline{C}_R(\underline{a})$: the set of elements of R that are regular modulo $\underline{a}$ (see page 125).

e(A) : the exponent of the algebra A (see page 48).

J(R) : the Jacobson radical of R.

k(R) : the Krull dimension of R.

$\ell_R(X)$: the left annihilator of X in R.

m(A) : the Schur index of the algebra A (see page 48).

Q(R) : the quotient ring of R, whenever it exists.

$\mathrm{rad}_R X$: see page 141.

$rg(X)$: the smallest subring of R (with 1) containing X.

R^* : the set of non-zero elements of R.

R_R : R regarded as a right module over itself.

$r_R(X)$: the right annihilator of X in R.

s_n : the standard polynomial identity of degree n.

U(R) : the group of units of R.

SPECIAL RINGS & FIELDS

$\mathbb{A}$: the algebraic closure of the rationals.

$\mathbb{C}$: the field of complex numbers.

$\mathbb{F}_p$: the field of p elements (p a prime).

GF(q) : the field of q elements (q a prime power).

$\mathbb{Q}$: the field of rational numbers.

$\mathbb{Q}_p$: the field of p-adic numbers.

$\mathbb{Q}(\zeta_m)$: the m-th cyclotomic field over the rationals.

$\mathbb{Z}$: the ring of rational integers.

$\mathbb{Z}_p$: the ring of p-adic integers.

RING CONSTRUCTIONS

A_P : see page 49.

A(L) : the universal enveloping algebra of the Lie algebra L.

(R,S,G,H) : the crossed product of S by H (see page 23).

Let S be a ring and G a group.

SG : the group ring of G over S.

S^*G : a skew group ring of G over S, some action of G on S being specified (see page 25).

If S is a subring of a ring R and G is a subgroup of U(R) normalizing S, then

S[G] : the subring of R generated by S and G.

S(G) : the division subring of R generated by S and G in situations where R and S are division rings.

FIELDS

Let F be a field.

$\underline{A}(F)$: the quaternion F-algebra (see page 49).
det(x) : the determinant of the matrix x.
deg(f) : the degree of the polynomial f.
$\tilde{F}$: the perfect closure of F.
Br(F) : the Brauer group of F.

Let F be a finite Galois extension of K.

$N_{F/K}(x)$: the norm of the element $x \in F^*$.
$N_{F/K}$: the subgroup of K^* of all $N_{F/K}(x)$ for $x \in F^*$.
Gal(F/K) : the Galois group of F over K.

Assume further that K is an algebraic number field and that $\mathfrak{P}$ is a non-trivial prime ideal of the ring of algebraic integers of K.

$e(\mathfrak{P},F/K)$: the ramification index of $\mathfrak{P}$ in F/K (see page 49).
$f(\mathfrak{P},F/K)$: the residue degree of $\mathfrak{P}$ in F/K (see page 49).
$\mathcal{O}_K$: the ring of algebraic integers of K.
$K_{\mathfrak{P}}$: the completion of K with respect to the valuation determined by $\mathfrak{P}$.

FUNCTORS

Let G be a group, V a G-module and S a ring.

Aut(G) : the group of automorphisms of G.
Aut(S) : the group of automorphisms of S.
Der(G,V) : the group of derivations of G into V.
End(G) : the endomorphism ring of G (whenever G is abelian).
$H^1(G,V)$: the first cohomology group of G with coefficients in V.
Ider(G,V) : the group of inner derivations of G into V.
Inn(G) : the group of inner automorphisms of G.
Inn(S) : the group of inner automorphisms of S.

Map(X,Y) : the set of all maps from the set X to the set Y.

SKEW LINEAR GROUPS

Let D be a division ring and n a positive integer.

D^n : row n-space over D.
$D^{n\times n}$: the ring of n×n matrices over D.
$\dim_D V$: the dimension of a D-space V over D.
$\mathrm{diag}(X_1,\ldots X_m)$: the set of diagonal matrices with diagonal entries from the appropriate sets X_i.
$(D:E)_r$: the right dimension of D over its division subring E.
GL(n,D) : the group of invertible n×n matrices over D.
SL(n,D) : the special linear group (see page 154).
$Tr_1(n,D)$: the subgroup of GL(n,D) of lower unitriangular matrices.
$Tr^1(n,D)$: the subgroup of GL(n,D) of upper unitriangular matrices.

Let G be a skew linear group and g an element of G.

G^- : see page 81.
G^+ : see page 81.
g_u, g_d : Jordan constituents of g (see page 83).
G_u, G_d : see page 85.
s(G) : see page 21.
u(G) : see page 21.
G^o : the connected component of the identity of the *linear* group G.

SPECIAL SYMBOLS

Let n be an integer.

[n] : the greatest integer less than or equal to n.
-[-n] : the least integer greater than or equal to n.
c(n,q) : see page 113.
e(n,q) : see page 112.
$\beta(n)$: a Jordan function (see page 151).
$\gamma(n,s)$: see page 47.

π'	: the set of primes not in the set π.
$\Phi(n)$	: the Euler function.
ω	: the first infinite ordinal.
$\varinjlim$	: the direct limit.
$\oplus$	: the direct sum.
$\otimes_R$	: the tensor product over a ring R.

AUTHOR INDEX

GENERAL INDEX

A

B

C

D

E

L

M

N

O

P